DECOUPLING CIVIL TIMEKEEPING FROM EARTH ROTATION

AAS PRESIDENT
    Frank A. Slazer   Northrop Grumman

VICE PRESIDENT - PUBLICATIONS
    Prof. David B. Spencer   Pennsylvania State University

EDITORS
    John H. Seago   Analytical Graphics, Inc.
    Robert L. Seaman   National Optical Astronomy Observatory
    Steven L. Allen   University of California Observatories,
                                                 Lick Observatory

SERIES EDITOR
    Robert H. Jacobs   Univelt, Incorporated

Front Cover Illustration:

The Decoupling of Civil Timekeeping from Earth Rotation shatters the astronomical paradigm of human timekeeping maintained since ages immemorial.

Image Credit: Artwork by Pete Marenfeld, courtesy of NOAO.

DECOUPLING CIVIL TIMEKEEPING FROM EARTH ROTATION

Edited by
John H. Seago
Robert L. Seaman
Steven L. Allen

Volume 113
SCIENCE AND TECHNOLOGY SERIES
A Supplement to Advances in the Astronautical Sciences

Proceedings of a Colloquium Exploring Implications of Redefining Coordinated Universal Time (UTC), held October 5–7, 2011 at Analytical Graphics, Inc., Exton, Pennsylvania.

Published for the American Astronautical Society
by Univelt, Incorporated, P.O. Box 28130, San Diego, California 92198
Web Site: http://www.univelt.com

Copyright 2011

by

AMERICAN ASTRONAUTICAL SOCIETY

AAS Publications Office
P.O. Box 28130
San Diego, California 92198

Affiliated with the American Association for the Advancement of Science
Member of the International Astronautical Federation

First Printing 2011

ISSN 0278-4017

ISBN 978-0-87703-575-6 (Hard Cover Plus CD ROM)
ISBN 978-0-87703-576-3 (CD ROM)

Published for the American Astronautical Society
by Univelt, Incorporated, P.O. Box 28130, San Diego, California 92198
Web Site: http://www.univelt.com

Printed and Bound in the U.S.A.

FOREWORD

On October 5 and October 6, 2011, a colloquium on *Decoupling Civil Timekeeping from Earth Rotation* was hosted in Exton, Pennsylvania by Analytical Graphics, Inc. (AGI), a leading software solutions company for the aerospace, defense, and intelligence industries. AGI's involvement with the topic of UTC redefinition began in 2003, when the company was asked to provide a position statement in preparation for the ITU-R *Special Colloquium on the Future of the UTC Timescale* at IEN in Torino, Italy. At the time, AGI's Chief Orbital Scientist did not expect UTC redefinition to impact its own internally developed software, yet AGI recognized that its own software products might experience indirect impacts from the adoption of third-party software models. AGI also realized that UTC redefinition "could adversely affect software applications or database definitions" as well as the "accuracy of certain low precision applications" throughout the astrodynamics community. AGI's position was that the company would necessarily adapt to changing standards to meet customer needs, but it was the company's "sincerest hope that the voice of the astrodynamics community is heard." Therefore, it seemed fitting that AGI should open its facilities to host a colloquium, co-sponsored by the American Astronautical Society (AAS) and the American Institute of Aeronautics and Astronautics (AIAA), to afford an opportunity for the astrodynamics community to be heard before a concluding vote to decouple civil timekeeping from Earth rotation by the Radiocommunications Assembly of the ITU-R in January 2012.

The colloquium was co-hosted by the National Optical Astronomy Observatory (NOAO) and the Virtual Astronomical Observatory (VAO). The opportunity for participation was not limited to astronomical and astronautical interests, however. Understanding that the implications of decoupling extend from technical infrastructure to legal, historical, logistical, sociological, and economic domains, technical and non-technical contributions were solicited from widely disparate fields. These fields included, but were not limited to: astronomy, astrodynamics and celestial mechanics, geophysical Earth-orientation, navigation (GNSS and celestial), remote sensing and space surveillance, spacecraft applications, sundialing, *etc*.

International interest in the colloquium was remarkable given that the meeting was announced with about four months' notice, with presenters having to provide original papers for the colloquium's archival proceedings. The meeting chairmen received many regrets from interested parties who were unable to contribute due to the constraints of personal schedule or economy. The number of contributed manuscripts was very strong relative to the efforts of similar meetings:[*] twenty-two (22) papers were eventually contributed on diverse topics related to a variety of timekeeping aspects. These were presented over two days to an audience of seventeen discussants representing opinions both favoring and opposing the decoupling of civil timekeeping from Earth rotation (Figure 1). Six contributing co-authors could not attend in person: Danny Hillis of Applied Minds, Inc., Denis Savoie of the Palais de la découverte and Observatoire de Paris, Florent Deleflie, Jérôme Berthier,

---

[*] Fourteen presentations were given at the 2003 ITU-R *Special Colloquium on the Future of the UTC Timescale* in Torino, and twelve presentations were given at the 2011 meeting *UTC for the 21st Century* sponsored by the Royal Society.

and Christophe Barache of the Observatoire de Paris, and Jon Giorgini of the Jet Propulsion Laboratory (JPL). Neil deGrasse Tyson of the Hayden Planetarium and the American Museum of Natural History, and Steven Slojkowski of NASA Goddard Space Flight Center, also participated as interested discussants.

In addition to the scheduled technical presentations and discussions, Frank Reed, a navigation instructor with Reed Navigation and Mystic Seaport Museum, provided a suppertime presentation entitled *GMT by Observation: The Historical Method of Lunars* on Wednesday, October 5. The "method of lunars" involves a determination of time using a hand-held sextant, as based on the angle of the Moon from other bright celestial objects. Mr. Reed's presentation featured excerpts from historic ship logbooks and plots of ocean voyages during the 18th and 19th centuries as determined from navigation fixes recorded in these logbooks. Blank pages from the backs of old logbooks often contained completely worked examples of the method of lunars, each requiring roughly 20 minutes of arithmetic to reduce sextant observations of the Moon to obtain the time. *The American Ephemeris and Nautical Almanac* continued to carry "lunar distance" ephemerides, used by practitioners of the method, through the 1911 edition. However, by the mid- to late-19th century, the method of lunars had become mostly obsolete, as it became more accurate and convenient to carry multiple ship-borne chronometers to determine and maintain time (and therefore position) at sea.

Mr. Reed offered attendees the opportunity to try their hand at determining time using the historical method of lunar-distance observation using a sextant. After supper, a handful of colloquium attendees gathered under darkness at a nearby hotel parking lot. Participants exercised their traditional navigation skills by measuring the arc between Jupiter and the lunar limb using sextants supplied by Mr. Reed. Using software, Mr. Reed reduced the limb measurements into a determination of time almost instantly. The best measurements obtained would have been good enough to fix a position to better than a nautical mile.

On Friday, October 7, nine colloquium attendees and their guests participated in a morning tour of the grand Analemmatic Sundial at Longwood Gardens in nearby Kennett Square, Pennsylvania. This event included a special orientation and behind-the-scenes technology tour of the Longwood Gardens facilities by staff historian Colvin Randall, and featured a short talk by P. Kenneth Seidelmann who redesigned the sundial in the early 1970s to keep accurate standard time. Dr. Seidelmann's manuscript on the analemmatic sundial adds to these proceedings a significant example of a timeless public apparatus that presumes civil timekeeping will remain accurately coupled to Earth rotation.

At the request of some colloquium attendees and other interested professionals, the co-chairs have included an extended introduction and summary of meeting's topic and technical points as a separate paper. This summary, which is an expression of the co-chairs, is no substitute for exploring the actual manuscripts and discussions, therefore readers are encouraged to consider the proceedings volume in its entirety.

John H. Seago
Robert L. Seaman
Steven L. Allen

Volume Editors

Figure 1. Colloquium Attendees.

Figure 2. The Analemmatic Sundial at Longwood Gardens.

Table 1. Colloquium Attendees (as ordered from left to right in Figure 1)

Name	Organization	Nationality
George H. Kaplan	IAU Commission 4	USA
Robert L. (Rob) Seaman	National Optical Astronomy Observatory (NOAO)	USA
Wolfgang R. Dick	IERS Central Bureau, Bundesamt für Kartographie und Geodäsie (BKG)	Germany
John H. Seago	Analytical Graphics, Inc. (AGI)	USA
Daniel Gambis	IERS Earth Orientation Center, Observatoire de Paris	France
Steven (Steve) Malys	National Geospatial-Intelligence Agency (NGA)	USA
Steven L. (Steve) Allen	UCO/Lick Observatory	USA
Neil deGrasse Tyson	Hayden Planetarium, American Museum of Natural History	USA
Paul Gabor	Vatican Observatory	Czech Republic
Mark F. Storz	United States Air Force Space Command (AFSPC)	USA
Dennis D. McCarthy	United States Naval Observatory (USNO, retired)	USA
Arnold H. Rots	Smithsonian Astrophysical Observatory (SAO)	The Netherlands
P. Kenneth (Ken) Seidelmann	University of Virginia (UVa)	USA
David G. Simpson	NASA Goddard Space Flight Center (GSFC)	USA
Steven Slojkowski	NASA Goddard Space Flight Center (GSFC)	USA
Frank E. Reed	Reed Navigation, Mystic Seaport Museum	USA
David L. Terrett	Rutherford Appleton Laboratory (RAL) Space	UK

CONTENTS

	Page
FOREWORD	vii
INTRODUCTION	1
The Colloquium on Decoupling Civil Timekeeping from Earth Rotation (AAS 11-660)	
John H. Seago, Robert L. Seaman and Steven L. Allen	3
SESSION 1: SETTING THE STAGE	13
Systems Engineering for Civil Timekeeping (AAS 11-661)	
Rob Seaman	15
Legislative Specifications for Coordinating with Universal Time (AAS 11-662)	
John H. Seago, P. Kenneth Seidelmann and Steve Allen	29
SESSION 2: THE PAST, PRESENT, AND FAR FUTURE	51
The Heavens and Timekeeping, Symbolism and Expediency (AAS 11-663)	
Paul Gabor	53
Leap Seconds in Literature (AAS 11-664)	
John H. Seago	65
Time in the 10,000-Year Clock (AAS 11-665)	
Danny Hillis, Rob Seaman, Steve Allen and Jon Giorgini	79
MIDDAY ROUND-TABLE DISCUSSION OF OCTOBER 5, 2011	97
Midday Round-Table Discussion (10-5-2011)	99
SESSION 3: EARTH ORIENTATION	103
Using UTC to Determine the Earth's Rotation Angle (AAS 11-666)	
Dennis D. McCarthy	105
The IERS, the Leap Second, and the Public (AAS 11-667)	
Wolfgang R. Dick	117
Results from the 2011 IERS Earth Orientation Center Survey about a Possible UTC Redefinition (AAS 11-668)	
Daniel Gambis, Gérard Francou and Teddy Carlucci	123

	Page
SESSION 4: TIME SCALE APPLICATIONS	**181**

Traditional Celestial Navigation and UTC (AAS 11-669)
    Frank E. Reed 183

The Consequences of Decoupling UTC on Sundials (AAS 11-670)
    Denis Savoie and Daniel Gambis 195

Time Scales in Astronomical and Navigational Almanacs (AAS 11-671)
    George H. Kaplan 201

Issues Concerning the Future of UTC (AAS 11-672)
    P. Kenneth Seidelmann and John H. Seago 215

SESSION 5: SPACE OPERATIONS — **235**

UTC and the Hubble Space Telescope Flight Software (AAS 11-673)
    David G. Simpson 237

Computation Errors in Look Angle and Range Due to Redefinition of UTC (AAS 11-674)
    Mark F. Storz 249

Proposal for the Redefinition of UTC: Influence on NGA Earth Orientation Predictions and GPS Operations (AAS 11-675)
    Stephen Malys 265

SESSION 6: GROUND OPERATIONS — **271**

UTC at the Harvard-Smithsonian Center for Astrophysics (CFA) and Environs (AAS 11-676)
    Arnold H. Rots 273

An Inventory of UTC Dependencies for IRAF (AAS 11-677)
    Rob Seaman 277

Telescope Systems at Lick Observatory and Keck Observatory (AAS 11-678)
    Steven L. Allen 289

MIDDAY ROUND-TABLE DISCUSSION OF OCTOBER 6, 2011 — **299**

Midday Round-Table Discussion (10-6-2011) 301

SESSION 7: CONTINGENCY PROPOSALS — **307**

Automating Retrieval of Earth Orientation Predictions (AAS 11-679)
    David L. Terrett 309

Dissemination of UT1-UTC through the use of Virtual Observatory (AAS 11-680)
    Florent Deleflie, Daniel Gambis, Christophe Barache and Jérôme Berthier . 317

Timekeeping System Implementations: Options for the *Pontifex Maximus* (AAS 11-681)
    Steven L. Allen 325

	Page
CONCLUDING ROUND-TABLE DISCUSSION	**335**
Concluding Round-Table Discussion (10-6-2011)	337
SPECIAL SESSION	**349**
The Longwood Gardens Analemmatic Sundial (AAS 11-682)	
P. Kenneth Seidelmann	351
APPENDICES	**361**
Appendix A: About the Contributors	363
Appendix B: Publications of the American Astronautical Society	367
Advances in the Astronautical Sciences	368
Science and Technology Series	378
AAS History Series	386
INDICES	**389**
Numerical Index	391
Author Index	393

INTRODUCTION

AAS 11-660

THE COLLOQUIUM ON DECOUPLING CIVIL TIMEKEEPING FROM EARTH ROTATION

John H. Seago,[*] Robert L. Seaman[†] and Steven L. Allen[‡]

On October 5 and October 6, 2011, the Colloquium on the *Decoupling Civil Timekeeping from Earth Rotation* was hosted in Exton, Pennsylvania by Analytical Graphics, Inc. (AGI). This paper highlights various technical perspectives offered through these proceedings, including expressions of concern and various recommendations offered by colloquium participants.

INTRODUCTION

In response to the requests of some colloquium participants and other professionals, the co-chairs of the colloquium on *Decoupling Civil Timekeeping from Earth Rotation*[§] have written an extended introduction and review for the proceedings. This summary does not intend to provide a complete, or even balanced, presentation of the issues surrounding the topic of decoupling of civil timekeeping from Earth rotation, but primarily serves to highlight some of the more significant technical perspectives made during the meeting according to the judgment of the co-chairs. It is hoped that the highlighting of these points encourages the reader to explore these proceedings thoroughly; this summary is no substitute for reading the actual manuscripts and discussions, as there is a wealth of detailed information to be discovered by careful review.

DECOUPLING CIVIL TIMEKEEPING FROM EARTH ROTATION

There is an old saying that "the man with one watch always knows the time, but the man with two watches can never be sure." Implicit to this proverb is the observation that timekeepers run at different rates. For practical reasons then, some ultimate timekeeper must be declared, with all others approximating it. For example, the *leap day* on February 29th keeps the two very different astronomical cycles of *day* and *year* synchronized. It does so by varying the number of days in the calendar year to keep up with the actual orbital year of approximately 365.242 days. The need for these adjustments is appreciated by a general public having learned it from a young age, and the requirement for leap days has been known since antiquity; for without leap-day insertions, the dates of a 365-day calendar would start to lose their familiar relationship with the seasons.

Until recently, our definitive timekeeper was the rotation of the Earth, or analogously, the motion of the sky as seen from the rotating Earth. The task of determining time was therefore left to astronomers, with whom Earth rotation came to be known in the 20th century as *Universal Time*

---

[*] Astrodynamics Engineer, Analytical Graphics, Inc., 220 Valley Creek Blvd., Exton, Pennsylvania 19341-2380, U.S.A.

[†] Senior Software Systems Engineer, National Optical Astronomy Observatory, 950 N Cherry Ave, Tucson, Arizona 85719, U.S.A.

[‡] Programmer/Analyst, UCO/Lick Observatory, ISB 1156 High Street, Santa Cruz, California 95064, U.S.A.

[§] http://futureofutc.org/

(UT), a measure which is synonymous with *mean solar time* at the meridian of Greenwich. *Mean solar time* is Earth rotation relative to the fixed stars (*sidereal time*) adjusted by approximately four extra minutes per day to keep up with the diurnal rising and setting of the Sun over the long term. The days of the calendar are thereby maintained with *mean solar days*.

The longstanding requirement to reconcile clocks with the mean-time rotation of the Earth has recently been questioned within communities that maintain and rely on ultra-precise frequency standards, sometimes known as "atomic clocks". Atomic resonators were invented in the 1950's, and by the 1960's ensembles of precise atomic-frequency standards were being averaged together to create laboratory time scales more uniform than the rotation of Earth. This led to a definition of duration within the metric system, known as the *Système International (SI) second*, in terms of a hyperfine transition frequency of cæsium. Although the *SI* second was calibrated to be close to $1/_{86400}$ of a mean solar day, the two measures of duration are fundamentally different, such that astronomical time slowly and irregularly drifts away from atomic time at an unpredictable rate. Like the man with two watches, a decision had to be made as to whether mean solar days, or atomic radiation, would be the ultimate timekeeper. Atomic frequency was important for broadcasting and telecommunications at the dawn of the space age, but astronomical time had its own technical and societal usages, with national laws and international regulations specifying that radio time-signal emissions should track Universal Time. Practically, Earth rotation cannot be adjusted; only the artificial atomic scale is adjustable—"how?" was the main question.

Atomic frequency standards enabled unprecedented synchronization of broadcasts with UT, informally known as "coordinated" Universal Time. Time had always been determined locally and astronomically; because the wireless transmission of time signals began before the International Bureau of Weights and Measures (BIPM) performed experimental work on the measurement of time or frequency, the regulation of time scales used for radio broadcasts landed with the Radiocommunication sector of the International Telecommunication Bureau (ITU-R). In 1972 regulations recommended that broadcast time signals should track the version of UT called UT1 using a new compromise which was by 1974 formally named Coordinated Universal Time (UTC).[*] Under this recommendation, a sequence of atomic seconds known as International Atomic Time (TAI) was continually maintained in the background by averaging the readings of many laboratory frequency standards. However, once Earth time (UT1) and atomic time (UTC) drifted about one-half second apart, the last day of a month was necessarily lengthened by one second by international agreement. This adjustment became known as the *leap second*, which serves the same purpose in clock time as a leap day serves calendar time.[1] Its scheduling is announced by the International Earth Rotation and Reference System Service (IERS) months in advance, and the notice is distributed electronically throughout telecommunications systems using services such as satellite navigation systems and via the internet using Network Time Protocol (NTP).

A proposal will come before the ITU-R Radiocommunication Assembly in January 2012 on whether to decouple civil timekeeping from Earth rotation by ceasing leap seconds from UTC after 2017. This proposal would allow UTC to diverge from UT1 without bound. The points of debate surrounding this proposal are quite numerous and now well-rehearsed from all sides.[2] The primary function of this colloquium, however, was to discuss the ramifications of such a proposal should it succeed, because the adverse impacts from redefining UTC have not been extensively researched. The colloquium chairmen were delighted with the breadth and quality of the contribu-

---

[*] Universal Time (UT) defines time-of-day on Earth. There are several realizations of UT, including UTC. UT1 best represents the instantaneous orientation of Earth about its rotational axis.

tions and the enthusiasm of the contributors, particularly through the thoughtful round-table discussions. It is therefore all the more remarkable that the general topic of civil timekeeping, and the specific issue of whether it should remain coupled to Earth rotation, have received so little airing outside of a few very narrow technical communities.

WHAT IS CIVIL TIMEKEEPING?

It would be misleading to now claim that civil timekeeping is based on the seasonally non-uniform time-of-day indicated by the length of shadows, also known as *apparent solar time*. Modern civilization moved from absolute reliance on sundials centuries ago to favor artificial timepieces that keep uniform time. Within the context of "decoupling civil timekeeping from Earth rotation," the technicalities of the meaning and measurement of Earth rotation as Universal Time (UT) were understood and appreciated within the technically inclined audience of the colloquium. Yet the more general topic of *civil timekeeping* is harder to compartmentalize.

In its broadest sense, *civil timekeeping* implies "time kept for civilization." However, this is much more than a standard to which the man-on-the-street sets his wristwatch. Timing signals are used by electronic devices which drive and control modern infrastructure, such as communication, transportation, finance, energy, and defense systems. Civil timekeeping involves both calendar-date stamping and high-precision frequency generation. The technological usage of civil time constrains its specification.

As civil timekeeping is considered within the service of today's culture, the most susceptible activity may be the programming of computing devices that drive the technology of modern society. This activity now has decades of legacy code deeply embedded, some of which operates as seemingly inaccessible "black boxes". There is a need to recognize that this is the way the world operates today, and that significant technology is built around the existing civil-timekeeping standard—both correctly and incorrectly. Although difficult, overcoming the "software inertia" of black boxes is possible, but it requires dedicated resources.

At the same time, there is also a need to recognize that computers work for people; people don't work for computers. Clocks have been invented and maintained by humanity for very specific reasons. Civil time cannot be defined arbitrarily and still be indefinitely useful; the rate of clocks must approximate mean solar time if date stamps from clocks are to remain useful for civil timekeeping purposes over the long term. The existing paradigm appears to operate well within our increasingly network-connected society, which raises doubts as to whether the continued coupling of civil timekeeping and Earth rotation presents significant or imminent problems.

THE CONCERNS OVER DECOUPLING

The topic of the colloquium relates to a very fundamental change, as this appears to be the first time in human history that civil timekeeping would be decoupled from Earth rotation *by design*. Historically, timekeeping has been principally perceived as an exercise in remaining faithful to what the sky was doing. Although timekeeping practices may have occasionally gone awry throughout recorded history, there was always willingness to make a correction once the errors became noticeable.

*Are we creating unknown problems for the **present** with the current proposal to decouple civil timekeeping from Earth rotation?* Based on limited declarations of ITU-R study groups, it is unclear how thoroughly certain technology domains have been represented and assessed by groups studying the issue of UTC redefinition. The technical discussions have been framed in terms of the elimination of leap seconds, which limits participation to those stakeholders who know about leap seconds, understand the difference between UT1 and UTC, and appreciate the implications of that relationship. The low number of responses to surveys, data calls, and other targeted inquiries suggests that the discussions have been too technical for most non-experts to appreciate. Many people outside the IERS especially lack awareness of this issue, with most astronomers not likely understanding the difference between UT1 and UTC. Unfortunately, the cessation of leap seconds would create complications for some organizations, notably those related to military defense, as well as astrodynamics and astronomy.

*Are we creating unknown problems for the **future** with the current proposal to decouple civil timekeeping from Earth rotation*? History suggests that astronomical timekeeping has always been specially regarded and there is no compelling evidence that it will be disregarded in the future. Rather, pressure may be eventually brought about from an innate principle of "astronomical conformity" which will lead to a desire for societies to recouple civil timekeeping and Earth rotation sometime in the future. How can we help future generations prepare for that recoupling now? Alternative schemes involving very infrequent intercalary adjustments (such as a leap minute or leap hour) cannot be credibly presumed to work once short-term adjustments are formally abolished. As technology becomes increasingly complex, it would seem extremely difficult to resynchronize civil time with Earth rotation if they were allowed to separate, and that the level of difficulty would grow with magnitude of the difference.

When our descendants ask why a fundamental change to human timekeeping was made in the early 21st century, the reasoning that was followed should be obvious, credible, and justifiable. Yet the study process in place is opaque and the motivations for UTC redefinition are not absolutely clear. For example, the ITU-R authorized a Study Question which decided that any recommendation for change should be based on a determination of user requirements, but that determination was never concluded. Some issues might never be acknowledged publicly; nevertheless, studies by the government agencies have generated tomes of public documentation for much less consequential questions. One would therefore expect voluminous documentation exploring the motivations, impacts, and repercussions across affected communities that one might normally expect out of a decision-making process. However, that documentation apparently does not exist, at least in the public domain, and therefore the information on which national governments are basing their decisions is unclear.

Because the existing process within the ITU-R has not clearly determined end-user requirements, organizations outside this process have attempted to poll for this information. Over the past decade, published survey results have expressed high favor for the *status quo* and rather low regard for the alternative of ceasing leap seconds. But survey attempts stimulate controversy as a means to collect opinions, inform people, and create awareness, and survey results have been largely dismissed by those who feel that such polling provides no useful information. Also, most surveys have been conducted "informally" to avoid the difficulty of gathering more official positions, so the opinions of decision-makers who must also consider the financial aspects may not be reflected. It is not clear that official decision-making processes within governmental organizations have been sufficiently transparent or that the most affected communities have registered substantive input.

THE CONSEQUENCES OF DECOUPLING

Impacts on Hardware

Celestial tracking and data acquisition are usually accomplished via multiple systems in which timekeeping signals are exchanged via complex messaging protocols. Telemetry is logged and maintained over many years, and provided to diverse users for different technical purposes which are often coordinated with other tracking systems. The relevant usable time intervals for such data may range from fractions of a second to decades (or longer). Tracking systems assuming UTC ≈ UT1 would need to be re-implemented to explicitly distinguish between UTC and Earth rotation (UT1), or isolated to receive a vetted UT1 input, or be retired or replaced. Systems already accommodating a UT1-UTC correction would have to be vetted for proper operation given values of |UT1-UTC| > 0.9s, possibly from a new source. Even systems unaffected by a decoupling of UTC and UT1 would have to be inventoried and assessed (with significant expense) to prove this.

Impacts on Software

Only a minority of software properly accounts for leap seconds, including software for real-time systems; nevertheless, it is still an outstanding question as to how much and to whom this matters. It is possible that the affected codebases of systems assuming UTC ≈ UT1 are much smaller than those that do not make such an assumption, and it is also possible that there are substantial costs for maintaining leap seconds. Yet, each of these possibilities argues for software inventories that have yet to happen. Some preliminary inventories (*e.g.*, for astronomical software systems) forecast major complications should UTC no longer closely approximate UT1.

There is agreement that timekeeping software must be more flexible than in the past, but what this means depends on one's viewpoint regarding UTC redefinition. For those who favor the decoupling from Earth rotation, "flexible software" should not assume that UT1-UTC is bounded in the future. For those respecting the current system, "flexible software" maintains the necessary infrastructure to accommodate small intercalary adjustments past, present, and future.

Would the decoupling of civil timekeeping from Earth rotation result in the development of more or less "flexible" software for the future? The cessation of future leap seconds does not mitigate the need to account for past leap seconds in software, so any argument that conventions must change to continue with badly written software is unpersuasive. Convenience to software programming is a weak reason for denying our progeny civil time linked to the astronomical day; programming mistakes will be made regardless of whether civil time is coupled or not to Earth rotation, so the future of civil timekeeping should not hinge on the anticipated incompetence of some programmers now or in the future.

Example Application: Impacts on GNSS

A convincing case can be made that existing global navigation satellite system (GNSS) programs would not be benefited, but would be financially impaired, by UTC redefinition. These systems already function through leap seconds and redefinition of UTC would require changes to operational software, established procedures, documentation, and testing. Such systems are relatively inflexible due to the criticality of their missions, such that seemingly inconsequential changes can be costly. Such costs would be unavoidable and borne by the taxpaying public.

Technical Confusion

Polling suggests that people are apprehensive about changing something that has worked for a very long time, and the one thing that is generally understood by people with any sensibility of timekeeping and time scales is that UTC is approximately UT1. A change in the meaning of time is therefore guaranteed to cause confusion, even in situations where systems do not malfunction. Service providers cannot control how users interpret standards or whether they are aware of any changes to implicit operating assumptions.

The proposed decoupling poses a special challenge to providing future data in software and almanacs that are a function of the rotational angle of the Earth. If data continue to be provided as a function of Universal Time, an intensive re-education strategy will be required. Users would thereafter need to clearly understand the difference between UTC and UT1 and know how to obtain and apply UT1-UTC from external sources. For example, even if the impact of dropping leap seconds were manageable for practicing celestial navigators through re-training and future almanac updates, a greater risk arises from the potential for confusion in literature, textbooks, and other educational materials already published for a subject that is a rarely exercised backup.

The creation of accurate technical documentation is a significant expense that is often neglected or deferred. The necessary documentation changes would disturb technology domains that are otherwise unaffected by a redefinition of UTC; therefore, the documentation aspects are likely to be very far reaching and extremely tricky to manage. A redefinition of a fundamental scientific time scale cannot avoid leaving permanent signatures in time-series taken before and after the change; this also would be a source of continuing confusion into the future. The technical definition of an atomic time scale called "UTC" first with, and then without, leap seconds will be complicated. Civil standards that do not approximate Universal Time would best avoid the label "Coordinated Universal Time" and its acronym "UTC", because these descriptions have always implied a technically useful realization of Universal Time.

Societal Confusion

The astronomical basis of timekeeping fulfills its social function even when it is not perfectly observed. This is because general perceptions are what count within the realm of symbolism, not just the facts. Many people would perceive a decision to decouple civil timekeeping from Earth rotation as an exercise of authoritarian power that no elite scientific or technical group should have been able to influence. Even if the effects of such a change were not immediately apparent or very gradual for most purposes, individuals would still be upset and would talk about the decoupling from Earth rotation as if timekeeping were already broken.

Educational Confusion

There is remarkable public fascination with the notion of leap seconds which can be leveraged to promote public awareness and emphasize the relevance of higher education in science, technology, engineering, and mathematics. Explanations of leap seconds often involve the Moon, tides, and gravity, making the topic a rewarding introduction into the domain of solar-system dynamics and its direct relevance to people on Earth. Leap seconds are an effect that an ordinary person can observe themselves, because a precise watch will reveal the one-second difference after a leap second compared to a GNSS receiver or a radio-controlled clock. Similarly, sundials are wonderful objects to teach the seasons or the inequality of the days, Kepler's laws, or the practical application of trigonometry. Therefore, from a pedagogical point of view, the decoupling of civil time from that kept by the Sun can be seen as an additional complexity that would not be easy to explain to pupils and the general public.

Legal Confusion

Some national governments recognize the astronomical basis of timekeeping explicitly, while others recognize it implicitly. The existing system is the only one known to be reconcilable with all existing national and international statutes, and a civil timekeeping standard decoupled from Earth rotation has no known precedent. Detailed consideration of various legal implications appears to be lacking; for example, future disruptions caused by adjustments larger than one second could spawn many legal complications.

DISCUSSED ALTERNATIVES

Alternatives to maintaining leap seconds have been proposed over the years, but options that keep civil time coupled to Earth rotation have been largely dismissed, offering no improvement over the existing functional system. Various options have included: relaxing the tolerance for |UT1-UTC|, redefining the *SI* second, and inserting required adjustments at less frequent but regular intervals.[3] Variations on these alternatives were discussed at the colloquium, although these variations do not represent all possible options.

Predictions from a Quadratic Model

A low-order (quadratic) model of the separation of Earth rotation (UT1) from uniform time could be used to make predictable adjustments to the calendar decades or centuries in advance. These corrections, based on a computational rule rather than empirical observation, would not necessarily be accurate enough to satisfy technical applications requiring UT1 from clocks, but such corrections would facilitate long-term software development and maintain a general perception of astronomical coupling. However, several concerns were noted with this approach.

1. Decadal variations can be significantly larger than the long term quadratic trend, making a quadratic model too inaccurate for the intended purpose in the near term.

2. Although extended predictions of UT1-UTC might help some software applications, software issues are primarily caused by complete ignorance of intercalary adjustments, rather than a need for their predictable forecast.

3. Having a predictable leaping rule for time of day still does not necessarily solve the software issues. (For example, programmers sometimes code the Gregorian calendar rules incorrectly.)

4. Loosening the existing ±0.9-second tolerance specification for UT1-UTC will still result in operational issues.

Regardless, such a proposed approach begs for continued geophysical studies to make Earth-orientation predictions more reliably accurate.

Use TAI Internally and Expose *Status-Quo* UTC Externally

It is clear that many astrodynamics, astronomical, navigational, and telecommunications applications already use this option operationally, or a variation of it (*e.g.*, GPS time in lieu of TAI seconds). It seems reasonable to promote a conventional sequence of TAI seconds without leap seconds as an internal scale, and as a source of determining precision time interval whenever necessary, as all modern scales are already functionally related to, or approximate the rate of, TAI. Prior to 2001, the internal use of TAI was recommended by the Director of the BIPM, the Consultative Committee on Time and Frequency (CCTF), and by the ITU-R through multiple Recommendations. The use of TAI-like time has been mainly hindered by the lack of transmission of DTAI = TAI–UTC, but the availability of GNSS scales through navigation signals may be overcoming this limitation. A technically interesting approach to implementing this method involves the synchronization of a computer's operating kernel to a uniform background scale and tracking the difference between UTC and the background scale (*e.g.*, TAI or GPS) using the widely deployed Time Zone Database.[*]

Increasing the Intercalary Interval

The idea of making the necessary adjustments at much longer intervals, for instance at the end of each century, is appealing in that most people do not live that long and therefore would have to

---

[*] http://www.iana.org/time-zones

deal with the problem either once or never. However, the following drawbacks of this proposal were noted:

1. New software is being written all the time so it is not simply a once-per-century matter. Software developers of the future are just as likely to focus on more immediate problems and may not expect their software to still be in use a few decades into the future, although such software might very well become entrenched as other code is built around it.

2. There is a real risk that any long-term adjustment scheme will not be reliably implemented when the declared time comes because long-term proposals such as a leap minute, leap hour, centennial adjustments, *etc.* push the technicalities sufficiently far into the future such that the recoupling would not be pragmatically addressed.

3. Future large adjustments still have the problem of labeling events during the adjustment interval in an atypical way, just as with leap seconds, except their introduction will be harder to ignore.

4. A change from the current standard will still result in immediate operational issues for some.

Perhaps the most surprising notion mentioned was whether the current proposal implied that humanity was supposed to wait until |UT1-UTC| approximated one day, at which point an unscheduled leap day would be introduced into the calendar. Such a scenario admits that there are really only two types of calendar adjustments presently, leap seconds and leap days, and abandonment of the smaller adjustment leaves humanity with only unrealistic and impractical options employing the larger, which is insufficient to address fundamental issues. For example, the current proposal to decouple creates a potential civil issue where a Saturday could turn into a Sunday; however, the seven-day week is a universally accepted convention based on mutually agreed tradition and is important to millions of faithful across the globe as well as the general populace.

Other Options Not Discussed

Feedback regarding adverse impacts from redefining Coordinated Universal Time was a primary goal of the meeting, which is a prerequisite to entertaining alternate options. However, it is not the only prerequisite, such that additional discussions, public meetings, and engineering-planning activities surrounding these issues should be encouraged for the future.

RECOMMENDATIONS FROM COLLOQUIUM DISCUSSANTS

No attempt was made to formalize any consensual recommendation(s) of the attending group because there was no specific professional, technical, or governmental entity requesting such recommendations from this gathering. Nevertheless, several suggestions were advanced by various colloquium attendees, some of which are now highlighted.

Reconsideration of the Issue

Action by the ITU-R at this time would seem unwise based on the potential widespread impacts of their decision and the lack of documentation supporting the need for a change. Therefore, a delay in any official decision is advisable. Consideration of civil timekeeping issues or adoption of UTC stewardship by other international entities or standards organization should be explored. The coherent collection of requirements and application of systems engineering best practices should provide a framework for reaching consensus.

Contact with National Governments

Colloquium participants recognized that a decision to change the *status quo* would impact various communities outside telecommunications. Responsible person(s) within government departments should be made aware that there is much more at stake than ramifications and consequences to telecommunications.

Transmission of TAI-UTC

DTAI (TAI-UTC) should be made more readily available. Transmission of DTAI would allow it to be received autonomously, avoiding haphazard changes by hand which is burdensome for systems where configurations are tightly controlled. Prior leap seconds are part of the historical record and whether or not UTC is redefined, a database of intercalary adjustments will be perpetually required for some applications. Access to the latest entry of this database is equivalent to "transmission of DTAI", and this requirement now persists however UTC is defined in the future.

Distributing the Offset between Civil Time and Earth Rotation

If Coordinated Universal Time is redefined to no longer be coordinated with Universal Time, the user base interested in UT1-UTC corrections would likely increase suddenly. For this reason, there must be some means of ensuring practical distribution of UT1-UTC, especially for machine usage. Suitable standards-based technologies exist but these must have a long expected lifetime, be practically implemented by both the producer and the consumer, and maintain system integrity and data security. An XML-based service is one possible approach.

Methods of distribution should start out as simple as possible, such as file access via http, and care must be exercised to avoid proposed solutions that cause unnecessary complications. For example, the idea of introducing an NTP service of variable frequency that attempts to track UT1 (as UTC did before 1972) might create unnecessary complications. Generally, it seemed that transmission of UT1-UTC corrections would be preferable. The IERS Directing Board should be involved in this issue, although the IERS is primarily a confederation of separately funded geophysical research centers rather than a transmission center for timing products. The recommendation to make UT1-UTC more readily available stands independently of whether civil timekeeping is decoupled from Earth rotation, because many operations need a realization of Earth rotation that is more precise than provided by civil clocks.

Extended Prediction of Leap Seconds and UT1-UTC

If leap seconds were retained, it seems that the IERS may already be able to confidently predict leap seconds two or three years in advance. Increasingly advanced prediction of UT1-UTC would be a good option for almanacs and other publications that are physically printed, regardless of whether civil timekeeping is decoupled from Earth rotation.

Development of Specialized Analysis Tools

If Earth rotation and civil time are decoupled, the development of specialized simulation tools would be necessary to provide some sense of the adverse operational impacts or to simulate the magnitude of symptoms caused by decoupling. There also seems to be a need for specialized diagnostic tools to assist in discovering where UTC is used as a realization of Earth rotation within complex operational systems and software.

Engagement of Additional Stakeholder Communities

Civil timekeeping is an issue of interest to all and efforts should be made to broadly engage possible stakeholder communities. If UTC is redefined, additional meetings should be organized for the future, some of which perhaps having narrower focus on more topical issues (astronomy, astrodynamics, *etc*.) to address the technical challenges that will result from the decoupling of civil timekeeping from Earth rotation.

CONCLUSION

While numerous points of view were well expressed by the contributors of this colloquium, a few points stood out in the opinion of the chairmen. Primarily, the motivations for decoupling civil timekeeping from Earth rotation are not entirely apparent, and the supposed advantages to making a change have not been shown to outweigh the supposed disadvantages both now and in the future. A coarser coupling between civil timekeeping and Earth rotation would result in adjustments of less frequency and larger magnitude, which would be more noticeable and less practical to maintain than the current system having predictable insertion points (at end of the month), constrained values (-1, 0, or 1), and a prescription for tagging events occurring during the adjustment (23:59:60). Because the existing UTC system with leap seconds is already implemented, it should be strongly preferred over alternative proposals or protocols that lack obvious advantages.

The historical tendency of timekeeping practices has been to move from empirically observed adjustments to predictable ones based on calculation. Perhaps the creation of a predictably accurate time-of-day adjustment algorithm, valid for at least a decade or more, might provide a viable alternative, but this has not been seriously studied up to the present. It appears that Earth rotation may now be predictably accurate to within one second out to three years with 99% confidence. If so, perhaps we should continue to seek advances that would improve the prediction of Earth rotation with high confidence out to one decade or better, with the goal of eventually replacing empirically predicted leap seconds with an algorithmic correction to time-of-day. Such an approach acknowledges that the coupling of civil timekeeping and Earth rotation is fundamentally an issue of calendar maintenance, and thereby satisfies public expectations of preserving astronomical relationships. Yet, because this approach may not be able to provide a sufficiently accurate realization of Earth rotation from clocks, it seems best to leave the current system in place until such time that Earth rotation models are improved, and for systems that prefer a uniform internal time scale to use IEEE 1588-2008 (Precision Time Protocol) with GNSS time or other time scale closely tied to TAI.

Finally, an outstanding question is whether "UTC without leap seconds" really represents a permanent scenario for decoupling. A persuasive argument can be made that the expectation of civil-timekeeping's astronomical basis is so deeply ingrained in our society that a departure from it is unlikely to last forever. If so, an attempted decoupling now may be the worst of all possible options, because it creates issues for both present users reliant upon the existing system and it causes unforeseeable complications for generations to come. The want of the current decoupling proposal to address the most fundamental issues erodes its favorability within many technical communities and the public, resulting in a lack of consensus.

REFERENCES

[1] Finkleman, D., S. Allen, J.H. Seago, R. Seaman and P.K. Seidelmann (2011), "The Future of Time: UTC and the Leap Second." *American Scientist*, Vol. 99, No. 4, p. 316.

[2] Finkleman, D., J.H. Seago, P.K. Seidelmann (2010), "The Debate over UTC and Leap Seconds". Paper AIAA 2010-8391, from the Proceedings of the AIAA/AAS Astrodynamics Specialist Conference, Toronto, Canada, August 2-5, 2010.

[3] McCarthy, D.D., W.J. Klepczynski (1999), "GPS and Leap Seconds—Time to Change?" *GPS World*, November, pp. 50–57.

Session 1:
SETTING THE STAGE

AAS 11-661

SYSTEMS ENGINEERING FOR CIVIL TIMEKEEPING

Rob Seaman[*]

The future of Coordinated Universal Time has been a topic of energetic discussions for more than a dozen years. Different communities view the issue in different ways. Diametrically opposed visions exist for the range of appropriate solutions that should be entertained. Rather than an insoluble quandary, we suggest that well-known systems engineering best practices would provide a framework for reaching consensus. This starts with the coherent collection of project requirements.

INTRODUCTION

One can speculate on the origins of human nature and our capacity for creative problem solving. Presumably there are strong evolutionary pressures as represented in well-worn phrases like "look before you leap" and "necessity is the mother of invention". The most fundamental observation about how humans address complex challenges is perhaps that we almost inevitably seek to think about potential solutions before completely analyzing the problem in front of us. This instinctual cognitive short circuit may have had survival value when evading predators on the savanna, but serves us less well in the setting of a complex technological civilization.[1]

Civilization's infrastructure has another widespread characteristic shared with biological evolution. Each generation of technology is derived from the one before. Our clocks are sexagesimal because our parents' and grandparents' clocks were, all the way back to the Sumerian[†] concepts of time 5,000 years ago. The combination of these two effects motivates technological solutions that are highly stable yet exhibit sudden phase transitions as the prior paradigm may be replaced entirely from one generation to the next.

Finding a robust consensual solution to an esoteric engineering problem thus requires navigating between the extremes of instinctual snap judgments and inherited tradition. *Systems Engineering* is the discipline that provides tools for managing this challenge, to reduce the brittleness of the decision-making process and to promote solutions that multiple stakeholders with diverse requirements and worldviews can jointly accept. Systems Engineering does not seek the "best" solution, rather it seeks a family of satisfactory solutions whose benefits and shortcomings can be traded-off against one another using unbiased techniques balancing cost, schedule, performance and not least, the risks of each proposal.

---

[*] Senior Software Systems Engineer, National Optical Astronomy Observatory, 950 N Cherry Ave, Tucson, Arizona 85719, U.S.A.

[†] http://www.nytimes.com/2010/11/23/science/23babylon.html

REQUIREMENTS VERSUS SPECIFICATIONS

One such esoteric engineering challenge is *civil timekeeping* – seemingly simple, but deceptively complex. Civil timekeeping provides another example of how humans grapple with problems. When contemplating an issue, we will often fixate on one small aspect of the situation and ascribe special importance to it. In the case of civil timekeeping this is the "leap second".

The problem of civil timekeeping is not leap seconds, rather the *problem* is how to meet a wealth of engineering requirements that includes aspects of atomic timekeeping and Earth orientation. The current *solution* to the problem of civil timekeeping happens to include the issuance of leap seconds. Ceasing leap seconds will not and cannot make the underlying engineering requirements go away. Requirements are just that – required. There are only three ways to handle any engineering requirement:

1. satisfy it,
2. issue an explicit lien against the requirement, or
3. reach consensus among the stakeholders that it is not, in fact, a requirement.

What is a requirement? It is not a specification. Requirements are inherent in the problem concept, whereas a specification describes some aspect of a proposed solution. A requirement is independent of human foibles and opinions (other than as humans themselves enter as part of the problem description). One **discovers** requirements. They exist independently of the observer. A requirement is also not a preference and is not something to be prioritized. Each solution that is proposed must address every requirement.

An example may help. This section is being written on an airplane. The problem concept may be stated as a flying vehicle. Two things among many that are required are lift and thrust. Wings, however, are not required. Wings are specified as a component of one class of solution, airplanes. But neither helicopters nor balloons have fixed wings. They do, however, develop lift. Jet engines are also not required. Helicopters can use their rotors to develop both lift and thrust, coupling those requirements in this class of solution, whereas hot-air balloons rely on the wind for their thrust. (A tethered balloon is not a vehicle, although its passengers may be said to fly.) That a dirigible blimp develops thrust reveals the separability of these two requirements in the more general lighter-than-air case. Lift is required else the vehicle could not be said to "fly", an element of the problem concept, but wings are only specified, not required. Note that this is not a value judgment – we passengers find wings very important nonetheless. But wings are important only because they provide a way for the vehicle to meet the requirements of air travel.

In the case of civil timekeeping, leap seconds have been specified as part of the currently codified and deployed solution.[2] What then is required? We will discuss this below.

SYSTEMS ENGINEERING BEST PRACTICES

The intent of this document is not to lecture on systems engineering. Rather, the goal is to highlight a few central concepts of systems engineering as they may apply to the discussion of the future of Coordinated Universal Time. In particular, it is to advocate for a more coherent community-wide examination of this issue and the implications of the different outcomes.

What is *systems engineering*? It is one organized method for reaching decisions when presented with complex challenges, convoluted environmental constraints, and diverse stakeholders. Other methods include such things as "Creative Problem Solving,"[3] "Principled Negotiation,"[4] and in a narrower context, software development processes (for one among many examples).[5]

More than a decision-making protocol, systems engineering works with project management techniques to actually accomplish the fruits of those decisions.

Systems Engineering Management Plans

Systems engineering is a type of planning discipline. In a formal exercise of these techniques that plan is always written down. The document format varies widely, but maps to the structure of the particular process being followed (Figure 1).

	Name	Order in which the documents are started	Order in which the documents are finished
Doc. 1	Problem Situation	1	8
Doc. 2	Customer Requirements	5	2
Doc. 3	Derived Requirements	6	4
Doc. 4	System Validation	7	5
Doc. 5	Concept Exploration	3	3
Doc. 6	Use Case Model	4	1
Doc. 7	Design Model	8	6
Doc. 8	Mappings and Management (schedules)	2	7

Figure 1. A Systems Engineering Management Plan (SEMP)[6] comprises multiple documents (sections) and is intended for diverse stakeholders.

In this process it is explicitly recognized that the order in which the documents are read differs from the order in which they were written, and in fact that the documents are started in one order but finished in another. One can quibble over the details of any formal process, but rather consider some of the general features expressed in this table. The first document to be started is a description of the *Problem Situation*, which is to say an expression of the underlying nature of the system in question. Perhaps surprisingly, this is the last document to be finished. The idea here is that throughout the engineering process – whatever process – the understanding of the problem space is extended and deepened. The Problem Situation is also the first document to be read. Whether one is conceptualizing and planning a system or rather investing in understanding someone else's plans, the place to start is with a description of the problem space.

At the far end of the writing process lies the *Design Model*, that is, an expression of the solution that has been refined and selected as proposed by the plan. This is also near the end of the plan documents as read by stakeholders – only the project schedules, budgets and other supporting documents follow, perhaps as an appendix. In short, the structure of the document formalism logically separates the characterization of the problem from its proposed solution(s).

In between the statement of the problem and the proposed design for a solution lie a variety of documents and activities that capture the conceptualizing invested in the engineering process. In the example of the specific process illustrated in Figure 1, these include descriptions of the *Customer* and *Derived Requirements* (sometimes called the *non-functional* and *functional requirements*) and *Use Cases* that capture the desired behavior of the system. *System Validation* is performed – does the system satisfy the requirements? Depending on the complexity and breadth of the problem being addressed, these documents may be quite extensive and detailed, in particular the *Concept Exploration* activities may involve significant research. But in even the simplest system the questions posed remain the same:

1. What is the problem at hand?
2. How must the system behave?
3. What do the stakeholders require?
4. How will compliance be ensured?

All Engineering Processes Share Fundamental Characteristics

Figure 1 expresses just one of many possible systems engineering processes and formal planning techniques. Perhaps there are others that proceed in a different order and comprise different sorts of documents and underlying activities? Rather, a broad survey of systems engineering practitioners has demonstrated that all widely used such processes are generally equivalent. The mnemonic that captures this notion is *SIMILAR*,[7] which maps well onto many other system development processes:

- **S**tate the problem
- **I**nvestigate alternatives
- **M**odel the system
- **I**ntegrate interfaces
- **L**aunch the project
- **A**ssess performance
- **R**e-evaluate your solution

For the purposes of this paper, we need not delve too deeply into the details of this equivalence, but note again the clear separation between 1) effort invested in understanding the problem space, and 2) work to design a solution that is compliant with the engineering requirements so discovered. This mnemonic captures another key aspect (in "Re-evaluate your solution") of systems engineering best practices, namely that engineering is an iterative exercise during all phases from design and development to operations and maintenance. In particular, an assertion that a particular solution will remain viable for all time is rarely appropriate. Operating on this assumption would guarantee widespread future disagreements and the frequent revisiting of supposedly settled civil timekeeping issues.

REQUIREMENTS OF CIVIL TIMEKEEPING

How can these techniques be applied to the problem of civil timekeeping? This brief paper cannot hope to capture the richness of an appropriately detailed Systems Engineering Management Plan. However, perhaps a bit of clarity can be achieved in the stating of underlying concepts to inform future discussions.

Consider again Figure 1, in particular, the order in which the planning documents (or sections or chapters) are finished. It was previously pointed out that the final document finished, the *Problem Situation,* is also the first one started, but what are the **first** documents finished? In this formalism it is the *Use Case Model* and the *Customer Requirements*, that is: What is the consensus among stakeholders for how the resulting system must behave and for how people will interact with it? In the case of the frequently zealous civil timekeeping discussions over the past several years, this is precisely the initial phase of the discussion that has been omitted. By focusing exclusively on proposed partisan solutions, opportunities for a durable consensus were missed. Time was lost that could have been better spent in engaging with stakeholders and in characterizing the common problem. It may even be that some variation of one of the proposed solutions could satisfy the engineering requirements—but how would this ever be discovered if the requirements are never written down, and if the stakeholders are never consulted for comment?

No solution will be proposed in this paper. The author is confident that a consensus solution can be found that is appropriate to our time and for our purposes. Instead of solutions, several requirements on the problem of civil timekeeping will be asserted. Indeed, a simple statement of faith in the existence of an appropriate solution to the problem leads immediately to discovering the first such engineering requirement:

> A. *Forward as well as backward compatibility must be preserved.*

Numerous discussions have pointed out the long-term quadratic behavior of the tidal slowing of the Earth's rotation.[8] No matter what mechanism is implemented now or in the future this will always be a characteristic behavior of civil timekeeping on Earth. The preservation of continuity over transitions from one civil timekeeping standard to the next is a clear engineering requirement. Remember, a requirement must either be satisfied – or an engineering *lien* must be placed explicitly on the requirement. A lien does not mean that the requirement can be ignored, rather the opposite. A lien means that an explicit mitigation plan exists to deal with the implications of not having met the requirement. Proposed solutions that are brittle over the long term are significantly disadvantaged by this requirement. In fact, implementing a lien is often more onerous and expensive than meeting the requirement in the first place, especially over the long term.

Engineering Vocabulary for Civil Timekeeping

The debate pertaining to a possible redefinition of Coordinated Universal Time (UTC) has been ongoing for more than a dozen years.[*] Unfortunately, these discussions have been less successful than they might have been. One reason is a lack of outreach – consider that the ITU-R has only held one public meeting on this topic the *Colloquium on the UTC Time Scale*[†] in Torino Italy, 28-29 May 2003. A more fundamental reason is a lack of shared vocabulary for constructing the discussions themselves. This needed vocabulary includes highly technical jargon, but starts with the most basic building blocks of the debate.

As this discussion continues, the language of systems engineering can make the relationships between frequently bandied terms clearer:

- *Problem statement*: civil timekeeping
- *Proposed solution*: UTC
- *Feature of solution*: intercalary steps

That is, we are asserting that timekeeping for civilian purposes comprises a single coherent problem space. Designing and deploying one-or-more timekeeping standards and systems will provide a solution to this problem. The current solution to this problem is Coordinated Universal Time. UTC meets the system requirements via intercalary steps,[‡] currently in the form of leap seconds (whether these are actual leaps or merely a representational overlay on TAI).

---

[*] http://groups.google.com/group/sci.astro.fits/msg/9326ad192333a560
[†] http://ucolick.org/~sla/leapsecs/torino/
[‡] Are intercalary steps *required*? No, they are an emergent property of the problem space. The current UTC standard happens to specify leap seconds, but if it were to be redefined to embargo these, the intercalary steps would continue to accumulate until some other accommodation is made to make the equivalent adjustment. It is the attempt to ignore this requirement that results in its being reasserted. If an alternate mitigation strategy is sought that has no intercalary steps, then the requirements must be faced head-on.

Two key vocabulary words may be understood through examples:
- *Requirement*: civil time approximates mean solar time (see below),
- as opposed to a *Specification*: leap seconds.

Systems engineering terminology can help to clarify talking points and avoid confusion and delay. This confusion started at the very beginning when the issue was characterized as the "problem with leap seconds." As has been explained, leap seconds are not the problem in the engineering sense. It goes further than this, however: the problem is not even UTC. Rather, the problem is *civil timekeeping* itself – that is, the worldwide system of timekeeping suitable for civilian purposes.

By focusing on leap seconds and UTC, the actual engineering requirements of civil timekeeping are obscured and become confused with requirements for other types of timekeeping. This is a second engineering requirement:

    B. *There is more than one kind of time and thus of timekeeping.*

Note an immediate benefit from clearly stating the problem and its requirements; namely, further discussions can steer clear of things that are not part of this particular type of timekeeping. Clocks come in many forms, but some kinds are more directly pertinent to civilian use than others.

The Stakeholders are Many

While the author strongly recommends systems-engineering best practices as a solid basis, readers serve themselves well by asking these three questions, whatever their decision-making process:

- What is civil timekeeping?
- What is required to implement a worldwide civil timekeeping system?
- Who are the users of civil timekeeping?

Let us focus now on the third question. The short answer is that there are seven billion civil timekeeping users. Focusing disproportionately on leap seconds serves to artificially narrow the list of stakeholders. Teasing this one thread out of its proper broader engineering context leads to the omission of affected communities, rather than their inclusion. User exclusivity simply does not arise when the true breadth of the problem space and its stakeholders is realized, leading to the next requirement:

    C. *Timekeeping users are ubiquitous.*

There is certainly not room at the table for seven billion seats. But there should be room for representatives from many more communities than currently consulted by the narrow interests of the Radiocommunication Sector of the International Telecommunication Union.[9]

This requirement warrants a bit more discussion. The short answer was that there are seven billion stakeholders in civil timekeeping. The longer answer is that very few will have enough familiarity with timekeeping issues to participate in detailed discussions. This is where systems engineering techniques truly shine. A formal but lightweight engineering process is particularly useful for structuring the gathering of requirements from diverse stakeholders with varying levels of knowledge and expertise. In fact, the alternative to using such techniques is precisely to exclude such stakeholders from participating in the decision-making process at all.

The Use Cases are Diverse

Civil timekeeping cannot be all things to all people, yet it must serve many purposes. The central problem with failing to follow a coherent process for collecting engineering requirements is that communication about the issues turns into an exercise in imagination. One party or another will indicate that it knows what is "best" for all purposes and all users. By treating requirements like preferences, some narrow aspect of the problem will be asserted to have a "higher priority" than all others. The reality is that the diverse and numerous requirements must all be satisfied at the same time.

A pertinent quote is attributed to Albert Einstein: "Make everything as simple as possible, but not simpler." To force a particular solution is to try to make things simpler than they really are. In the case of civil timekeeping, sometimes clock readings are needed to measure time intervals, and sometimes clock readings are needed to indicate time of day. Both aspects are inherent in the many use cases that pertain to civil time.

What are *use cases*? They are a broad selection of necessary behaviors that capture a sufficiently complex picture of how the system behaves across the community of stakeholders. This doesn't mean that all possible usages of clocks need be captured or should be given equal weight if that were even possible. But it does mean that any overly narrow list of clock functions will fail to meet the challenge.

Paraphrasing Einstein, the requirement on the complexity of timekeeping is:

> D. *Make clocks as simple as possible, but not simpler.*

In modern engineering processes it is predominantly use cases that are used to drive the discovery of requirements. A broad universe of civil timekeeping behaviors will inform a lengthy, but by no means vast, list of use cases. These use cases will include technical, scientific, cultural, legal, religious, philosophical, historical, logistical, economic and other aspects. With a system as widespread and fundamental as civil timekeeping, the list is indeed long. Similarities will become evident among diverse classes of usage. The major underlying engineering requirements will be relatively few in number—this paper makes a start.

We Live on a Planet Lit by the Sun

Some facts are so well known that they become invisible; in that case we may say that we "can't see the forest for the trees." On planet Earth nothing is more blatant than the Sun in the sky.[*] We know it is there even on a cloudy day. Somewhat more shy is the Moon, but we could infer its existence from the tides even without the ability to make a single celestial observation. It is unsurprising that the star that we orbit and the satellite that orbits us both affect the nature of timekeeping on Earth.

---

[*] The presentation corresponding to this paper at the colloquium *Decoupling Civil Timekeeping from Earth Rotation* included comparisons between timekeeping on the surface of the Earth and on Edgar Rice Burroughs' worlds of Pellucidar (at the Earth's core) and Barsoom (Mars). The issue is not science fiction – the issue is how humans would behave under those circumstances.

The alternative is to suggest that our clocks could be set to any hour of the day whatsoever. In this case the phrase "hour of the day" loses all meaning and the date itself becomes wholly arbitrary, rather than something delimited by intervening periods of nighttime. Readers might ask themselves why, exactly, the SI-second was chosen not only to have the same *name* as the second of sexagesimal time and angle representations – but also was specified to *approximate* the fraction of 1/86,400th part of a mean solar day. If the SI-second had been chosen even so slightly different, as to be the 1/86,399th part or the 1/86,401th part of a day, there would have to be a leap second every day or a leap hour every decade.[*]

The question to ask here is not "How would we solve this?" The first question to ask is "What is the underlying engineering requirement?" That requirement is:

> E. *Civil time must closely approximate mean solar time.*

Understanding what it means to say this is a requirement is the heart of the exercise in reaching consensus on how to move forward with civil timekeeping. A more accurate statement is: "Civil time *is* mean solar time", because this is really just a definition of terms. However, the notion of approximation is inherent in all engineering trade-offs. Remember, systems engineering does not seek the *best* solution, it seeks a family of *satisfactory* solutions whose cost, schedule, performance and risks can be traded off against each other. Trading off, however, does not mean discarding requirements, but rather a rebalancing of a proposed solution's accommodation of the requirements.

Technology Evolves Rapidly

By focusing on putative solutions rather than actual engineering requirements our options narrow to overemphasize currently perceived hurdles, especially of technology. For instance, the jargon word "POSIX"[†] is very rare in everyday speech, but very frequently heard on the so-called LEAPSECS[‡] mailing list (so-called because its name would more accurately reference UTC, rather than the narrow issue of leap seconds). Timekeeping standards are – or at least, should be – much longer lived. Greenwich Mean Time[10] (GMT) extends back to the Nineteenth Century[11] before computers and airplanes, automobiles or artificial satellites. Coordinated Universal Time[12, 13] (UTC) as an approximation to GMT[§] has been a standard for four decades, predating the Global Positioning System (GPS) and smart phones. The corresponding engineering requirement might be stated as:

> F. *Civil timekeeping must support technologies not yet invented.*

A corollary to this is that we should seek instances of long-lived technologies as especially valuable exemplars when considering similarly long-lived standards. In the case of civil timekeeping one such technology is precisely that of the astronomical telescope, which has recently celebrated its 400th anniversary. One should view the undisputed fact that redefining UTC would require rewriting large amounts of astronomical software as the "canary in the coal mine" for civil timekeeping.[14]

---

[*] By contrast, if the SI-second had been selected to be a value completely disjoint from this fraction of the mean solar day, the underlying laws of physics would retain exactly the same form. The overriding engineering requirement is supplied by those facts of nature that cannot be easily modified (mean solar time), not by those things that can (physics works out the same in the end).
[†] http://standards.ieee.org/develop/wg/POSIX.html
[‡] http://six.pairlist.net/mailman/listinfo/leapsecs/, messages from 2000-2007 at http://www.ucolick.org/~sla/navyls/
[§] Per CCIR Recommendation 460-4, "GMT may be regarded as the general equivalent of UT."

History is long

Astronomy is an ancient pursuit, but so are clocks and calendars.[15, 16] Any proposal to redefine the practice of civil timekeeping should be informed by related historical events. One particularly apt analogue was the Gregorian calendar reform.[17] Programmers will attest that calendar-related bugs continue to turn up in their software. This is not an argument to eliminate the Gregorian calendar as a common standard spanning all the world's cultures from centuries past to centuries yet-to-come. History does extend into the future.[18] Current timekeeping deliberations should take special care to properly value long-term requirements. This leads us to the final requirement to be asserted here:

    G. *We cannot un-ring the bell.*

Another way to say this is the Hippocratic dictum "*First, do no harm.*" A re-engineering effort that properly acknowledges the underlying system requirements retains the freedom to evolve further in the future. By choosing a new solution that meets the same coherent set of requirements as the original solution, a lightweight transition is effected that would also permit transitioning back to the *status quo* as necessary. Since the same underlying conceptual model is implemented, it preserves a "paper trail" for interpreting dates and times across the transitions.

On the other hand, if an abrupt transition is made to a completely different conceptual model then the signature of this discontinuity will be preserved forever. The interpretation of dates and times will require software and hardware that encapsulates both the old (*UTC is time-of-day*) and new (*UTC is atomic time*) rules. Since the new model is overly simplistic, any future transitions (guaranteed to occur by the ever-accruing embargoed leap seconds) will involve a similarly large **conceptual leap**. Which is to say that a proper systems engineering plan includes a discussion of risks and how these might be mitigated.

WHEN IS A SOLUTION "GOOD ENOUGH"?

This paper has served not only as a brief introduction to high-level systems engineering concepts, but also as an exercise in requirements discovery for civil timekeeping. The question naturally arises of what to do with the requirements so discovered:

    A. Forward as well as backward compatibility must be preserved.
    B. There is more than one kind of time and thus of timekeeping.
    C. Timekeeping users are ubiquitous.
    D. Make clocks as simple as possible, but not simpler.
    E. Civil time must approximate mean solar time.
    F. Civil timekeeping must support technologies not yet invented.
    G. We cannot un-ring the bell.

First note again that **requirements are not specifications**. Release any notion of a one-to-one correspondence. Requirement E is not an attempt to handcuff the process to "require" leap seconds. Leap seconds are not and cannot be required – they have rather been specified for several decades as an international standard. But other features of other possible solutions could serve to satisfy this requirement.

The existence of a requirement, however, ensures that a proposal to simply eliminate leap seconds will cause other behaviors to pop up somewhere else. Requirements exist separately from our comprehension of them – they are inherent in the problem space, in the behavior of the universe. Even if we don't recognize a particular requirement, it will exist to pester human systems.

Suppressing future leap seconds does indeed cause widely recognized issues. The embargoed leap seconds continue to accrue. At some future date, a larger intercalary adjustment will become necessary. This is the behavior of an engineering requirement spurned.

Well, okay – some may think – we will deal with the larger adjustment at some future date. The issue here is that a problem like civil timekeeping is characterized by many requirements simultaneously. An attempt to slay one requirement at a time is like playing *Whac-a-Mole*.* As each mole is pounded down another jumps up – in fact, more than one jumps up. In the case of civil timekeeping, when Requirement E is pushed down, not only does Requirement A jump up, but also F, and E itself refuses to stay down. Suddenly one is left looking three moles in the eye.

Which is to say that relaxing an engineering tolerance is one thing—and is a normal activity for a trade-off study. However, increasing a tolerance by three or four orders of magnitude is not relaxing it, but rather eliminating it. The precise requirement (to "approximate") one tries to eliminate reasserts itself because it is inherently required by the problem situation itself. And issues of continuity arise (Requirement A) as well as brittleness toward future technologies (Requirement F) and undoubtedly other requirements not listed here.

It is not so easy to completely redefine a problem space as to avoid all the pitfalls:

A. Sweeping issues under the rug of how the embargoed leap seconds will be accommodated in the future does not mitigate our responsibility to plan for this inevitable result—and it does not address short-term technical issues brought on by the decoupling of civil timekeeping from Earth rotation.

B. Precision time-interval and time-of-day are not the same things.

C. Focusing narrowly on high-technology users ignores a vastly larger population of more general users—especially considering that eliminating leap seconds does not address all high-technology use cases.

D. An oversimplified replacement standard does not address the inherent complexity of the underlying problem space.

E. Selective consideration of use cases that accommodate hour-long excursions from mean solar time does not address those use cases for which closer approximations are needed.

F. Paying too much attention to the limitations of current technologies and projects risks falling short in recognizing future trends and accommodating new opportunities.

G. Time and date standards are extremely long-lived and their impacts are broadly exposed to the lay public. A proposal that would redefine the essence of what timekeeping means must meet a very high level of due diligence.

CONCLUSION

Individual readers may disagree with the author's list of requirements. In fact, this is a central point of the exercise. By focusing on discussing use cases and requirements, consensus can be reached before any solution is proposed whatsoever. These requirements, and any others that may suggest themselves, should be debated. We should construct a coherent conceptual model for the

---

* http://en.wikipedia.org/wiki/File:Whackamole.jpg

problem space. We should seek solutions that satisfy all aspects of that model – or that dispose of shortcomings explicitly through liens. We should engage with all the stakeholders – certainly not just including the ones from our own communities. Building consensus and understanding of the problem space is not only the best way to conduct an exercise in systems engineering – it is the fastest way to identify and validate possible solutions.

REFERENCES

[1] Gilbert, D., "Buried by bad decisions", *Nature*, vol. 474, pp. 275-277, 2011.

[2] ITU-R, "Standard-frequency and time-signal emissions", Recommendation TF.460-6, 2002. (URL http://www.itu.int/rec/R-REC-TF.460/en)

[3] Treffinger, D. J., et. al., *Creative Problem Solving: An introduction*, 4th Ed., Prufrock Press, Inc., ISBN 1-59363-187-1, 2006

[4] Fisher, R., Ury, W., & Patton, B., *Getting to Yes: Negotiating agreement without giving in*, Houghton Mifflin, ISBN 0-395-63124-6, 1981, 1991.

[5] Rosenberg, D. & Scott, K., *Use Case Driven Object Modeling with UML*, Addison-Wesley, ISBN 0-201-43289-7, 1999.

[6] Chapman, W., Bahill, A.T., & Wymore, A.W., *Engineering Modeling and Design*, CRC Press LLC, ISBN 0-8493-8011-1, 1992.

[7] Bahill, A.T. & Gissing, B., "Re-evaluating systems engineering concepts using systems thinking", *IEEE Transactions on Systems, Man, and Cybernetics, Part C: Applications and Reviews*, vol. 28, no. 4, pp. 516-527, 1998.

[8] Morrison, L. V. & Stephenson, F. R., "Historical values of the Earth's clock error ΔT and the calculation of eclipses", *Journal for the History of Astronomy*, vol. 35, part 3, no. 120, pp. 327-336, 2004.

[9] Beard, R. L., "Role of the ITU-R in time scale definition and dissemination", *Metrologia*, vol. 48, no. 4, pp. S125-S131.

[10] US Dept. of State, *International Meridian Conference: Protocols of the Proceedings*, Gibson Bros, Washington, D.C., 1884.

[11] Royal Astronomical Society, "Report of the Council to the Eleventh Annual General Meeting of the Society", *Monthly Notices of the Royal Astronomical Society*, vol. 2, pp. 11-12, January 1831.

[12] CCIR Recommendation 460-2, "Frequency and Time-Signal Emissions", *Recommendations and Reports of the CCIR (XIVth Plenary Assembly)*, vol. VII, 1978.

[13] CCIR Recommendation 460-4, "Frequency and Time-Signal Emissions", *Recommendation and Reports of the CCIR, 16th Plenary Assembly (Dubrovnik)*, vol. 7, p. 12, International Telecommunications Union, Geneva, 1986.

[14] Seaman, R., "An Inventory of UTC Dependencies for IRAF", Paper AAS 11-677 from *Decoupling Civil Timekeeping from Earth Rotation-A Colloquium Exploring Implications of Redefining UTC*, American Astronautical Society Science and Technology Series, vol. 113, Univelt, Inc., San Diego, 2012.

[15] Sobel, D., *Longitude*, Walker Publishing, Inc., ISBN 0-8027-9967-1, 1995.

[16] Steel, D., *Marking Time: The epic quest to invent the perfect calendar*, John Wiley & Sons, Inc., ISBN 0-471-29827-1, 2000.

[17] Finkleman, D., et. al., "The Future of Time: UTC and the Leap Second", American Scientist, vol. 99, no. 4, pp. 312-319, 2011.

[18] Hillis, W.D., et. al., "Time in the 10,000-Year Clock", Paper AAS 11-665 from *Decoupling Civil Timekeeping from Earth Rotation-A Colloquium Exploring Implications of Redefining UTC*, American Astronautical Society Science and Technology Series, vol. 113, Univelt, Inc., San Diego, 2012.

DISCUSSION CONCLUDING AAS 11-661

Mark Storz noted a list of various alternative options for redefining UTC, ranging from maintaining the *status quo* to discontinuing leap seconds. One of the proposed options mentioned "periodic adjustments of leap seconds." Storz asked how that option differed from the *status quo*. Dennis McCarthy replied that this describes a situation where an indeterminate time interval would be periodically inserted into the calendar, say every four years. Storz noted that this situation is therefore different from the *status quo* in that the insertion point is regular but the amount of time inserted is variable. McCarthy said that some people like that idea because then they know when there will be leap seconds, but David Simpson said that they will not necessarily know how many. McCarthy agreed, but said that they can plan for inserting leap seconds, say, every fourth year or so. Seidelmann also said that there would be no bound on |UT1-UTC| in that situation. Storz said that perhaps the description of this option should be changed, because the current description reads much like the *status quo*. Seaman said that the option and its phrasing originated from a paper of McCarthy's.[1] McCarthy said that some software people especially like the idea. John Seago commented that this option is really not different than the *status quo* because the insertion points of leap seconds already occur at a known, predictable epoch: the end of each month. Under the present system, the value of the leap second is constrained to -1, 0, or +1, whereas under the alternative the exact value would not be constrained. Daniel Gambis noted that inserted values would be predicted in advance.

With regard to the IAU and American Astronomical Society (AAS) not having taken a position on this issue, Steve Malys asked if there would be any discussion regarding which organizations have taken a position. Malys believed that governmental entities would be involved in making the decision, and that most of the colloquium attendees do not represent governmental entities. McCarthy replied that the USA formulated a position in response to an ITU-R questionnaire. The response was handled by the US State Department because the ITU is an international treaty organization and therefore the actions of the ITU-R Radiocommuncation Assembly take on the status and force of an international treaty agreement. McCarthy described the ITU-R questionnaire as being rather simple minded.[2,3]

Regarding the AAS and the IAU, McCarthy was chairman of the IAU working group and participated in the AAS working group on leap seconds. Both of those organizations did not "forcefully come out and say 'Yes we need to get rid of leap seconds' or 'Yes we want to keep leap seconds.' The problem was that there was a big yawn on the part of both the IAU and the AAS on this." McCarthy said that he experienced difficulty getting interest within these organizations. The AAS working group sent letters to every conceivable astronomical institution and received limited feedback, such that "there was no consensus that anybody cared." Thus within the AAS it was felt that the astronomical community could live with whatever the ITU-R decided. Seaman commented that *"absence of evidence* is not *evidence of absence."* McCarthy continued that the IAU essentially had the same sort of response as the AAS, except that within the IAU working group there were perhaps more people favoring the abolition of leap seconds.

Malys said it seemed that only two of the five notional solutions are being considered by the ITU-R: *status quo* and "discontinue leap seconds." McCarthy said this situation exists "because of the way the question is phrased within the ITU." The proposal forwarded to the Radiocommunication Assembly by Study Group 7 seeks to amend Recommendation TF.460-6 to define Coordinated Universal Time so that it no longer has leap seconds. George Kaplan noted that framing the discussion as one related to the elimination of leap seconds limits participation only to those stakeholders who know about leap seconds, understand the difference between UT1 and UTC, and appreciate the implications of that relationship. Kaplan guessed that the vast majority of astronomers in particular would not understand the difference between UT1 and UTC.

REFERENCES

[1] McCarthy, D.D., W.J. Klepczynski (1999), "GPS and Leap Seconds—Time to Change?" *GPS World*, November, pp. 50–57.

[2] Timofeev, V. (2010), "Questionnaire on a draft revision of Recommendation ITU-R TF.460-6, Standard-frequency and time-signal emissions." ITU-R Administrative Circular CACE/516, July 28, 2010.

[3] Racey, F. (2011), "Questionnaire on a draft revision of Recommendation ITU-R TF.460-6, Standard-frequency and time-signal emissions." ITU-R Administrative Circular CACE/539, May 27, 2011.

AAS 11-662

LEGISLATIVE SPECIFICATIONS FOR COORDINATING WITH UNIVERSAL TIME[*]

John H. Seago,[†] P. Kenneth Seidelmann[‡] and Steve Allen[§]

The abolition of intercalary (leap) seconds within Coordinated Universal Time (UTC) would create a new civil timekeeping standard fundamentally different from solar timekeeping or Earth rotation. Such a standard has no known civil precedent and would present national governments with certain legal, technical and philosophical questions brought by the abandonment of the long-standing mean-solar-time standard. This paper elevates awareness of some of these questions; specifically, the laws of some nations and international unions require time based on the mean solar time at the meridian of Greenwich, or, if one prefers, Universal Time (UT). Since statutory specifications have not demanded ultra-precise uniformity of rate for civil timekeeping based on mean solar time, the continued synchronization of atomic UTC with Universal Time has allowed UTC to proliferate as a legally acceptable world standard. It is presumable that some nations promoted the legal status of "UTC" in the belief that a time scale named "Coordinated Universal Time" might remain coordinated with Universal Time in perpetuity. For this reason, a civil broadcast standard no longer coordinated with UT might not be easily reconciled with existing national statutes, thus requiring changes to statues or exceptional broadcast realizations. Civil broadcast standards failing to approximate Universal Time would best avoid the label "Coordinated Universal Time" and its acronym "UTC", since these descriptions have always implied a realization of Universal Time, in title and purpose, both inside and outside statutory and regulatory prescriptions.

INTRODUCTION

Coordinated Universal Time (UTC) establishes a base for the coordinated distribution of standard frequencies and timing signals per ITU-R Recommendation 460.[1] UTC has the same rate as International Atomic Time (TAI) maintained by the International Bureau of Weights and Measures (BIPM), but UTC epochs are infrequently adjusted relative to TAI by inserting (positive) or neglecting (negative) intercalary ("leap") seconds to assure its rough concordance with Universal Time. These adjustments are announced by the International Earth Rotation and Reference Systems Service (IERS), a scientific analysis organization established by the International Astronomical Union (IAU) and the International Union of Geodesy and Geophysics (IUGG).

---

[*] This paper updates the unpublished manuscript "National Legal Requirements for Coordinating with Universal Time" written in 2003 by John H. Seago and P. Kenneth Seidelmann.

[†] Astrodynamics Engineer, Analytical Graphics, Inc., 220 Valley Creek Blvd., Exton, Pennsylvania 19341-2380, U.S.A.

[‡] Research Professor, Astronomy Department, University of Virginia, P.O. Box 400325, Charlottesville, Virginia 22904, U.S.A.

[§] Programmer/Analyst, UCO/Lick Observatory, ISB 1156 High Street, Santa Cruz, California 95064, U.S.A.

There are at least two concepts that define the modern duration of time known as the second. *Universal Time* (UT or UT1) is a precise astronomical measure of the rotation of the Earth on its axis, synonymous with *mean solar time at the meridian of Greenwich*, sometimes known simply as *Greenwich mean time* (GMT).[2,3] The mean astronomical second is $1/_{86400}$ of the mean solar day.[4] The more recent *Système International d'Unités* (*SI*) second is based on 9192631770 periods of the radiation emitted from cesium 133 at a temperature of 0 K.[5] This atomic second was calibrated against the theoretically uniform "Ephemeris time"—the pre-relativistic independent variable of solar-system ephemerides based on astronomical observations from the 18th and 19th centuries.[6] Astronomical time serves as the basis for civil time, but Ephemeris time was never intended or designed to represent mean solar time exactly.[7,8] Rather, at its inception, Ephemeris time was considered to be a specialty time scale "for the convenience of astronomers and other scientists only," whereby it "seemed logical to continue the use of mean solar time […] for civil purposes."[9] Clock adjustments are therefore needed if civil time (based on Earth rotation) is also expressed in terms of *SI* seconds (calibrated by Ephemeris time).[10]

UTC is a continuous[*] time scale having duality of purpose: it completely preserves the ultra-precise uniformity of atomic frequency standards while maintaining a system of labeling epochs that remains in close proximity to Universal Time. Specifically, UTC has always respected national statutory requirements for mean solar time at Greenwich to better than one second. The *Conférence Générale des Poids et Mesures* (CGPM) endorsed the usefulness of UTC as a basis of civil time only after "considering that […] UTC is […] an approximation to Universal Time, (or, if one prefers, mean solar time)."[11] Prior to CGPM endorsement, few (if any) countries had recognized UTC by statute. Almost all countries now acknowledging UTC as a legal or regulatory standard adhered to a UT standard previously, with UTC also being a realization of Universal Time in title and purpose at the time of adoption. Consequently, a requirement for mean solar time is reflected in all time-keeping law today; legal time is explicitly referenced to Earth rotation in some countries, in others it is based on atomic time adjusted for Earth rotation, but in no country is legal time known to disregard Earth rotation.

Because the global transmission of UTC began wirelessly, international responsibility for its definition historically landed with the Radiocommuncation Bureau of the International Telecommunication Union (ITU-R). The ITU-R is an administrative body chartered under the United Nations, providing regulatory recommendations that affect information and telecommunication technology.[12] After

> *considering... that UTC is the legal basis for time-keeping for most countries in the world, and de-facto is the time scale used in most others...*

via Study Question ITU-R 236/7 "The Future of the UTC Timescale" (2001), the ITU-R Assembly agreed that the following should be studied: [13]

> *What are the requirements for globally-accepted time scales for use both in navigation and telecommunications systems, and for civil time-keeping?*
>
> *What are the present and future requirements for the tolerance limit between UTC and UT1?*

---

[*] The practice of coercing mean-solar-time clocks to display UTC results in the unfortunate mischaracterization of the atomic UTC as a time scale lacking sequence or coherence (*e.g.*, is "discontinuous"). UTC is completely sequential and coherent within the prescriptions of the UTC standard.

Implied by the ITU-R Study Question is that the timekeeping requirements for "navigation and telecommunications systems" may be distinct from those of "civil time-keeping", although requirements may overlap. Within the context of the ITU-R's purview, *civil time-keeping* is, arguably, time kept for purposes beyond navigation and telecommunications systems. As implied by the Study Question, civil timekeeping requires "legal" context for precise definition and assessment of "present and future requirements."

ITU-R Working Party 7A appointed a Special Rapporteur Group (SRG) on the future of UTC in October 2000 to address its Study Question.[14] One of the earliest public reports of the SRG's activities noted that consideration of legal aspects were highly relevant to its decision making:

> ...the second meeting of the ITU-R Special Rapporteur Group met in Paris 21-22 March 2002. The Special Rapporteur Group has converged to the opinion of freezing the present difference between UTC and International Atomic Time (TAI) at the current value of 32 seconds. It was decided at the Paris meeting that it would be necessary to retain the name "Coordinated Universal Time" and the abbreviation (UTC) to avoid potential problems regarding the definition of national time scales. UTC is the legal basis for time in many countries. Consequently, many laws might have to be rewritten to account for this change.[15]

The practical effect of halting leap seconds is to change the basis of civil time-keeping from mean solar time and/or Earth rotation. Thus, national laws might have to be rewritten regardless; jurisdictions that recognize Universal Time or mean solar time at Greenwich as an explicit legal standard instead of UTC (United Kingdom, Ireland,[16] provincial Canada, *etc.*) cannot be spared from addressing the legality of a fundamentally new civil standard like UTC without leap seconds. Only very recently have some industrialized nations come to recognize UTC explicitly by statute (*e.g.* provincial Australia by 2005, the United States of America in 2007). For nations that have legally recognized "UTC" for many decades (France, Germany, the Netherlands, Switzerland, *etc.*), there may be a legal question of whether the meaning of the time scale entitled "Coordinated Universal Time," once legally adopted should be changed without further legislative endorsements. Legal and technical considerations may therefore necessitate a uniquely different name for such a fundamentally different standard, contrary to the reported opinion of the study group.

WHAT TIME IS "IT"?

Civil Time

The ultimate nature of time is a deep and uncertain philosophical question.[17] In the 15th century, Leonardo da Vinci noted that time-keeping involved concepts of position and extension or duration, the former responding to the question of "When?" while the latter responding to the question of "How long?"[18] When atomic time scales became available, philosophical arguments flourished about the fundamental distinction between the astronomical practice of *dating events*, versus the atomic measurement of *time interval*, because "the difference between these concepts of date and time interval is important and has often been confused in the single word *time*."[19] One argument for having civil atomic time closely coordinated with Universal Time was that atomic oscillators, generating ultra-precise frequency, only provided for a measure of "time interval" from an arbitrary epoch but not a measure of "time" in the traditionally understood sense of date epoch.[20] This appears to be one of the motivating reasons for the UTC time scale as it exists today with leap seconds and why this practice has served as a legally acceptable standard internationally.

Another motive for the current definition is continuity with solar time-keeping going back to antiquity. Historically, civil time has been recorded by recurring astronomical phenomena, these being moderately verifiable by the general public. According to Newcomb (1906), the distinction between "day" (as a calendar date) and "time of day" had resulted in two divisions for time expression:[21]

> The main purpose of a measure of time is to define with precision the moment of a phenomenon. The methods of expressing a moment of time fall under two divisions: one relating to what in ordinary language is called the "time of day," and depending on the earth's rotation on its axis; the other on the count of days, which leads us to the use of years or centuries. In any case, the foundation of the system is the earth's rotation. The time of this rotation we are obliged, in all ordinary cases, to treat as invariable, for the reason that its change, if any, is so minute that no means are available for determining it with precision and certainty. [sic] There are theoretical reasons for believing that the speed of rotation is slowly diminishing from age to age, and observations of the moon make it probable there are minute changes from one century to another. If such is the case the retardation is so minute that the change in the length of any one day cannot amount to a thousandth of a second. Yet, by the accumulation of a change even smaller than this through an entire century, the total deviation may rise to a few seconds and, in the course of many centuries, to minutes.

Of Newcomb's two divisions for time expression, recurring "time of day" appears less fundamental, being a subdivision of the fundamental unit of astronomical civil and religious calendars— the "day". Newcomb's basic understanding of time is also interesting because the non-uniformity of the mean solar day was already presumed, even before it was widely adopted as a statutory basis for legal and civil times.

Mean Solar Time

By precisely measuring the duration of the year in terms of sidereal days (*i.e.*, 366.242 transits of the vernal equinox), and by recognizing that there is one less solar day per annum than sidereal days, *mean solar time* is realizable with a clock whose diurnal rate of operation exceeds the rate of sidereal time by approximately 3^m56^s (1/366.242) per sidereal day.[22] The rates of mean solar time and mean sidereal time are thereby proportional to Earth rotation which is measured by observing cataloged celestial objects beyond the solar system.[23] Historically this was done from fixed observatories such as the Airy transit circle at Greenwich.[24] Today, this is done with Very Long Baseline Interferometry observing extra-galactic quasi-stellar radio sources. As a uniform time scale, mean solar time remained basic to both civil and scientific time keeping into the 20th century, but mean solar time is especially useful for civil timekeeping purposes, being the form of sidereal time that keeps pace with the synodic day on average.

While mean solar time had been used for centuries, the introduction of Newcomb's *Fictitious Mean Sun* enhanced the practicality and accuracy of this standard. Explicit almanac references to the Mean Sun were used until the official implementation of Ephemeris time in 1960.[25] Today, Universal Time is not explicitly based on an analytical Mean Sun, but it is defined to be linearly proportional to Earth rotation angle using a constant of proportionality traceable to Newcomb's determination of the mean motion of the apparent Sun.[26] This makes UT1 a very close approximation to the mean diurnal motion of the Sun, such that the rates of "Universal Time" and "mean solar time" are practically equivalent in the very long-term.[27]

Mean Solar Time at Greenwich

Earth-rotation angle provides a sequentially increasing continuum that is everlasting and widely apparent, and its rate of uniformity is far superior to most mass-produced clocks and computers in use today.[28] To define a global time scale based on Earth rotation, the specification of a standard reference meridian on the Earth is necessary. The International Meridian Conference of 1884 resolved to adopt "the meridian passing through the center of the transit instrument at the Observatory of Greenwich as the initial meridian for longitude."[29] Their decision was facilitated by the fact that many nautical charts and almanacs were referenced to the Greenwich meridian and mean solar time at Greenwich was already commonly used for navigation.

Often described as *Greenwich Mean Time* (GMT), mean solar time at Greenwich is the legally recognized basis for civil timekeeping of most nations now or historically. The acronym GMT survives as a common navigational synonym for UT1 despite admonitions from as far back as 1928 that "astronomers are advised not to use the letters GMT in any sense for the present."[30, 31] The decrease in GMT's technical usage over time has caused its description to suffer in more general literature, with some dictionaries and encyclopedias now vaguely defining Greenwich mean time or GMT as simply as "the local time at Greenwich, located on the 0° meridian."[32] Such imprecise descriptions are sometimes coupled with factually incorrect statements: one example that has now multiplied into the definitions of some dictionaries and technical glossaries is the unusual claim that "GMT was replaced by UTC in 1986."[*]

While the rates of Universal Time and "mean solar time" have been made practically indistinguishable by virtue of definition, the specification of what is meant by "at Greenwich" may be a different question. In 1925 astronomical and navigational almanacs in the USA and Great Britain changed from "astronomical days" which began and ended at noon to adopt "civil days" beginning and ending at midnight. Thus, the epoch previously referred to as 31.5 December 1924 in pre-1925 editions became known as 1.0 January 1925 starting with the 1925 edition. However, the British *Nautical Almanac* continued to label this new convention as GMT, resulting in the recommended use of the term "Universal Time" by the IAU.[33] Nevertheless, this 12 h ambiguity posed no apparent issues in the context of legal statutes, as *Greenwich mean time* was already codified before 1925, and in legal contexts it unambiguously meant civil time of day beginning and ending at midnight.

A more subtle matter might be the precise location of the prime meridian from which to reference Earth orientation. The reference meridian of the modern global terrestrial reference frames (*e.g.*, ITRF, WGS-84) no longer passes directly through the Transit Circle Room at Greenwich, but is presently located about 100 m to the east, or, advanced by roughly one-third of a second of Earth rotation (Figure 1). Without elaborating on the very many details, this situation is mainly due to the fact that terrestrial reference frames are now maintained using different techniques, and are also affected by gradual geophysical effects such as polar motion and plate tectonics.[34] The celestial reference frames from which Earth orientation is determined have also changed substantially in terms of accuracy and observing techniques since the Airy transit circle was commissioned in the mid-19th century.[35]

Because *Greenwich mean time* is no longer maintained according to the same methods as when it was adopted into law, these technicalities can lead to questions over its legal relevance. For example, in response to concerns of legality, the Report of the International Union of Radio Science (URSI) Commission J Working Group on the Leap Second (2000) parenthetically suggested that Greenwich mean time "has not existed for thirty years."[36] The question of whether

---

[*] Time-keeping nomenclature is often inaccurate in general literature, but the origin of this incorrect claim is particularly unclear. The significance of "1986" might refer to CCIR Recommendation 460-4 (1986), which was the defining document for UTC in force when the IERS was officially established.

"Greenwich mean time is ambiguous" has also been debated within the UK Parliament's House of Lords.[37] However, these concerns may be assuaged as one considers the different levels of legality that may address such.

Figure 1. Greenwich Transit Circle Room v. the IERS Reference Meridian.

"Official Time"

Within the context of this discussion, *legal time* is that prescribed by the law or decree of a sovereign national authority within its own jurisdictional boundaries. The word *legal* means "of or relating to law," where *law* implies imposition by a sovereign authority and obligation of obedience by all subject to that authority.[38] The matters of "imposition" and "obligation" become complicated when internationally recommended practice conflicts with national legality. Furthermore, there are various classes of national law: *statute law* is that prescribed by legislatures, *case law* is established by judicial decisions (sometimes owing to an interpretation of ambiguous legal nomenclature or obsolete terminology), and *regulation* deals with details or procedure by one so legally authorized.[39] Related to case law is presumptive *common law*, the body of law "that derives its force and authority from the universal consent and immemorial practice of the people," sometimes unwritten in statute or code, and constituting the basis of English legal systems.[40] When statutes and regulations conflict, they are often resolved by changes in regulation or through the consideration of case law and/or common law, rather than by revision of statute.

The large majority of statutory citations of legal time seem concerned with the management and application of time as a commodity, rather than the basis of its definition.[41] For example, the specification of official holidays, so-called "blue laws" restricting commercial activity or labor to certain days or times, *etc.* all make use of timekeeping at a level where the use of UTC versus UT1 has no practical consequences. This is not unexpected however, for…

> With very limited exceptions, however, common law legal systems have long reckoned periods of legal significance by the calendar, not by the clock. See *Mason v. Bd. of Educ.*,

> 375 Md. 504, 826 A.2d 433, 435 (2003) ("[A] day is usually considered by the law to encompass a single, indivisible moment in time."); *State v. Stanley,* 67 S.W.3d 1, 3 (Tenn.Crim.App.2001) ("[T]he general rule for computation of time is that the law knows no fractions of a day." (internal quotation marks and citation omitted)); 2 WILLIAM BLACKSTONE, COMMENTARIES ON THE LAWS OF ENGLAND 141 (1769) ("In the space of a day all the twenty four hours are usually reckoned; the law generally rejecting all fractions of a day, in order to avoid disputes.").[42]

Indeed, common law pressed into wide use the time span called "a year and a day" that should include the date of an event (such as an offense) without regard to the time of day of its occurrence, to avoid calculating precise time intervals.[43]

Statutes often defer the details of official realizations to an authorized regulatory agency or another responsible entity; however, there is some question of how well time of day realized by national regulatory authorities remains true to the intentions of its legal prescription. There is little choice but for regulators to suppose that national statutory specifications are important, if only because of their existence, yet because of potential differences, it may be prudent to give this kind of time a different label, such as *official time* or *regulatory time*. For common-law legal systems, one might offer:

> *Official time* or *regulatory time* is a realization of the legal time legislated by a sovereign authority which intends to satisfy public expectations for civil time based on historical, philosophical, religious, and technological prejudices, precedents, and requirements.

Within the context of Question ITU-R 236/7, "civil time-keeping" presumably refers to legal time enforced and maintained—that is, *official* or *regulatory time* at a national level. UTC is primarily a regulatory time (defined by what is foremost a regulatory body, the ITU-R), the legal status of which has been codified in many, but not all, jurisdictions.

THE NEED TO CONSIDER NATIONAL TIME-KEEPING LAWS

International treaties serve to establish and harmonize standards insofar as the civilian user communities recognize, and their local governments enforce, those definitions and recommendations through national legislation (where "enforcement" might include the disbursement of public monies for the national maintenance and distribution of so-called official or regulatory time). Because national governments maintain their own realizations of UTC for official use in real time, the Study Question "What are the requirements… for civil time-keeping?" seemingly urges careful consideration of national statutes establishing time standards for individual governments. However, this viewpoint may not be shared by all who study this issue. For example, Nelson *et al.* (2001) suggest "should the definition of UTC be revised, the effect on legal codes may need to be investigated."[44] This implies that the laws of individual nations are an afterthought when revising the definition of UTC; the rationale may be that few national laws numerically stipulate a required proximity between official time and astronomically based legal concepts like mean solar time or Universal Time.[45]

Consideration of what might be satisfactory to national authorities contributed to a "consensual opinion" reached within the SRG and presented to interested and representative parties at the ITU-R *Special Colloquium on the Future of UTC* in 2003.[46]

> Serious consideration was given to a contribution proposing that the maximum tolerance of DUT1, the difference between UT1 and UTC, be increased to one hour. This alternative was based on a similar concept of daylight saving time. This modification of standard time used by nations that is determined by national civil authority appeared to satisfy all civil requirements and concerns.[47]

However, that the leap-hour proposal "was based on a similar concept to daylight saving time" should not suggest that it satisfies "all civil requirements and concerns" in the sense of "globally-accepted requirements" for civil timekeeping. This becomes evident through consideration of national legislation.

Presently, the majority of the world does not practice summer-time, or, daylight-saving-time clock adjustments, whether measured by population, number of countries, or regional land mass (Figure 2). In tropical regions where the duration of daylight is less variable, summer-time adjustments are generally unwarranted and undesired.[48] Nations that still practice summer-time adjustments do so at different times of the year, with the northern and southern hemispheres exercising seasonal clock adjustments out of phase with each other. In the history of civil timekeeping, daylight-saving adjustments are relatively new and its long-term practice is unclear.[*]

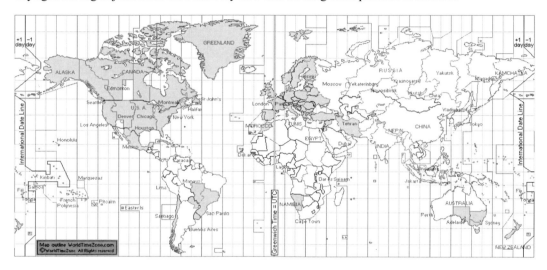

Figure 2. Daylight-Saving Time Practice (shading indicating current legislated practice).
Source: WorldTimeZone.com, used with permission.

Also, national statutory prescriptions for standard (zone) time and daylight-saving time describe two distinct concepts. One concept is the local or regional time indicated by official clocks; the other concept is the coordinated background or *basis time* from which all official clocks are offset, such as mean solar time or UTC. Some (but not all) nations allow for summer-time adjustments of local civil clocks, but no nation is yet known to express legal tolerance for significant adjustments to the coordinated background or *basis time* which underlies and regulates civil time-keeping.

Finally, while the proposed leap-hour adjustment attempts to manage statutory expectations that UTC must remain coordinated with Universal Time, there is no evidence that coordination to the nearest hour (15 degrees) has any technically useful purpose. Therefore, there is no technical

---

[*] The eight members of the SRG, and its three special representatives, were from nations where daylight-saving time was practiced at the time of the proposal.

basis by which regulatory authorities should interpret the statutory specifications for Universal Time or Greenwich Time so inaccurately.

The leap-hour proposal was questioned by attendees of the ITU-R Special Colloquium; because the label "Universal Time" has always been a term reserved for time linked to Earth rotation, continued use of the terms "Coordinated Universal Time" and "UTC" seemed inappropriate for atomic timekeeping uncoordinated with Earth rotation.[49] Any small problems now associated with leap seconds would be greatly amplified by larger adjustments in the future. The need for a leap-hour adjustment would likely not occur for another six to eight centuries, such that it appeared presumptuous to codify such adjustments now.[50] Rather, the cessation of leap seconds simply relinquishes the long-standing mean solar day, supplanting it with a "metric" day of exactly 794,243,384,928,000 cycles of cesium-133 radiation, or 86400 *SI* seconds.[51]

LEGAL REQUIREMENTS FOR MAINTAINING UNIVERSAL TIME

Statutory standards for timekeeping have been historically expressed and understood in navigational or astronomical terms (*e.g.*, "longitude", "meridian", "Greenwich", *etc.*).[*] Many timekeeping statutes still reflect some expectation that Earth rotation regulates the civilian notion of Time (including the meanings of commonly understood concepts and words such as "day", *etc.*). Even time legislation explicitly based on "UTC" often includes navigational or astronomical elements or terminology (time zones, "antemeridian", *etc.*)

Today, the navigational use of "Greenwich mean time" still implies UT1 as the instantaneous orientation of Earth determined by the IERS. Yet the uniformity of UT2—UT1 corrected for seasonal variations of Earth rotation (measured in milliseconds)—also made UT2 the basis for standard-time broadcasts for many years.[52] However, the original (legal) concept for mean solar time predates the various other realizations of Universal Time (*e.g.*, UT0, UT1, UT2, UTC). Universal Time is also something to be observed and extrapolated, or reduced after the fact, and different people may acceptably observe and reduce Universal Time differently (the methods of which are also unspecified under law). This further implies that the law does not place extremely rigid statutory prescriptions on the realization of Universal Time for legal purposes; rather the meaning of any technical term must be assessed within the context of the state of the art.

For these reasons, one cannot definitively assign a specific realization of Universal Time within most legal contexts. This may also be why UTC has endured as a legally acceptable proxy for Universal Time; atomic UTC is a technically useful realization of Universal Time. Certainly, the unchallenged juxtaposition of UTC for "mean solar time at Greenwich" in many applications suggests that a fraction of a second may already be a legally allowable level for civil-time ambiguity, but a redefinition of UTC that exceeds this tolerance should necessitate consideration of national statutes. Technically "day" and "year" are non-*SI* units and the status of a calendar maintained with metric days versus Earth rotations also becomes a potential legal question of the future. Hence, consideration of national laws seems necessary to ensure that internationally broadcast time standards remain acceptably legal across all jurisdictions.

Standard Time of the United States of America

Statutory authority over standard time in the United States of America resides with its Congress. When the US Congress first enacted the Standard Time Act of 1918, it legislated

---

[*] For example, Argentina's final summer-time declaration in 2009 references the "meridiano de Greenwich." http://www.hidro.gov.ar/Noticias/RENoticias.asp?idnot=197

"That, for the purpose of establishing the standard time of the United States, [...t]he standard time [...] shall be based on the mean astronomical time of [...] longitude west from Greenwich."[53]

At this time there were only two concepts that could be interpreted as "mean astronomical time": mean sidereal time and mean solar time. Both are defined by Earth rotation, the rate of one proportional to the other, and it was already thought that the length of the mean solar day was increasing at a rate of many seconds per century.[21] Newcomb had suggested that "astronomical mean time" technically described the day starting at noon, but the difference of twelve hours between civil and astronomical time was as apparent as night and day and there was little concern over the need for more specialized legal nomenclature. So within the technical and historical context of long-standing civil conventions, the phrase "mean astronomical time" was all that Congress needed to convey a precise legal notion of the mean solar day beginning at midnight relative to Greenwich.

By 1958 however, the IAU had defined the more uniform astronomical time scale known as Ephemeris Time, the rate of which was adopted by the CGPM in 1960 to define an *SI* second.[54, 55] The specification of another uniform, yet fundamentally different, astronomical time scale approximately one-half minute from mean solar time perhaps rendered the previous legal descriptor "mean astronomical time" ambiguous. When Congress passed the Uniform Time Act of 1966, language was clarified by replacing the phrase "mean astronomical time" with "mean solar time" which ensured that standard time would be regulated by the astronomical concept of Universal Time rather than the astronomical concept of Ephemeris Time.[56]

Such Congressional action afforded unambiguous legal protection for mean solar time when a more uniform (but secularly deviating) time scale was available and might have been interchanged owing to the obsolete wording of law. In hindsight, this action might also suggest a low legal tolerance for a basic time standard differing more than several seconds from what was legally intended or required at the time of adoption. But perhaps just as important, the distinctions in the realization of broadcast Universal Time (*i.e.*, UT2), and even the more astronomically precise term "Universal Time" itself, went unrecognized under the Uniform Time Act. This further suggested that legislation was not only tolerant of subtle ambiguities in the realization and legal meaning of "mean solar time" (all being well below one second), but statutes chose to emphasize the conceptual aspect of "solar" time in an astronomical standard.

While Congress left standard time defined in astronomical and navigational terms, addenda to federal code acknowledged UTC as an acceptable regulatory proxy for mean solar time where limited to most practical purposes associated with radio regulations and telecommunication.[57] Specifically

> *Coordinated Universal Time (UTC)*. Time scale, based on the second (SI), as defined in Recommendation ITU-R TF.460–6...
>
> Note: For most practical purposes associated with the ITU *Radio Regulations*, UTC is equivalent to mean solar time at the prime meridian (0° longitude), formerly expressed in GMT.

Thus "UTC" in this context still refers to an atomic time scale that remains within ±0.9 seconds of Earth rotation, per the Federal Radionavigation Plan (an official US policy published jointly by the US Departments of Transportation* and Defense).[58] But as a legal basis for regulating civil time, US code is not known to have otherwise acknowledged or supported the use of Ephemeris Time, or any parallel variants of its successors such as TAI. Instead, US code legally authorized the use of the *SI* second as a measure of time interval as part of the metric system.[59] Mean solar time measured in *SI* seconds thereby appears to be legal in the US. This, of course, describes UTC with leap seconds.

In 2002, the *National Institute of Standards and Technology* (NIST) *Authorization Act of 2002* was introduced into the US Senate.[60] Primarily an appropriations bill for NIST, Section 207 of the bill proposed to amend the Uniform Time Act of 1966 by changing the basis of standard (zone) time from the "mean solar time" of standard meridians west of Greenwich to "Coordinated Universal Time" retarded by specific numbers of hours. It also added the following statutory definition for UTC:

> Coordinated Universal Time Defined—In this section, the term 'Coordinated Universal Time' means the time scale maintained through the General Conference of Weights and Measures and interpreted or modified for the United States by the Secretary of Commerce.

The 2002 Senate bill did not pass, but the language reappeared within the so-called America COMPETES Act of 2007, which became public law on August 9, 2007.[61] In the final version the statutory definition for UTC was amended with "…in coordination with the Secretary of the Navy."† That UTC needed a statutory definition (whereas the astronomical concepts of "mean solar time" and "mean astronomical time" did not) implies there might be greater legal uncertainty or ambiguity as to its meaning without such a definition.

The statutory change from "mean solar time" to UTC was offered as a technical amendment to revise "Outdated Specifications" associated with use of the metric system. Within this context, the General Conference of Weights and Measures (CGPM) is the international delegation under the Treaty of the Meter that manages arrangements for sustaining and improving the metric system, including major decisions concerning the organization and development of the International Bureau of Weights and Measures (BIPM). However, there is no BIPM service by which one can obtain official UTC time signals or otherwise set a timekeeping device.[62] Officially, UTC is evaluated in arrears by the BIPM through published corrections to the emissions of primary frequency standards via *Circular T* a month or more after the fact.[63] However, the multitudes of ordinary users require an instantaneous realization for legal purposes. Moreover, the CGPM does not define UTC; UTC is a real-time broadcast specification defined by ITU-R Recommendation 460 which is not under the direct purview of the CGPM.

While the accuracy of civilian timepieces has greatly improved to a point where a leap second might be detectable, very few timepieces support (display) leap seconds. Arguably, clocks of such manufacture are still generic mean-solar-time clocks. Such anecdotal evidence might further imply that generic Universal Time is the standard being upheld, employed, or expected. Therefore, it might be argued that, after more than a century of statutory recognition and civil usage, time based on Earth rotation is anticipated by custom and precedent. Congress is presumed to legislate against the background of the common law, but Congress can override any common-law pre-

---

* The responsibility for standard time and time zones in the USA resides with the Secretary of Transportation, per US Code Title 15, Chapter 6, Subchapter IX, Sec. 262.
† The Secretaries of Transportation and the Navy are named because NIST and the US Naval Observatory respectively operate under their authorities.

sumption with express language.[64] Yet, because UTC was a realization of mean solar time at Greenwich at the time of its adoption, it could be suggested that Congress intended to substitute one expression for mean solar time at Greenwich for another having more regulatory precision.

Standard Time of the United Kingdom

When *Greenwich mean time* became a legal standard across Great Britain in 1880, there was no other civil meaning associated with it beyond mean solar time at Greenwich.[65] Today, one confusing aspect is that some civilian applications casually describe UTC broadcasts as "GMT". Furthermore, Parliamentary law did not specify a legal title for British Summer time (which is GMT plus one hour); this too results in occasional descriptions of British summer time as GMT.[*]

However, that Greenwich mean time has come to be recognized as being casually synonymous with UTC in ordinary language does not further imply that these two concepts are to be permissibly interchanged. Rather, the presence of leap seconds makes UTC synonymous with *Greenwich mean time* and justifies the practical use of UTC as a proxy for GMT wherever GMT is prescribed today. In 1978, Donald Sadler commented that "the two forms of [atomic and solar] time-scale are fundamentally different; both are essential [...] and it would seem important to ensure that no unnecessary confusion between them is introduced."[66] Yet in those countries where UTC has not been made explicitly legal, one may conclude that Earth rotation, and not atomic frequency, was intended as the ultimate basis for civil time. This intention seems most explicit within the UK, where bills attempting to replace GMT with UTC have been debated yet failed to overcome Government neutrality.[67, 68] This is because English speaking countries tend to make statutory distinctions between UTC and GMT; for example, New Zealand amended its Time Act of 1974 effective 30 March 1987 to henceforth reference UTC in place of GMT.[69]

Standard Time of Canada

Canada is an example of a nation where both UTC and Greenwich mean time (or, simply "Greenwich time") are simultaneously legislated by different provinces. Québec's recognized standard has been UTC since 2006,[70] while other provinces such as Alberta,[71] Manitoba,[72] Ontario,[73] Saskatchewan,[74] *etc.*, recognize Greenwich mean time. Therefore, legal consistency between the Canadian provinces suggests a need for continued coordination of Coordinated Universal Time and Greenwich mean time.

Standard Time and the European Union

The directives of European Parliament reconciling the application of summer-time adjustments across the European Union is an example of an instruction applicable to EU member nations having different bases for national time.[75] Of the twenty-two (22) versions of the directive available in their original languages, twelve (12) cite *Greenwich mean time*, *Greenwich time*, or *GMT*, six (6) cite *Universal Time*, three (3) cite *world time*,[†] and two (2) cite *UTC* or *Coordinated Universal Time* explicitly (Table 1). Thus, 59% make explicit reference to GMT or "Greenwich" by name in their original language versions, while only 9% of the versions unambiguously declare UTC. An interesting question might be whether the dominance of *Greenwich time* within translations because the original text was expressed that way. Regardless, it appears that legal

---

[*] A similar situation exists in the USA, where the term *Daylight-Saving Time* is not explicitly codified and the exact wording of statute implies that *Standard Time* changes according time of year. Nevertheless, in practice the time in effect designated either *Standard Time* or *Daylight-Saving Time*.
[†] Traditionally, "world time" is synonymous with and translated as either *Universal Time* or GMT, although the Danish version annotates that UTC was specifically meant.

consistency between member countries of the European Union may require close coordination of Coordinated Universal Time and Greenwich mean time.

**Table 1. Specifications of National Time Bases
in EU Directive 2000/84/EC on Summer-Time Arrangements**

	Language	Original Language Quotation	Translation
BG	Bulgarian	*по Гринуич*	Greenwich time
EL	Greek	*ώρα Γκρίνουιτς*	Greenwich time
EN	English	Greenwich Mean Time	Greenwich mean time
ET	Estonian	*Greenwichi aja järgi*	Greenwich mean time
LT	Lithuanian	*nakties* GMT *laiku*	Greenwich mean time
LV	Latvian	*pēc Grīnvičas laika*	Greenwich mean time
HU	Hungarian	*greenwich-i idő*	Greenwich mean time
MT	Maltese	Greenwich Mean Time	Greenwich mean time
SK	Slovakian	*greenwichského času*	Greenwich mean time
SV	Swedish	*Greenwichtid* (Greenwich Mean Time, GMT)	Greenwich time (Greenwich Mean Time, GMT)
FI	Finnish	(GMT)	(GMT)
CS	Czech	*světového času* (GMT)	universal time (GMT)
PL	Polish	*czasu uniwersalnego* (GMT)	universal time (GMT)
ES	Spanish	*hora universa*	universal time
FR	French	*temps universel*	universal time
IT	Italian	*ora universale*	universal time
PT	Portuguese	*tempo universal*	universal time
RO	Romanian	*timp universal*	universal time
DE	German	*Weltzeit*	"world time"
NL	Dutch	*wereldtijd*	"world time"
DA	Danish	*verdenstid* (UTC)	"world time" (UTC)
SL	Slovenian	*univerzalnem koordinirane času* (UTC)	coordinated universal time (UTC)

Historic Universal Time (GMT) in France

Prior to its complete legal adoption of UTC on August 9, 1978,[76] France decreed its legal standard for Greenwich mean time as "Paris mean time, retarded by 9 minutes and 21 seconds."[77] This previous decree specified a level of precision of one (1) second for GMT or Universal Time. By design, UTC has never deviated from France's originally designated legal resolution for Universal Time; France thereby adopted a UTC time scale historically compatible with, if not identical to, Universal Time as previously acknowledged under French law.

There are two types of solar time, mean and apparent, which are unbiased with each other and differ ±15 minutes annually. It has been suggested that civil authorities are likely to be tolerant of large differences between UT and UTC, insofar as they remain at levels near the difference of mean and apparent solar times.[78] However, it is noteworthy that the static difference between Paris and Greenwich local time is smaller than the periodic difference between mean and apparent solar time, and much smaller than the eventual static offset implied by "leap hours." Paris and Greenwich time differed to such a degree that France found it necessary to legally account for the difference, suggesting another precedent for legal intolerance of standard-time differences beyond one second of what may have been intended or required.

Legal Time of All Nations

As mechanical time pieces flourished, civil conventions for uniform time became almost exclusively expressed with relation to mean solar time. Even after uniform atomic time became available as a broadcast standard, its civil and legal acceptability was secured through leap-second adjustments for the sole purpose of emulating the mean-solar-time standard. The International Radio Consultative Committee (CCIR)[*] affirmed that the establishment of UTC was to provide a realization of Universal Time and mean solar time at Greenwich, *e.g.*,

> The CCIR, *considering*… the continuing need of many users for Universal Time (UT)… *unanimously recommends*… that the transmitted time scale should be adjusted when necessary in steps of exactly one second to maintain approximate agreement with Universal Time (UT)…[79]

and "GMT may be regarded as the general equivalent to Universal Time."[80] It is reasonable to presume that some—perhaps most—countries elevated the legal status of UTC understanding that a time scale named "Coordinated Universal Time" would remain coordinated with Universal Time in perpetuity.

POLITICAL CONCERNS FOR UTC WITHOUT LEAP SECONDS

Even among experts in horology, precision time-scale definitions have not come easy.[81, 82] UTC without leap seconds would be a time scale equal to International Atomic Time (TAI) plus a static offset. It is therefore interesting to speculate about whether the formal definition of TAI may be insufficiently terse for legal purposes, owing to the complexity of the subject. In 1971, the CGPM requested the *Comité International des Poids et Mesures* (CIPM) give a definition to International Atomic Time already in use,[83] the tenuous submission of the *Comité Consultatif pour la Définition de la Seconde* (CCDS)[†] having been

> International Atomic Time (TAI) is the time reference coordinate established by the Bureau International de l'Heure[‡] on the basis of readings of atomic clocks operating in various establishments in accordance with the definition of the second, the unit of time of the International System of Units.[84]

Later, it became necessary to burden the definition of TAI with General Relativity Theory. By 1980 its definition was reportedly "completed" in this sentence:[85]

> TAI is the coordinate time scale defined in a geocentric reference frame with the *SI* second as realized on the rotating geoid as the scale unit.[86]

---

[*] The CCIR was the predecessor of the ITU-R
[†] The CCDS was predecessor of the CCTF, *Comité Consultatif du Temps et des Fréquences*
[‡] This responsibility is now with the BIPM.

But to astronomers understanding that atomic resonators can only define time as an interval relative to some (arbitrary) epoch, this definition was still insufficient. A further clarification was advanced by the IAU in 1991:[*]

> TAI is a realized time scale whose ideal form, neglecting a constant offset of 32.184s, is Terrestrial Time (TT), itself related to the time coordinate of the geocentric reference frame, Geocentric Coordinate Time (TCG) by a constant rate.[87]

The IAU resolution implies that the origin of TAI is ideally defined in terms of TCG, although practically speaking, TAI is the realization based on the *SI* second (accurate to the level of the frequency standards) and TCG is practically realized by an ideal mathematical prescription relative to it. To complicate matters a bit more, the IAU further refined the definition of TT in 2000.[88]

The realization of TAI (or, UTC without leap seconds) is more involved than these definitions indicate.[89] TAI is a "paper clock" determined from the weighted average of contributions from frequency standards in many countries. Some less accurate national standards are included with minimal weight, mainly for political reasons. There are different realizations of TAI determined on a monthly basis as well as after the end of the calendar year.

The global establishment of mean solar time at Greenwich overcame many political obstacles and took several decades. Placing atomic UTC without leap seconds in the legal foreground may place additional political or regulatory demands on atomic civil time that background TAI has beneficially avoided. So far TAI continues to be practically defined through BIPM edict, unfettered by national legislation. National law may do well to avoid the subject, but it may not be able should the basis of official time seem arbitrary. US Code has already assigned UTC a legal definition that can differ from what the BIPM prescribes. Without international unanimity on the subject of UTC redefinition, there remains a possibility that different nations could slip onto differing time bases, depending on how closely they preferred these bases to be aligned to Earth rotation.

CONCLUDING SUMMARY

Two legal bases for timekeeping are codified globally. Mean solar time at Greenwich (also known as Greenwich mean time, GMT, or Universal Time, UT1) is an astronomical measure of Earth rotation referenced to the international reference meridian. Coordinated Universal Time (UTC) is a precise timescale based on atomic frequency and used for broadcasting and telecommunications. Because atomic resonators maintain a rate different than mean solar time, an atomic realization of mean solar time must be adjusted; for UTC this adjustment is made to the length of the "UTC day" (the day being a non-*SI* base unit) using occasional *leap seconds*. UTC is therefore an atomic realization of Universal Time in title and practice, which is, in turn, the modern-day complement of Greenwich mean (solar) time.

By international agreement, UTC broadcasts have never differed from UT1 by more than ±0.9 seconds. This is a tolerance which appears to satisfy most legal requirements for civil time scales defined in relation to the mean solar day at Greenwich as stipulated under law by most nations (now or historically), and allows UTC to be used in jurisdictions where astronomical time is (or was) legally prescribed. Allowable deviations larger than this have no known legal precedent in modern times and do not appear to have been tested or reviewed by national judicial or legislative systems.

---

[*] As far as the authors have been able to discover, this clarification has yet to be recognized by the CGPM.

A proposal will come before the ITU-R Radiocommunication Assembly in January 2012 on whether to create a new time scale uncoordinated with Universal Time. This scale will still be called *Coordinated Universal Time* but it will lack UTC's original duality of purpose to provide both time of day and fundamental frequency. The abandonment of the long-standing mean-solar-time standard will present governments with certain legal, technical and philosophical questions of which this paper attempts to elevate awareness. While the discussion is not intended to be complete or authoritative, it suffices to illustrate that UTC without leap seconds may not be easily reconciled with some existing national statutes. Legislation endorsing a time scale called "Coordinated Universal Time" that is no longer coordinated with Universal Time will remain legally and technically confusing. The potential for legal complications in national courts is conceivable given the complexity of national legal systems, even in countries that acknowledge "UTC" by name as a legal standard. Therefore, fundamental changes to the UTC standard would require explicit changes to national laws.

UTC without leap seconds also creates a new question: is a precisely maintained, indefinite sequence of synthetically generated time intervals a sufficiently viable concept to permanently displace humanity's long-standing precepts of what ultimately regulates Time? This question is not related to satisfying the technical conveniences of today's telecommunication and navigation systems, but is more of a moral, philosophical, or historical question to be thoughtfully upheld by civil law. Similar questions were already considered somewhat by the CCIR and the IAU four decades ago, and the answer was UTC with its existing system of leap seconds.

As currently defined, the existing UTC system appears capable of uniquely tagging any event that may possibly occur during the next 1000 years with full atomic accuracy, and at this time, there does not appear to be any legal requirement for ultra-precise uniformity in civil time beyond what is already being supplied with existing UTC. For decades, applications with very stringent timing requirements (including the US Global Positioning System) have continued to operate successfully despite the existence of leap seconds. Almost all applications non-compliant with UTC came into existence within the last four decades—well after today's UTC standard was established. With the proposal now under consideration, national governments may do well to investigate why certain modern-day applications are still either unwilling or unable to comply with international time-keeping standards while others are functionally compliant. National legislative investigations would help discover what, if any, changes to UTC are warranted. This could avoid unnecessary changes to, or tests of, existing (inter)national legislation, and avoid unnecessary burden to systems, applications, and industries already compliant with current standards.

REFERENCES

[1] Recommendation ITU-R 460-6 (2002), Standard-frequency and time-signal emissions (Question ITU-R 102/7), in ITU-R Recommendations: Time Signals and Frequency Standards Emission, Geneva, International Telecommunications Union, Radiocommunication Bureau.

[2] Seidelmann, P.K. (ed., 1992), *Explanatory Supplement to the Astronomical Almanac*. University Science Books, Mill Valley, CA, p. 6.

[3] Stratton, F.J.M. (ed., 1929), Transactions of the IAU Vol. III, Proceedings of the IIIrd General Assembly, Leiden, The Netherlands, July 5 - 13, 1928, Cambridge University Press, p. 224.

[4] Chadsey, H., D. McCarthy (2000), "Relating Time to the Earth's Variable Rotation." Proceedings of the 32nd Annual Precise Time and Time Interval (PTTI) Systems and Applications Meeting, Reston, Virginia, November 28-30, 2000. pp. 237.

[5] 13th CGPM (1967-8), Resolution 1, (CR, 103). Also, *Metrologia*, 1968, Vol. 4, p. 43.

[6] Markowitz, W., R.G. Hall, L. Essen, J.V.L. Perry (1958), "Frequency of Cesium in Terms of Ephemeris Time." *Physical Review Letters*, Vol. 1, pp. 105-7.

[7] McCarthy, D.D. (1991), "Astronomical Time." *Proc. IEEE*, Vol. 79, No. 7, pp. 915-920.

[8] *Explanatory Supplement to The Astronomical Ephemeris and The American Ephemeris and Nautical Almanac* (1961). Her Majesty's Stationery Office, London, p. 66-7.

[9] Clemence, G.M (1948), "On the System of Astronomical Constants," *Astronomical Journal,* Vol. 53, No. 6, Issue 1170, pp. 171–72.

[10] McCarthy, D.D. (2000), "Future Definition of UTC", in *Proceedings of IAU Colloquium 180, Towards Models and Constants for Sub-microarcsecond Astrometry*, K.J. Johnston, D.D. McCarthy, B.J. Luzum, G.H. Kaplan, eds., U.S. Naval Observatory, Washington DC, USA, pp. 363-371.

[11] 15th CGPM (1975), Resolution 5 (CR, 104). Also, *Metrologia*, 1975, Vol. 11, p. 180.

[12] Finkleman, D., S. Allen, J.H. Seago, R. Seaman and P.K. Seidelmann (2011), "The Future of Time: UTC and the Leap Second." *American Scientist*, Vol. 99, No. 4, p. 316.

[13] Jones, R.W. (2001), Question ITU-R 236/7, "The Future of the UTC Time Scale." Annex I of ITU-R Administrative Circular CACE/212, March 7, 2001.

[14] Document CCTF/01-33, "Report of ITU-R Working Party 7A in the Period 1999 to 2001 to the 15th Meeting of the CCTF (Sevres, 20 – 21 June 2001)."

[15] Capitaine, N., Chapront, J., Hadjidemetriou, J. D., Jin, W., Petit, G., and Seidelmann, K. (2003). "Division I: Fundamental Astronomy (Astronomie Fondamentale)," in *Reports on Astronomy 1999-2002, Transactions of the International Astronomical Union*, Vol. 25A, edited by H. Rickman (San Francisco: Astronomical Society of the Pacific), p. 8.

[16] Houses of the Oireachtas, Ireland, Standard Time (Amendment) Act, 20th July 1971: "An Act To Provide For A Period Which Shall Be Known As Winter Time And During Which The Time For General Purposes In The State Shall Be Greenwich Mean Time, And To Provide For Other Matters Connected With The Matter Aforesaid."

[17] Green, R.M. (1985), *Spherical Astronomy*. Cambridge University Press, London. p. 236.

[18] Klein, H.A. (1988), *The Science of Measurement: A Historical Survey*. Dover Publications, Inc. New York, p. 169.

[19] Blair, E.B., (ed.), NBS Monograph 140, Time and Frequency Fundamentals, p. 4.

[20] Sadler, D.H. (1978), "Mean Solar Time on the Meridian of Greenwich." *Quarterly Journal of the Royal Astronomical Society*, Vol. 19, p. 307.

[21] Newcomb, S. (1906), *A compendium of spherical astronomy*. Macmillan Company, p. 114-7.

[22] Seidelmann, P.K., J.H. Seago (2011), "Time Scales, Their Users, and Leap Seconds." *Metrologia*, Vol. 48, pp. S186–S194.

[23] *Explanatory Supplement to The Astronomical Ephemeris and The American Ephemeris and Nautical Almanac* (1961). Her Majesty's Stationery Office, London, p. 67.

[24] Lewis, T., (1885), "The Greenwich system of sympathetic clocks and the distribution of time-signals." *The Observatory*, Vol. 8, p. 364.

[25] Seidelmann, P.K. (ed., 1992), *Explanatory Supplement to the Astronomical Almanac*. pp. 78-9.

[26] Capitaine, N., Guinot, B., and McCarthy, D.D., "Definition of the Celestial Ephemeris Origin and of UT1 in the International Celestial Reference Frame," *Astronomy &. Astrophysics*, Vol. 355, 2000, pp. 398-405.

[27] Seidelmann, P.K. (ed., 1992), *Explanatory Supplement to the Astronomical Almanac*. p. 51.

[28] Finkleman, D., J.H. Seago, P.K. Seidelmann (2010), "The Debate over UTC and Leap Seconds". Paper AIAA 2010-8391, from the Proceedings of the AIAA/AAS Astrodynamics Specialist Conference, Toronto, Canada, August 2-5, 2010.

[29] US Government, International Conference held at Washington for the purpose of fixing a prime meridian and a universal day, October 1884: Protocol of the Proceedings. pp. 199-201.

[30] McCarthy, D.D. (2011), "Evolution of timescales from astronomy to physical metrology." *Metrologia*, Vol. 48 p. S134.

[31] IIIrd General Assembly - Transactions of the IAU Vol. III B Proceedings of the 3rd General Assembly Leiden, The Netherlands, July 5- 13, 1928 Ed. F.J.M. Stratton Cambridge University Press, p. 224.

[32] *The Macmillan Encyclopedia 2001* (2000), Market House Books Ltd.

[33] Sadler, D.H. (1978), "Mean Solar Time on the Meridian of Greenwich." p. 300.

[34] Guinot, B. (2001) "Solar time, legal time, time in use." *Metrologia*, Vol. 48, S181-S185.

[35] McCarthy, D.D. (2011), "Using UTC to Determine the Earth's Rotation Angle." Paper AAS 11-666, , from Seago, J.H., et al., *Decoupling Civil Timekeeping from Earth Rotation—A Colloquium Exploring Implications of Redefining UTC*. American Astronautical Society Science and Technology Series, Vol. 113, Univelt, Inc., San Diego, 2012

[36] Matsakis, D., et al. (2000), Report of the URSI Commission J Working Group on the Leap Second, July 2, 2000. (URL http://www.ietf.org/mail-archive/web/ietf/current/msg13828.html)

[37] *Lords Hansard*, Parliament of the United Kingdom, 27 Nov 1996: Columns 257-60.

[38] Merriam-Webster's Collegiate Dictionary 10th Ed. (1998), Merriam-Webster, Inc.

[39] *Ibid.*

[40] Bouvier, J. (1856), *A Law Dictionary, Adapted to the Constitution and Laws of the United States.*

[41] Rakoff, T.D., (2002), *A Time for Every Purpose: Law and the Balance of Life*. Harvard.

[42] *Lagandaon v. Ashcroft*, 383 F.3d 983, 985 (9th Cir. 2004),

[43] *Ibid.*

[44] Nelson, R.A., D.D. McCarthy, S. Malys, J. Levine, B. Guinot, H.F. Fliegel, R.L. Beard, T.R. Bartholomew, (2001), "The leap second: its history and possible future." *Metrologia*, Vol. 38, p. 524.

[45] Levine, J., (2001) "GPS and the legal traceability of time." *GPS World*, Vol. 12, No. 1, pp. 52-58.

[46] Press Release "UTC Timescale Conference", *The Institute of Navigation (ION) Newsletter*, Vol. 12, No. 4 (Winter 2002-2003)

[47] Beard, R. (2011) "Role of the ITU-R in time scale definition and dissemination." *Metrologia*, Vol. 48, p. S130.

[48] Steel, D. (2000), *Marking Time – The Epic Quest to Invent the Perfect Calendar*. John Wiley & Sons, New York, pp. 276-77.

[49] CCTF (2004), Consultative Committee for Time and Frequency (CCTF) Report of the 16th meeting to the International Committee for Weights and Measures (April 1–2, 2004), Bureau International des Poids et Mesures. p. 17.

[50] Stephenson, F.R., L.V. Morrison (1995), "Long-term fluctuations in the Earth's rotation: 700 BC to AD 1990." *Philosophical Transactions of the Royal Society of London, Series A*, 351 pp. 165-202.

[51] Taylor, B.N. (ed., 2001), *The International System of Units (SI)*, NIST Special Publication 330, p. 16.

[52] Sadler, D.H., (ed., 1960), Transactions of the IAU Vol. X, Proceedings of the Xth General Assembly, Moscow, Russia, August 12 - 20, 1958, Cambridge University Press, p. 489.

[53] S. 1854 [Public, No. 106.] 40 Stat. 450, March 19, 1918.

[54] Sadler, D.H., (ed., 1960), Transactions of the IAU Vol. X, Proceedings of the Xth General Assembly, Moscow, Russia, August 12 - 20, 1958, Cambridge University Press, p. 500.

[55] 11th CGPM, 1960, Resolution 9, (CR 86) "definition of the unit of time (second)."

[56] S. 1404 (Public, No. 89-387), 80 Stat. 109, April 13, 1966.

[57] US Code Title 47, Chapter 1, Part 2, Subpart A, Section 2.1, Terms and Definitions.

[58] US Departments of Transportation and Defense, 2001 Federal Radionavigation Plan, DOT-VNTSC-RSPA-01-3 / DOD-4650.5, January 2000-December 2001, Appendix A (Definitions), p. A-2.

[59] US Code Title 15, Chapter 6, Subchapter I, Sec. 204

[60] S.3177—National Institute of Standards and Technology Authorization Act of 2002.

[61] 110th Congress Public Law 69—America Creating Opportunities to Meaningfully Promote Excellence in Technology, Education, and Science Act.

[62] Arias E F, G. Panfilo, G. Petit (2011), "Timescales at the BIPM." *Metrologia*, Vol. 48, pp. S145– S153

[63] Arias, E.F. (2005), "The Metrology of Time." *Phil. Trans. R. Soc. A* Vol. 363, p. 2291.

[64] *Astoria Fed. Sav. & Loan Ass'n v. Solimino,* 501 U.S. 104, 108, 111 S.Ct. 2166, 115 L.Ed.2d 96 (1991).

[65] "An Act to remove doubts as to the meaning of Expressions relative to Time occurring in Acts of Parliament, deeds, and other legal instruments." 2nd August 1880. Statutes (Definition of Time) Act, 1880 (43 & 44 Vict. c. 9)

[66] Sadler, D.H., "Mean Solar Time on the Meridian of Greenwich." p. 291.

[67] Lords Hansard, Parliament of the United Kingdom, 27 Nov 1996: Columns 257-60.

[68] Lords Hansard, Parliament of the United Kingdom, 11 Jun 1997: Columns 964-76.

[69] General Assembly of New Zealand in Parliament, Time Act of 1974, (1974 No. 39, "An Act to consolidate and amend the law relating to fixing the time for general purposes in New Zealand"), substituted by No. 57 of 1987.

[70] *Legal Time Act* (States of Québec, 2006, c. 39)

[71] Daylight Saving Time Act, *The Revised Statutes of Alberta*, 2000, Chapter D-5.

[72] *The Official Time Act*, C.C.S.M. c. O30

[73] Time Act, 1990, Revised Statutes of Ontario, Chapter T.9, last amendment: 2009.

[74] *Time Act*, Chapter T-14 of *The Revised Statutes of Saskatchewan*, effective February 26, 1978.

[75] Directive 2000/84/EC of the European Parliament and of the Council of 19 January 2001 on summer-time arrangements, *Official Journal of the European Communities* (English ed.), 02/02/2001, L 31 , p. 21.

[76] Décret no. 78-885 du 9 août 1978, loc. cit. 3080.

[77] Howse, D. (1997), *Greenwich Time and the Longitude*. Philip Wilson Publishers Ltd., London.

[78] Nelson *et al.* (2001), "The leap second: its history and possible future." p. 525, 526.

[79] CCIR, "Standard frequency and time-signal emissions." (Recommendation 460), in *XIIth Plenary Assembly CCIR*, (New Delhi 1970), Vol. III, p. 227 (ITU, Geneva, Switzerland, 1970).

[80] CCIR, "Standard frequency and time-signal emissions." (Recommendation 460-2), in *XIVth Plenary Assembly CCIR*, (Kyoto 1978), Vol. VII, p. 9 (ITU, Geneva, Switzerland, 1978)

[81] Guinot, B., P.K. Seidelmann, "Time scales: their history, definition and interpretation." *Astronomy and Astrophysics*, Vol. 194, pp. 304-308.

[82] Seidelmann, P.K., T. Fukushima, "Why new time scales?" *Astronomy and Astrophysics*, Vol. 265, pp. 833-8.

[83] 14th CGPM, 1971, Resolution 1 (CR, 77). Also, Metrologia, 1972, 8, p. 35.

[84] CIPM Recommendation S2, PV, 38, 110 (1970). Also, Metrologia, 1971, 7, p. 43.

[85] Taylor, B.N. (ed., 2001), The International System of Units (SI), NIST Special Publication 330, p. 54.

[86] Declaration of the *Comité Consultatif pour la Définition de la Seconde*, 1980, 9, S15. Also, *Metrologia*, 1981, 17, p. 70.

[87] Bergeron , J. (ed., 1992) Transactions of the IAU Vol. XXIB, Proceedings of the XXIst General Assembly, Buenos Aires, Argentina, July 23 - August 1, 1991, Kluwer Academic Publishers, Dordrecht.

[88] Seidelmann, P.K., (2000), "Summary and Conclusions", in *Proceedings of IAU Colloquium 180, Towards Models and Constants for Sub-microarcsecond Astrometry*, K.J. Johnston, D.D. McCarthy, B.J. Luzum, G.H. Kaplan, eds.,

U.S. Naval Observatory, Washington DC, USA, pp. 411-12. (adopted by the XXIVth General Assembly of the IAU, Manchester, August 2000).

[89] Guinot, B. (1986), "Is the International Atomic Time TAI a Coordinate Time or a Proper Time?" *Celestial Mechanics*, Vol. 38, pp. 155-61.

DISCUSSION CONCLUDING AAS 11-662

With regard to the Canadian provinces, Mark Storz asked what is meant by "GMT" when it is prescribed by legislation. John Seago replied that UTC is broadcast and used in practice. Storz asked if difference was merely semantics; Seago said that the legal specifications were different but UTC as presently defined appeared to satisfy legal specifications for GMT, as they represent the same thing to within one second. Referring to the earlier comment by George Kaplan that many astronomers don't understand the difference between UT1 and UTC, David Terrett remarked almost no lawyers would understand the difference!

Frank Reed suggested that the differences in Canadian provinces was "obviously" a case of French versus English language, and "obviously the English provinces prefer something that has the name Greenwich in it." Seago replied that was possible, but added that Québec had changed from Greenwich time only recently (2006). With regard to the table regarding different European language specifications within the EU summer-time directive, Reed offered that Slovenia had only been an independent country for about 20 years and they were trying very hard to be Middle European rather than Eastern European, and trying to appear modern by choosing UTC. That the Maltese cite GMT is due to their very long association with the UK. Reed's point was that one should not read too much into these citations, as they simply reflect history. Seago replied that he was simply presenting the language of the EU directive.

Paul Gabor said that European regulatory texts tend to be translated by people trying to do their best but are not experts in astronomy. The translations are then debated and voted upon as they were presented. Gabor said he would like to inquire as to how these were translated, a question that may be more interesting that the EU directive language itself, because there is an official vocabulary dictionary used for translations, and if that dictionary is in error then those errors would propagate through to the legislative language.

George Kaplan asked about a slide that briefly appeared with regard to the standard time of the USA. Seago said that slide appeared in the presentation by mistake, as he had intended to omit that slide simply because of a lack of presentation time. He said that the slide's purpose was to present some historical background on US standard time. Originally, the US codified "mean astronomical time" in 1918, but that specification was changed to "mean solar time" in 1966 after the introduction of Ephemeris Time in the 1950s and the SI second in 1960. Ephemeris time was a type of "astronomical time" that differed from mean solar time by about half a minute. The point was that the legislative standards seemed to care about differences between a time scale linked to Earth rotation versus one that was more uniform. Another point was that the US went off the mean-solar-time standard in 2007, but codified a definition for UTC which allows the USA to interpret or modify whatever it thinks UTC is. This was likely for practical reasons because the USA maintains its own realization, but legally the USA could do what it wanted.

Rob Seaman said that the notion of leap hours had been tossed around, as well as the notion of hiding hourly adjustments within the daylight-saving time or time-zone systems. He commented that there are very interesting holes in the map of daylight-saving time practices, including Arizona, and that the changes in summer-time between the northern and southern hemisphere in opposite direction gets to be very confusing. Terrett said that these systems show that societies can indeed cope with shifting clocks by an hour; it is something than *can* be done, even if it is not practiced in every country now. Seago responded that, while people can tolerate such practices, this doesn't mean that they prefer it. He had also found a map which showed that most of the world has experimented with daylight saving time at one point but no longer. Seago said the trend seems to be that more and more nations are getting away from making seasonal clock adjustments. At this point, summer-time adjustment is pre-dominantly a North-American and a European phenomenon, and these nations dominate the ITU-R study groups that came up with the leap-hour proposal.

Session 2:
THE PAST, PRESENT, AND FAR FUTURE

AAS 11-663

THE HEAVENS AND TIMEKEEPING, SYMBOLISM AND EXPEDIENCY

Paul Gabor[*]

Timekeeping has always followed the heavens for reasons of practicality and symbolism. These two motivations can have conflicting implications for the concrete implementation of timekeeping mechanisms. Even in this eminently pragmatic age we have little control over the power of temporal symbols. This paper focuses on the deeper dynamics underlying timekeeping, proposing several notions which can serve as tools in examining the past and the present of chronology: continuity, timelessness, inertia, expediency, empirical and calculated schemes, symbolism and reality. We show that the principle of astronomical conformity in timekeeping seems to fulfill its social function even when it is not perfectly observed. In the realm of symbolism, what counts is the general perception, not the fact. Throughout history, expediency dictates a general trend away from empirical timekeeping to calculated schemes. These do not follow the principle of astronomical conformity strictly, rather, they respect it on average, aided by empirical correction mechanisms. In the present case, debating the relative merits of UTC as defined in 1972 and the proposed uniform civil time is very much an instance of this general shift away from empirical timekeeping. Calculated timekeeping does not actually agree with the Heavens at any given moment but this does not seem to jeopardize the general perception of its astronomical conformity. The general perception seems to follow the intention rather than the fact. This paper proposes to examine the dynamics of symbolism and expediency in the long history of timekeeping compromise, and apply these findings to the proposed decoupling of civil time from Earth's rotation.

INTRODUCTION

This conference focuses on issues raised by a proposal submitted to the upcoming Radiocommunication Assembly of 2012. The matter is primarily technical but it also touches upon a principle which has been regarded as inviolate and undisputed for millennia: that civil timekeeping needs to follow astronomical events. Let me state my purpose already at the outset. I will not be calling for an unconditional preservation of the UTC as defined in 1972, nor will I be advocating a concrete proposal for its revision.

What I would like to recall, however, is the depth of meaning that timekeeping represents. I would like to get in touch with the symbolic substrata and other social dynamics of timekeeping. However, I shall also point out that these dynamics exert conflicting influences upon the development of timekeeping schemes throughout history. By consequence the actual schemes are never perfect.

The principle of astronomical conformity possesses great symbolic power. And yet, as I shall try to show, this does not imply that timekeeping schemes ought to follow the Heavens as accurately as possible — just accurately enough. Symbolism can suffer some astronomical inaccuracy. In fact,

---

[*] Vatican Observatory, Department of Astronomy, University of Arizona, Tucson, Arizona, 85721-0065, U.S.A.

my overall message is that of reassurance. I believe that even if a decoupling of civil time from Earth's rotation becomes a reality in the near future, it is unlikely to have lasting consequences.

To outline the contents of this paper briefly, Section 1 examines some of the purposes of timekeeping, in order to show the depths of motivation for the astronomical conformity of timekeeping which is a principle deeply ingrained in its history. Section 2 discusses various aspects of the social dynamics underlying timekeeping, including interplay of astronomical conformity with other forces and principles. The purpose of this Section will be to gather notional tools, (hopefully) useful in approaching the subject at hand, and Section 3 applies them to several concrete cases.

Most (if not all) of the historical examples do not concern the diurnal rhythm but other elements of timekeeping. I believe this is legitimate because the issues are clearly analogous and the history of the definition of the day does not provide sufficient material.

THE HIDDEN DEPTHS OF TIMEKEEPING

Purposes of timekeeping: practical and otherwise

Timekeeping (closely followed by navigation) is doubtless the most ancient mission of astronomy. It is an eminently practical endeavor. Already in 360 BC, Plato mentions this evidence.

> The observation of the seasons and of months and years is as essential to the general as it is to the farmer or sailor. (*Republic*, 527c)[1]

This is all the more true today when, assisted by omnipresent accurate timepieces, our lives are meted out by the second under the dictate of schedules, agendas, timetables, milestones, and deadlines. We tend to forget that apart from these most prosaic and pedestrian of utilitarian purposes, there is more to timekeeping than practicality, and how ever inane these other purposes may seem, they are much more profound and pervasive.

I believe that the practical purposes of timekeeping will receive quite extensive attention from many of the speakers at this conference. What I feel needs mentioning, however, are its purposes which may not seem all that practical. That is why I will leave the former to others and concentrate on the latter.

The purpose of this conference is to ponder the consequences of "decoupling civil timekeeping from Earth's rotation". If our deliberations are to proceed with clarity and balance, I believe we need to remember that timekeeping is not just a practicality. It touches upon some of the basic, and indeed, most primitive aspects of civilization and culture. What I will try to bring forward in this Section are the results of reflections of anthropologists and historians of religion, and I have to ask your indulgence, because I am no expert in either of these fields.

Regeneration of Time

Timekeeping satisfies a considerable number of social needs, and apart from the obvious practical ones, there are many of a different sort: the need to come together to celebrate or to deal with the past, to mark a new beginning or to help bring about closure and healing, to feast and to fast. I will limit myself to an example of a social ritual we all know and participate in: the New Year's Eve.

Why do we celebrate a randomly chosen moment of the civil year? I am sure you have asked yourselves this question before. To my utter horror and consternation, I find that many undergraduates today have an answer ready. They tell me, "It is a convention," and consider the matter closed. If pressed, they explain that since it is a convention, it is just a random accident of historical evolution which could have gone another way easily. Therefore, the subject is of no deeper relevance. (This does not stop them from wholeheartedly indulging in the celebration of this "mere convention".)

The precise moment of this social ritual is indeed a convention. But the ritual itself? Is it a celebration of a convention? Is it simply that the transition from one calendar year to the next is a good pretext to indulge in some merrymaking? Or is there a more profound meaning behind it?

Mircea Eliade identifies the fundamental motivation behind the New Year's celebrations with a very archaic operation which he calls the Regeneration of Time.[2] How ancient is this procedure? Since it is clearly present in the culture of Australian Aboriginals, as well as in the cultures of the Fertile Crescent, we can derive a lower limit on the age of the Regeneration of Time by looking into the DNA chronology of prehistoric migrations (unless this is a case of "convergent evolution" which is unlikely). A recently published result allows us to deduce a lower limit of about 60,000 years![3]

The exact purpose of these rites may be debated but it is clear that they are related. According to Eliade, the Regeneration of Time is a way of dealing with the past and establishing a new beginning. This is achieved by a ritual cosmogony, a ritual re-creation of the world, which is preceded (and perhaps symbolically provoked) by a return to the primordial Chaos through another set of rituals. There are numerous variations in these but mostly they try to effect this abolition of time and return to a chaotic timelessness by suspending social and moral codes, e.g., slaves and their masters exchanging roles for the day, the whole community drinking to excess. Generally speaking, the more confusion the better.

Some elements of these rituals remain with us to this day as somewhat puzzling customs devoid of their symbolic power: the public and generalized ebriety of the New Year's Eve or the custom still alive in some schools of the youngest student becoming the headmaster for a day. It is worth remarking that these remnants of ancient rituals are not practiced only on the New Year's Eve but can be found in conjunction with other important dates public and private: the carnival before Lent, the bachelor party before the wedding, etc.

I believe we need to realize how far-reaching Eliade's insight really is. What he is saying is that human societies need a periodic abolition of "time" itself during an immersion into primordial Chaos so that Cosmos, the ordered world, may arise anew. And this implies that those who are in charge of timekeeping are in fact in charge of ritual cosmogony!

Timekeeping schemes: artifacts with symbolic power

Timekeeping schemes (calendars, UTC, the seven-day week...) are artifacts devised by individuals and societies. Once they become symbols they gain a life of their own. Regardless of their authors and their intentions, timekeeping schemes come to possess symbolic powers. They have the ability to evoke mysteries and depths which hold a powerful subliminal sway over us. They draw us in and produce diverse emotional responses. The seven-day week has a link to Creation and thus to the Creator. It is a symbol of the Creator Himself. And, according to Paul Tillich, a symbol "participates in the power of that which it symbolizes".[4] The lunar month (which has been effectively left out by Julius Cesar's calendar reform) has a clear link to the feminine with all the richness and depth of meanings this signifies. A few decades ago the Platonic year's link to the nearly forgotten idea of secular change and the ages of world has been brought forth from the collective subconscious as the supposed advent of the "Age of Aquarius" excited many minds and hearts.

I would like to stress that even those institutions which have the authority to modify timekeeping schemes themselves, have little control over the symbolic significance of what they regulate. Even though the results are mostly out of the hands of individuals and institutions, it is nonetheless infinitely preferable to meddle in such matters with at least some idea of the underlying dynamics.

UNDERLYING DYNAMICS AND ASTRONOMICAL CONFORMITY

Principle of Astronomical Conformity

The issue under consideration at this conference is an instance of the astronomical conformity of timekeeping schemes. How does astronomical conformity impact the symbolic power of timekeeping schemes? One might argue that for symbolic and ritual purposes, astronomical conformity is of no import. Eliade's Time Regeneration festivals could be declared at any point of the year. Let us examine the historical record. On the one hand, we must acknowledge that not only they can be declared without regard to Earth's rotation or orbital position, but also that sometimes this indeed was the case. On the other hand, however, we can point out that *although astronomical accuracy was often lacking in fact, what is universally present is the principle that feasts should be celebrated in conformity with natural rhythms in general, and astronomical phenomena in particular.*

In the 1st Century BC, Geminos examining the history of the Greek calendar expresses this quite clearly as the general opinion of Antiquity:

> "Accuracy in observing feasts pleases the Gods." (*Isagoge*, VIII, 6-9)[5]

The Chinese Empire was very distant from the cultures of the Fertile Crescent and the Mediterranean basin, and yet even there the principle of calendar accuracy was an undisputed goal. Between 104 BC and AD 1644 the Chinese government reformed the calendar about 50 (fifty!) times precisely because they wanted the calendar to conform to the heavens.[6] This principle was expressed in the official astronomical canons for the calendar:

> "The verification of the principles of ancient astronomical canons has to be sought in the Heavens." (*Hanshu*, j. 21A)[6]

> "It is good to conform to the Heavens in order to bring about the accords [between observation and calculation]." (*Jinshu*, j. 18)[6]

Coupling timekeeping schemes with celestial phenomena has been, for the most part, a very practical arrangement. Although I do not have an exhaustive data set I venture to claim that the evidence at hand suggests that the astronomical conformity was universally accepted as a principle, although it was not always adhered to in fact.

How can something be *universally accepted but not respected in fact*? Certain moral principles are violated by many or most people, and yet they represent a foundation of very important social institutions. The institution of marriage, for instance, is founded upon a number of principles, one of which is fidelity. This principle is a part of what defines marriage as an institution in our society. It may seem strange but the principle of fidelity "works", i.e., provides the society with an idea of what marriage is, even when some people, sometimes, are not faithful. This example shows that a principle can fulfill its function even when its observance is not perfect.

Similarly, *the principle of astronomical conformity in timekeeping seems to fulfill its social function even when it is not perfectly observed.*

This discrepancy was sometimes caused by simple lack of astronomical knowledge (e.g., in the above-mentioned Chinese case), but more often by a conflict with other principles.

Continuity and Timelessness

Continuity is a very practical aspect of any timekeeping scheme, and it can gain a special symbolic significance. A continuity since "time immemorial" lends the timekeeping scheme a certain air of timelessness which makes it sacrosanct and thus immune to change. Social institutions enjoying an air of timelessness are often accepted as sacred rather than profane, as *contemporary with the foundation myths* of the given civilization.

> "Calendars are resistant to change. The Kings of Egypt had to swear before they took office that they would not change the calendar..."[7]

The Egyptian calendar was perceived as a divine gift and thus it was unthinkable to change it even though non-conformity with the seasons soon became apparent (the year was 365 days long with no leap years). Let me stress that *timelessness is not just another name for continuity or inertia.*

Our own timekeeping still contains one "timeless" element: the 7-day week. It has been around literally since "time immemorial" because its early Mesopotamian origins are shrouded in mystery. It is clearly older than the Hebrew Scripture which enshrines it within the Creation narrative of Genesis 1, linking it to the beginning of time itself. In the context of 20th Century debate on the reform of the calendar, the Second Vatican Council maintained that the perennial status of the 7-day week needs to be safeguarded:

> "Among the various systems which are being suggested to stabilize a perpetual calendar and to introduce it into civil life, the Church has no objection only in the case of those systems which retain and safeguard a seven-day week with Sunday, without the introduction of any days outside the week, so that the succession of weeks may be left intact, unless there is question of the most serious reasons." (Constitution *Sacrosanctum Concilium* of the Second Vatican Council, Appendix *Declaration on the Revision of the Calendar*)[8]

This example demonstrates the difference between timelessness and continuity. Whereas continuity is obviously important for practical purposes, it is by no means a sufficient reason for maintaining the highly impractical continuous cycle of 7-day weeks. This particular institution can be explained by its timelessness: It cannot change because it has been in place since "the beginning of time".

It is important to note that *what counts is the general perception, not the historical fact.* In other words, although most of us know that the cycle of 7-day weeks cannot possibly have its origin at the beginning of time, this fact is overshadowed by *the feeling of timelessness surrounding this particular institution. As long as this feeling remains, the institution remains and retains its symbolic power* of evoking the Creation narrative.

Inertia

Inertia is another property of social institutions which often prevented astronomical conformity. The overdue reform of the Julian calendar is a well-known example. The reasons why the change was so difficult to tackle ranged from simple procrastination to fear caused by previous traumatic experiences. Relevant authorities did not wish to rekindle old conflicts or cause new divisions in Christendom. The long struggle to unify the celebration of Easter (2nd–6rd Century) was still vividly present in historical memory. Indeed, it is there even now.

In the mid-1920's, the League of Nations established a committee to study calendar reform, including fixing the date of Easter (e.g., on the first Sunday in April). The Committee diligently launched a comprehensive inquiry, requesting official positions of governments and other bodies, including ecclesial authorities. What is quite remarkable is that all major Christian Churches responded in the same way: There are no doctrinal objections to such a proposal but we do not want this to cause further division among Christians.[9]

Gregory XIII's reform of the calendar was a courageous move, which did indeed lead to considerable tension which still remains to be resolved with the Orthodox Churches. *Inertia in such sensitive matters as timekeeping can be a force to be reckoned with.*

Expediency: shift from empirical timekeeping to calculated schemes

Last but not least there is a general trend to replace empirical timekeeping with calculated schemes which do not follow the actual astronomical phenomena but mean parameters. As an undesired but inevitable consequence a certain discrepancy between the Heavens and civil timekeeping is introduced.

This tendency has been evident throughout the history of timekeeping. Let me recall the example of Jewish timekeeping which, until AD 70, was strictly empirical. The months were meticulously observed according to the phases of the Moon and solar years were kept according to the Metonic cycle. After the Jewish-Roman wars, however, Hillel II, president of the Sanhedrin (AD c. 320-385), enacted a transition to a calculated calendar scheme. I find this particularly revealing because of the contrast with Judaism's insistence on accurate observance of the ritual rules and regulations: even the Great Sanhedrin capitulated under expediency's inexorable pressure, abandoning the strict, empirical coupling of Jewish feasts and festivals with the Heavens.

What speaks in favor of the calculated schemes is expediency. It is much easier to follow computational rules than to maintain an astronomical vigil over the celestial phenomena relevant for timekeeping. Empirical schemes need protocols and networks in order to announce the observed parameters, making it difficult to follow the empirical time. Calculated schemes can be extrapolated into the future (which empirical ones cannot), and into the past (for which empirical ones require an access to records).

In the present case, debating the relative merits UTC and TAI is very much an instance of this general shift away from empirical timekeeping.

By construction, *calculated schemes do not follow the principle of astronomical conformity strictly, rather, they respect it on average, minimizing long-term drifts.* Since computational rules cannot predict astronomical phenomena with perfect accuracy, *calculated timekeeping needs some correction mechanisms based on observation.* The major drawback is that *often no such mechanism is defined at the outset.* As a matter of course Clavius says (*Romani calendarii... explicatio*, V, 17)[10] that the Gregorian calendar should be tweaked after several millennia by judiciously omitting a leap day to keep it synchronized with the tropical year. Yet there is no mention of such a mechanism in the Papal Bull *Inter gravissimas*:

> "Then, lest the equinox recede from XII calends April [March 21st] in the future, we establish every fourth year to be bissextile (as the custom is), except in centennial years ; which always were bissextile until now; we wish that year 1600 is still bissextile; after that, however, those centennial years that follow are not all bissextile, but in each four hundred years, the first three centennial years are not bissextile, and the fourth centennial year, however, is bissextile... and the same rule of intermittent bissextile intercalations in each four hundred year period will be preserved in perpetuity." (Gregory XIII, *Inter gravissimas*, 1581)[10]

Despite this drawback, it is clear that the calculated schemes are highly attractive because of their expediency. They are always mere tools allowing to conform with the Heavens with a minimum of empirical input in day-to-day timekeeping. What is also crucial to note is that *in general, calculated timekeeping at any given moment does not actually agree with the Heavens. This, however, does not seem to jeopardize the general perception of the astronomical conformity of such schemes. The general perception seems to follow the intention rather than the fact.*

Symbols and Reality

Why is astronomical accuracy such a central principle (although poorly observed) not just for the everyday and practical but also for the symbolic and ritual purposes of timekeeping? I believe the fundamental reason has to do with the relationship between symbols and reality. The relationship is rather complicated. Not so long ago, all human societies would have understood that the eternal realm of symbols and myths, values and virtues, Gods and heroes, was the ultimate reality, and our lives were a part of this true world inasmuch we imitated these models, i.e., inasmuch we participated in the timeless and therefore permanent (principally through ritualized behavior). Plato's ontology is a philosophical reflection of this relationship of symbol and reality.

Our society today is somewhat at odds with this view, and a different ontology rules. We say that real is what we see, touch, measure. At the same time, a substratum of our collective consciousness remains faithful to the previous view. I shall refrain from developing this topic any further. The only purpose of these terse remarks was to indicate why I think the relationship between symbols and reality is far from straightforward in our culture.

Be it as it may, *symbols are instinctively perceived as at least grounded in reality if not indeed true and real.* A good timekeeping example being noon and its dark twin, midnight. There is something magical about noon and midnight. Historically, high noon was regarded as the symbolic axis of the day, a stepping stone in the river of time. Similarly midnight represented the essence of darkness. Hours are labeled with respect to noon as hours "ante" and "post meridiem". Noon was announced (and in many places still is) by the tolling of bells, and in the days of unreliable mechanical clocks, noon was the privileged moment of clock synchronization. This was understandable as the culmination of the Sun was very easy to observe accurately. By a certain effect of inertia, even after a century of time zones people think of 12 o'clock as the moment when the Sun is the highest. *It is the symbolic significance that dominates over factual trivia.*

In the event of decoupling the civil day from Earth's rotation, the inertia of general perception will maintain all the symbolic significance of the diurnal cycle for a long time. *Symbols work as long as they are perceived as grounded in reality: timekeeping symbols work as long as they conform to astronomical phenomena at least in some way that would allow the general perception to persist.*

Perceived astronomical conformity thus lends force to timekeeping symbolism. But this relationship is mutual. *The symbolic value of timekeeping exerts a pressure on timekeeping schemes until they conform to astronomical Heavens.*

APPLICATION

We can now apply the notions outlined above to several instances of timekeeping, primarily those under consideration at this conference.

UTC Scenario 0: No change

In this scenario UTC as defined in 1972 remains the basis of civil timekeeping. The system of how the public is informed about the leap seconds may be improved perhaps.

From the point of view of the underlying dynamics as described in the previous Section, this scenario abides by the principle of astronomical conformity but is contrary to the expedient trend of abandoning empirical timekeeping for calculated schemes. I believe this discrepancy will exercise a pressure on the UTC, sooner or later bringing about a change.

UTC Scenario 1: A new calculated rule

In this case, the empirical mechanism included in the 1972 definition of UTC would be replaced by a new calculated rule which would allow the UTC to conform with Earth's rotation to within half a second.

Such a rule would satisfy all the underlying social dynamics of timekeeping. The problem is that, at this point in history, it is unrealistic. We do not possess a sufficiently long series of observations to be able to formulate such a rule.

UTC Scenario 2: Decoupled "forever"

The second option is the adoption of a uniform time (TAI) as the sole basis of civil timekeeping. For specialists, navigators, astronomers, etc., UT1 might become available in some form or another. The general public, however, will use TAI exclusively. At the moment of transition to this system, there will be no mention of ever re-coupling civil timekeeping to Earth's rotation.

Whereas such a reform maximizes expediency, it also destabilizes the diurnal symbolism, creating a growing discrepancy between the symbolic significance and the reality upon which it is perceived to be grounded. I would estimate that this instability will be very small in the beginning. It will not take too long (maybe a century), however, before the principle of astronomical conformity reasserts itself.

UTC Scenario 3: To be re-coupled

From the point of view of the underlying dynamics, the main problem with the previous scenario is that it pretends to decouple timekeeping from Earth's rotation once and for all. First of all, I do not believe that is possible because in the long run astronomical conformity will prevail. And secondly, it is just the pretense of Scenario 1 which makes it incompatible with the powerful principle of astronomical conformity; the actual measures themselves do not have to alter the general perception of timekeeping and its symbolism.

I would suggest, therefore, that if Scenario 2 is to be adopted, it would be wise to include provisions for eventual re-coupling of civil time to Earth's rotation. Such re-coupling, however, should not be a return to the existing empirical protocol, rather it should be a rule for leap seconds in conjunction with periodic reviews of the rule (Scenario 1).

Maintaining the diurnal rhythm in Space

Finally, let us apply our notions of the social dynamics underlying timekeeping to the problem of timekeeping in Space. Obviously, maintaining the diurnal rhythm will be a physiological necessity.* It is also quite clear that in the era of Space exploration, the most reasonable scheme will simply follow Earth time. But when colonies are established on other bodies, following the Earth may be impractical and it may even become a political issue. The principle of astronomical conformity will assert itself, and may prevail in the long run, although it would be struggling against the confusion of time conversions between the Earth and the various colonies. The colonies on Mars will thus adopt the Martian diurnal rhythm (with the Martian solar day of 24h 40 min); assuming that the human body will be adapt, adjusting its internal clock. Similarly the colonies on the Moon or on Ganymede might choose their diurnal rhythms as a simple fraction (1/27 or 1/7, respectively) of their orbital periods (leading to a day respectively of 24 h 17 min and 24 h 32 min).

CONCLUSION

To conclude, let me summarize my position. The principle of astronomical conformity is crucial in civil timekeeping for reasons that go far beyond the purely practical. However, precisely because this principle is so deeply ingrained in our society, a departure from it is unlikely to last.

The drawback of the 1972 protocol is that it is empirical, depending upon observations, whereas the evolution of timekeeping in general has been moving away from empirical protocols to calculated schemes. Several decades of observations should allow to predict the average spin-down rate of Earth's axial rotation with sufficient accuracy so that leap seconds could be introduced according to a simple predefined rule (with an occasional empirical tweak).

Although timekeeping enjoys the mysterious air of timelessness, its history demonstrates a surprising level of flexibility if the underlying symbolism can be maintained. Symbolism depends upon general perception, and is somewhat vague in its relationship to exact measurement.

I would prefer that civil timekeeping remains coupled to Earth's rotation. Yet, if it is to be decoupled, I suggest that the solution most in line with the long-term social dynamics described in this paper, will simply make it clear from the outset that the decoupling is temporary, and that there is a clear intention to introduce, in a few decades, a new and improved coupled scheme, based on a computational rule rather than directly on observation.

REFERENCES

[1] B. Jowett (transl.), *The Dialogues of Plato translated into English with Analyses and Introductions by B. Jowett, M.A. in Five Volumes.* Oxford University Press, 3rd ed., 1892.

---

*In the 1997 movie "Men in Black", the eponymous agency maintains Centaurian time with 37-hour days. As the character Zed says, "Give it a few months. You'll get used to it... or you'll have a psychotic episode."

[2] M. Eliade, *Le mythe de l'éternel retour: archétypes et répétition*. Gallimard, 1969.

[3] M. Rasmussen, X. Guo, Y. Wang, K. E. Lohmueller, S. Rasmussen, A. Albrechtsen, L. Skotte, S. Lindgreen, M. Metspalu, T. Jombart, T. Kivisild, W. Zhai, A. Eriksson, A. Manica, L. Orlando, F. De La Vega, S. Tridico, E. Metspalu, K. Nielsen, M. C. Ávila-Arcos, J. V. Moreno-Mayar, C. Muller, J. Dortch, M. T. P. Gilbert, O. Lund, A. Wesolowska, M. Karmin, L. A. Weinert, B. Wang, J. Li, S. Tai, F. Xiao, T. Hanihara, G. v. Driem, A. R. Jha, F.-X. Ricaut, P. d. Knijff, A. B. Migliano, I. Gallego-Romero, K. Kristiansen, D. M. Lambert, S. Brunak, P. Forster, B. Brinkmann, O. Nehlich, M. Bunce, M. Richards, R. Gupta, C. D. Bustamante, A. Krogh, R. A. Foley, M. M. Lahr, F. Balloux, T. Sicheritz-Pontn, R. Villems, R. Nielsen, W. Jun, and E. Willerslev, "An Aboriginal Australian Genome Reveals Separate Human Dispersals into Asia," *Science*, Vol. 333, 2011, pp. 1689–1691.

[4] P. Tillich, *Systematic Theology, Volume 1*. University of Chicago Press, 1973; p. 240.

[5] K. Manitius (ed.), *Des Geminos Isagoge [Geminos, Eisagoge eis ta phainomena]*. Leipzig, 1898.

[6] J.-C. Martzloff, *Le Calendrier Chinois: Structure et calculs (104 av. J.C. - 1644)*. Paris, 2009.

[7] E. G. Richards, *Mapping Time: the Calendar and its History*. Oxford University Press, 2000; p. 110.

[8] A. P. Flannery (ed.), *Documents of Vatican II*. Eerdmans, 2nd ed., 1975.

[9] *Rapport relatif à la réforme du calendrier présenté à la Commission consultative et technique des communications et transit de la Société des Nations par le Comité spécial d'étude de la réforme du calendrier*. League of Nations, Geneva, 17 August 1926.

[10] C. Clavius, *Romani Calendarii a Gregorio XIII. P. M. Restituti Explicatio. Clementis VIII. P. M. iussu edita*. Aloysius Zannettus, Roma, 1603.

DISCUSSION CONCLUDING AAS 11-663

Neil deGrasse Tyson said many interesting points were made by Paul Gabor, but there appeared to be a lack of distinction between celebrations that might be made for the "regeneration of time" versus those celebrations that are more explicitly linked to astronomical phenomena. Tyson noted that the January 1st celebration has no astronomical significance yet that does not diminish the revelry that occurs in Western cultures. Tyson asked why there should be concern about the decoupling of timekeeping devices from an astronomical measure given that many annual celebrations are already unassociated with astronomical phenomena. Noting that there are other societal, cultural, and economic factors that drive such celebratory activity, should a distinction at least be made between these two types of celebrations? Paul Gabor replied that his New Year's Eve example was to illustrate the forces behind such things, not necessarily an example of how they end up within the timekeeping schemes we have. It was just an example of an existing widespread ritual and the dynamics behind quite a number of things. The subject of the "regeneration of time" doesn't directly address the issue of decoupling civil timekeeping from Earth rotation and its mention did not intend to address that issue; rather, the point was that symbolic sub-strata are present in timekeeping. Gabor said that it may not have been the most direct example but it was the example that he felt was perhaps most telling when it comes to these more "mystical aspects".

Tyson noted that the seven-day week has no astronomical relevance yet there seems to be agreement that there is strong inertia for that to never change. The week offers its own level of "regeneration of time"—each week is a fresh start. Tyson did not necessarily see a strong urge to recouple back to a cosmic rhythm, given that societies are perfectly comfortable celebrating events in time that are non-cosmic. Gabor responded that Tyson's point was now clearer to him. He replied that while it is true that we quite enjoy the celebration of the New Year, the symbolism is no longer fully there. New Year's Day is an example of a celebration that became badly decoupled from reality and the symbolism no longer works quite properly. Tyson responded that the celebration seemed as popular as ever, to which Gabor replied that it definitely isn't. It was celebrated much longer and in much more complicated ways—in the Middle Ages the festivities continued for twelve (12) days—and today only very tiny vestiges of those revelries still exist. Terrett said that January 1 may be an arbitrary date but the interval between successive New Year's Days is not arbitrary. Seaman added that "rhythm" was mentioned by Gabor and this seems to be a more important aspect than precise solar position.

McCarthy thought that the discussion was "missing the point here." He said that the most ancient things we can find in timekeeping regard the month and year as symbols of harvest (for the year) and cycles of extra light and dark (for the month). He added that the oldest bit of timekeeping paraphernalia discovered so far kept track of the days of the month. There is always a question of why that was done, but "as near as we come up with" it was done because "an extra bit of light" was available for hunting or whatever else needed to be done. Gabor replied that was "a very utilitarian attitude;" the lunar cycle has been primarily and obviously linked with the feminine and that many of these things seem to be purely ritualistic. McCarthy countered that these are not so much rituals but are conceived only after a practical need has been established, *e.g.*, when to plant and when to harvest, with a celebration developing around a successful harvest. Gabor responded that New Year's celebrations (regeneration of time) were apparently done long before the development of agriculture. McCarthy said he would have to see that [proven].

George Kaplan asked if significant celebration occurred when the New Year was maintained closer with vernal equinox. Gabor said no, adding that the establishment of New Year's Day close to the vernal equinox was purely administrative and that celebrations of a change of year have been traditionally linked with the winter solstice. Nevertheless, celebrations at other times of year have been linked with the New Year; for example, the harvest festival is a type of New-Year's celebration as well.

AAS 11-664

LEAP SECONDS IN LITERATURE

John H. Seago[*]

The advent of electronic textbooks and the digitization of less-recent scholarly documents has resulted in significantly increasing amounts of archived information searchable via computer networks. Internet search-engine technology can therefore be used to casually discover hundreds of archival references that reference *status-quo* UTC with leap seconds. While searchable electronic documents represent only a fraction of literature actually published, such reviews suggest a range of technical fields that may rely on UTC, the literature of which would be need to be revised should UTC be redefined.

INTRODUCTION

UTC is a broadcast standard for coordinating the distribution of standard frequencies and timing signals per ITU-R Recommendation 460.[1] By international agreement, UTC represents a sequence of SI seconds progressing at the same rate as International Atomic Time (TAI) maintained by the International Bureau of Weights and Measures (BIPM), except that the epochs of UTC are infrequently adjusted relative to TAI by inserting or neglecting so-called leap seconds to assure its concordance with Universal Time (or, mean solar time at Greenwich) to within ±0.9 seconds. These adjustments are announced in advance by the International Earth Rotation and Reference Systems Service (IERS). The adjustments have been necessary because Universal Time is the astronomical basis of civil timekeeping and ensembles of atomic clocks forming TAI run at a different rate than Universal Time.[2]

The cessation of leap seconds has been discoursed without consensus for more than a decade.[3] The functional definition of UTC has remained largely unchanged since the early 1970's, so changes to the UTC standard now are expected to impact many technical operations of unknown scope. One issue contributing to this debate is the degree of expense. The absence of accurate information regarding cost is understandable because highly reliable, official cost-estimation is typically expensive to generate and approve; thus, organizations and businesses are not motivated to start financial impact assessments until they absolutely must.

Technical cost assessments must consider the labor needed to initially identify and report compulsory modifications—that is, the cost of estimating the cost. For many systems that are thought to be unaffected by UTC redefinition, methodical investigations will still be needed to conclusively prove that no modifications are in fact necessary. Once initial investigations are funded and accomplished, the next level of necessary expenditures for affected systems might involve the development of requirements or specifications (planning meetings, regulatory paperwork, *etc*.), software and hardware development, testing and benchmark development, implemen-

---

[*] Astrodynamics Engineer, Analytical Graphics, Inc., 220 Valley Creek Blvd., Exton, Pennsylvania 19341-2380, U.S.A.

tation and installation. Accurate cost estimation is also complicated by the uncertain value to be placed on lost productivity or data missed during outages caused by system upgrades and testing. Government programs may experience additional managerial and regulatory expenses to oversee and approve changes to government systems.

Personnel will also need to be re-trained or re-educated at uncertain expense, where personnel training and education are directly related to quality documentation. Much existing documentation would be invalidated by a change in the definition of UTC. Without dedicated outlays to remedy this across all technology fields reliant on UTC, technical confusion could ensue, which has its own financial repercussions.

The absence of accurate cost information is exacerbated by the ubiquitous use of UTC. Fortunately, surveys of digitized documentation might provide insight into affected technology domains. This paper experiments with an internet search engine to categorize potential areas of technical specialization that might warrant careful examination if UTC is redefined. As a check on the results, this approach is also applied to a set of documentation collected by the study groups responsible for studying and recommending a redefinition of UTC.

Figure 1. Title Search Using the Google Books Internet Search Engine.

THE EXPERIMENT

The advent of electronic publications and the digitization of older scholarly work has significantly increased the amount of archived information searchable via computer networks. Straightforward internet search-engine technology can be used to casually discover hundreds of archived references that refer to *status-quo* UTC with leap seconds. While searchable electronic documents represent only a fraction of literature actually published, queries of searchable documentation might suggest the varying range of technical fields that may care about the definition of UTC, or at the very least, identify literature that would need to be revised should UTC be redefined.

Approach

The approach of this limited study was to simply execute the following query using the Google Books internet search engine using a web browser (Figure 1):[*]

(UTC OR GMT OR "universal time") ("leap second" OR "leap seconds")

This particular query can find documents categorized as "books" that at least includes "UTC" *or* "GMT" *or* the phrase "universal time", and also includes the phrase "leap second" or its plural. The query on *universal time* will also discover *Coordinated Universal Time*.

The query for this analysis was performed in mid-September 2011 and declared about 2,400 matches and returned approximately 370 viewable outcomes. The search was intentionally limited to books because books are a relatively expensive method of archiving and disseminating information; therefore, it might be presumed that the identification of *status-quo* UTC in a book implies that the definition of UTC is not irrelevant to the book's subject or audience. Another practical reason to limit searches to books is that a general internet search query[†] results in over 200,000 reported discoveries, which was unmanageable for this exercise. Finally, a search using the Google Scholar search engine[‡] provided up to 1000 viewable results out of a reported 1,730 available returns (neglecting patents and legal papers), but many of these results were topical papers about the leap-second controversy, which were less useful for the immediate purpose of discovering potentially unknown or overlooked stakeholders in the definition of UTC.

Computer Network Time Synchronization: The Network Time Protocol

Time: From Earth Rotation to Atomic Physics

Satellite Orbits: Models, Methods, and Applications

Global Positioning Systems, Inertial Navigation, and Integration

Reference Data for Engineers: Radio, Electronics, Computer, and Communications

Figure 2. Top Five Results Searching for
(***UTC* or *GMT* or *universal-time***) & (***leap-second* or *leap-seconds***)

The five book results ranked most relevant by Google's search technology at the time of the query included books about computer network-time synchronization,[4] horology,[5] satellite orbit determination,[6] navigation (including GPS),[7] and a reference book for electrical and communications engineers[8] (Figure 2). The top results of such a search are not definitive or terribly significant, as the search rankings can vary somewhat from day by day; nevertheless these outcomes present a nice cross section of some of the various technology fields recognized by the search process.

As a point of reference, a more general search criterion was also tried without requiring the presence of "UTC" *or* "GMT" *or* the phrase "universal time", namely

---

[*] URL http://books.google.com/
[†] URL http://www.google.com/
[‡] URL http://scholar.google.com/

"leap second" OR "leap seconds"

The more general query declared more than twice as many matches (5,700) but only about one-dozen more viewable returns. The more general query also ranked less-technical books higher in the search outcomes. This is evidenced by looking at the top five search results for this more general search criterion, which included a history book and a general interest science book in the third and fourth positions. The general-interest science book contains non-technical treatments of recent science news—in this case, the controversy over abolishing leap seconds.[9] The astronomical history book includes a page-length discussion explaining the need for leap seconds to the layman.[10] This latter book does not appear to mention UTC, GMT, or Universal Time at all, as therefore it did not appear in the more-limited search requiring the presence of "UTC" *or* "GMT" *or* the phrase "universal time".

| Time: From Earth Rotation to Atomic Physics | Computer Network Time Synchronization: The Network Time Protocol | Eclipse: The Celestial Phenomenon That Changed the Course of History | The Why? Files: The Science Behind The News | Reference Data for Engineers: Radio, Electronics, Computer, and Communications |

Figure 3. Top Five Results Searching for
leap-second **or** *leap-seconds*

Caveats

Unfortunately this experiment is fraught with limitations that affect the completeness and conclusiveness of the results. Several shortcomings are now mentioned, which should be considered when trying to interpret these data.

Extremely Limited Discovery. The methods of discovery only identified titles written in English; also, searchable electronic documents represent only a fraction of published material. Searchable titles are sometimes scanned imperfectly, inhibiting recognition of the search keyword(s). Therefore, the results presented here must represent an extremely limited sample. Because the discover methods are limited, one cannot make firm conclususions about the full size and scope of fields interested in UTC redefinition.

Degree of Consequence. The context of most search outcomes is lacking. Often only a few lines of text surrounding the search words are exposed. The relevance of the definition of UTC is not always apparent, so it is hard to draw conclusions about the implications of consequences of a possible redefinition. The mention of UTC and leap seconds within a publication does not guarantee that there will be consequences within the author's profession, excepting the matter of obvious documentation revisions.

Subjectivity of Categorizations. The identification of potential stakeholder groups requires classification of references according to topical domains. This process is somewhat subjective and likely imperfect. Many titles are multi-discipline, so a determination was made as to which category the title best fits. More than one copy or edition of the same title can occasionally appear within the outcome of searches, the duplication which may result in a slight increase in the size of some categories. Because of these reasons, one cannot make firm conclusions about relative sizes of fields identified.

Despite these qualifications, there appears to be some merit to the search method as a means for exploring potential areas that could be investigated more deeply as to their reliance upon and usage of UTC. The process is also useful for illustrating that documentation revisions may not be a trivial or inexpensive issue if UTC is redefined. Search results are nonetheless informative, but not conclusive.

Outcomes

Each viewable title was assigning to one of 23 topical categories. These categories were developed based on what was discovered. The categories and their contents are now generally described, with the number assigned outcomes specified by a parenthetical number.

Technical Reference (43). This category included the largest numbers of entries and includes encyclopedias, dictionaries, handbooks, and reference data related to computer science, engineering (general), audio engineering, environmental engineering, geophysics, astrophysics, astronomy, radio standards, electronics, telecommunications, broadcasting, scientific units, government and military terminology, mapping, charting and geodetic terminology, physics, fiber optics, scientific guides, economists desk reference, tables and formulae, science and technology.

General Reference (3). This category included general encyclopedias and dictionaries like Britannica.

Astronomy – professional and student (31). This category included titles aimed at professional astronomers, with topics related to fundamental astronomy, observational astronomy, spherical astronomy, radio astronomy and interferometry, astrometry, cosmology, high-energy astrophysics and astrophysical quantities, asteroseismology, relativity, and astronomy of ancient cultures. It also included titles related to astronomical methods, such as data analysis and reduction, including analysis software and information handling. The references mentioning *status-quo* UTC were predominately textbooks, but also included a few meeting proceedings, doctoral dissertations, and professional notices. This category included published proceedings related to International Astronomical Society (IAU) assemblies, meetings and colloquia.

Astronomy – amateur (15). This category included titles aimed at amateur or advanced amateur astronomers, including telescope usage, astronomical computing, observing methods and observing technology (notably eclipse and comet observation), and star atlases.

Computing (25). This category included titles related to computing technology, multimedia computing, program and system information protocols, (distributed) operating systems and OS timekeeping, system administration, embedded and ubiquitous computing, and fault-tolerant computing. This category attempted to collect titles on computer science not obviously dedicated to programming.

Database Technology (9). This category included titles related to databases and database architectures, database management, query languages and programming, and metadata tools.

Information Technology (5). This category included titles related to information technology / information services and its systems, applied informatics, local networking, web-based energy information and control systems.

Software (13). This category included titles related to commercial software, libraries, software languages and programming, web application design, and spatial and geographic visualization (KML), Unix programming, object-oriented programming. Programming or scripting languages specifically notably cited included C, C++, R, Ada, awk, gnu, Haskel, ActionScript, Matlab, Standard ML (SML), and Java.

Metrology (28). This category included titles related to metrology, including units and fundamental constants, measurement and control instrumentation, standards and guides, data processing, the metric system, historical surveys, government-issued publications, proceedings, transverse disciplines, both scientific and of general interest.

Almanacs, Atlases, and Yearbooks (25). This category included titles related to almanacs, explanatory supplements, or similar reference information, aimed at either professionals (such as astronomers or navigators) or at the general public, including some calendars and national yearbooks. Because some general-interest almanacs are annual publications, some slight duplication was noticed, but no attempt was made to track and remove any suspected duplications from the sample.

Navigation and Surveying (22). This category included titles related to general navigation technology, perhaps including but not specific to GNSS, such as inertial navigation and integration, avionics, celestial navigation, radio navigation, surveying, transport geography and spatial systems, emergency navigation, piloting, sailing, and seamanship, This category included mostly books but also a few journal articles and published proceedings.

GNSS Technology, Applications, and Practice (19). This category included titles predominately devoted to the theory, technology and practice of global navigation satellite systems (GNSS), including GPS, GLONASS, Beidou / Compass, and Galileo, GNSS-based geodesy and surveying techniques, software and data processing, receiving equipment, and some geographic information systems (GIS) reliant on GNSS.

Earth science (18). This category included titles related to hydrography, oceanography, seismology, (geometrical and satellite) geodesy, geodynamics, geophysics, geological site surveys, polar motion monitoring and nutation studies. This category also included one IUGG proceedings.

Spacecraft (18). This category included titles related to the technology of space-based systems, including spacecraft operations and communications, deep-space networks, spacecraft technology, space vehicle design, spacecraft guidance, remote sensing, and interplanetary missions. It also included titles related to the theory and application of celestial mechanics, dynamical astronomy, astrodynamics, and related areas, including planetary or satellite motion, satellite orbit determination, geostationary orbits, and satellite navigation. It included some published government reports and journals.

Science – general interest (16). This category included titles related to scientific presentations aimed mostly at the general public, including space science, timekeeping issues, history of scientific technology, nature of light, critical thinking, published magazine volumes (American Scientist, Nature, Science, Popular Science), and books on current events.

Telecommunications (17). This category included titles related to telecommunication technology, including digital radio and audio broadcasts, fiber optic communication, real-time systems, telemetering, wireless internet, cellular networks, signals, network synchronization, radio relay, future of UTC, television, and amateur radio. It included mostly books but some published government reports and documents. This category also included one proceedings related to PTTI.

Network time transfer (7). This category included titles related to network time protocol (NTP), signal communication architectures and protocols, network clock synchronization, and government services related to network time distribution.

Horology (16). This category included titles related to timekeeping technology aimed at either technologists or the general public, including historical surveys of timekeeping technology, calendrical computations, clock hardware, astronomically-based time, time-scale determination, clock instrumentation and control, and video time encoding. It included three wikipedia-based print-on-demand e-books discussing leap seconds, UTC, and TAI.

Physics, Science, and Math (12). This category included a variety of titles related to physics and science (and to a much lesser extent mathematics), including student textbooks, instructional materials for teachers, physical science, scientific computation, quantum mechanics, gravitation and relativity, and Royal Society proceedings.

Applied engineering (8). This category included titles related to miscellaneous engineering applications, such as mechatronics, cryptography, industrial research, estimation theory, manufacturing, quality control and applied statistics, and machine design.

Electronics (4). This category included titles related to information about electronics, general textbooks, instrumentation, power system management, and popular electronics.

Economics (4). This category included titles catering to economic applications, such as global trade issues, research and development, and the technology of high finance.

General Interest (11). This category included titles related to information to popular culture, trivia, pseudoscience, science fiction, human philosophy, metaphysics, psychology, science words, history, literature, (some of which may only reference UTC and leap seconds in their glossaries).

Table 1. Title Count by Topical Category

Topical Category	#	Topical Category	#
References (Technical + General)	46	Spacecraft	18
Astronomy	46	Telecommunications	17
Computing	25	Time Transfer	7
Databases (+ IT/IS)	14	Horology	16
Software	13	Science (general interest)	16
Metrology	28	Physics, Science, Math	12
Almanacs, Atlases, and Yearbooks	25	Applied engineering	8
Navigation and Surveying	22	Economics	4
GNSS	19	Electronics	4
Earth science	18	General Interest	11
		Total	*369*

A larger number of smaller categories is most useful for pinpointing potential domains that might be affected by UTC redefinition; however, for a concise tabular presentation, it was useful to combine a few related categories (Table 1). Databases and IT/IS are arguably related, as is professional and amateur astronomy, and general and technical references.

Although we can draw no strong conclusions regarding the relevance of UTC to certain fields based on the numbers of counts per category, it may not be surprising to see that most references citing leap seconds occur in what is described here as "technical reference" literature (encyclopedia, dictionaries, handbooks, atlases, almanacs, *etc.*), astronomy, and computing.

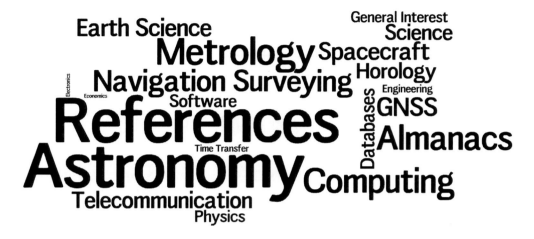

Figure 4. Word Cloud of Assigned Book Categories.

To get a visual sense of the categories and their relative sample sizes, we employ the use of a "word cloud".[*] A *word cloud* is a semi-random arrangement of the words whereby the height of the word correlates to word's usage frequency. For this application, we express a consolidated category as a single word or word pair, with its height being in proportion to the number of samples assigned to that category. Figure 4 presents a word cloud for the book data consolidated in Table 1. With the word cloud, it is visually apparent that *astronomy*, *references*, *metrology*, *almanacs*, and *computing* are the top five domains in terms of frequency of mentioning UTC and leap seconds.

EXPLORING INCORRECT NOMENCLATURE

This activity can be extended to discover other interesting results. For example, the transposed phrasing "Universal Coordinated Time" is sometimes observed in literature; it is conjectured that people mistakenly use this rearrangement because the acronym "UTC" suggests that the adjective "Universal" should be first. One might speculate that such usage reflects a lack of familiarity with UTC as a realization of Universal Time, which could be useful in prioritizing outcomes to be investigated.

A search on the term "Universal Coordinated Time" used in combination with "UTC" displayed approximately 450 books, the largest percentage appearing in texts related to computer science and programming, general interest science and technology, navigation, and communica-

---

[*] http://www.wordle.net/create

tions. Of these, approximately 90 books actually used the erroneous term "Universal Coordinated Time" while also mentioning leap seconds. Other incorrect usages and their frequency of occurrence are noted in Table 2.

Table 2. Transposed Nomenclature Used with "UTC" Appearing in Books

Searched Criterion	Declared Matches	Displayed Results
Coordinated Universal Time	9,160	426
Universal Coordinated Time	2,440	446
Universal *Coordinate* Time	65	65
Coordinate Universal Time	24	24
Coordinated Time Universal	5	5

CATEGORIZATION OF STUDY DOCUMENTS

For the purposes of comparison, categorizations were also applied to a list of 43 documents collected by the ITU-R for supporting a redefinition of UTC. In April 2008, a drafting group chaired by Dr. Elisa Felicitas Arias of the BIPM was formed to document and summarize the data and materials considered by the ITU-R Working Party 7A and its Special Rapporteur Group during its ten (10) year study period 1999-2008.[11] The drafting group developed a "final" report (not publicly released) that included an overall summary and a number of derived conclusions. The drafting group believed that the collection of documents "offered a full and comprehensive perspective of the overall [study] effort" with the caveat that the available material was still "not necessarily a complete compilation."

Outcomes

The drafting group reportedly processed 26 contributed documents collected over a ten-year study interval, in addition to 17 reports and statements generated within the ITU-R. Most of these were received following the 2005 leap second, and several were received in 2008 especially for the "final" report. Because the details of the original documents are not publicly available, the classifications are loosely assigned according the area of expertise of their originators.

Network time transfer (6). Three letters from NIST personnel, one report from Time Dissemination Working Group, PTB, one report from the USNO, one report from the Internet Engineering Task Force.

Metrology (2). One letter from Bundesamt für Eich- und Vermessungswesen (BEV, Austria), one letter from Time Section of the *Real Observatorio de la Armada* (ROA, Spain).

Astronomy (5). Two reports from the American Astronomical Society, two reports from the IAU, and one communication from the Royal Astronomical Society.

Navigation and Surveying (2). One letter from Ephemeris Section of the *Real Observatorio de la Armada* (ROA, Spain), one communication from the Royal Institute of Navigation.

GNSS (3). One letter from the Galileo Project Office, European Space Agency (ESA), one general service bulletin from SAAB, one general service bulletin from ACR Electronics.

Earth Science (1). One letter from the International VLBI Service for Geodesy and Astronomy.

Spacecraft (5). One survey report from the French Space Agency (CNES) and Paris Observatory, one (1) letter from EUMETSAT, one letter from JSAT Corp., two conference papers from this author.[12, 13]

Applied Engineering (1). Report of the International Union of Radio Science (URSI) Commission J Working Group on the Leap Second (2000).[14]

Horology (1). One journal article.[15]

Telecom (17). ITU-R reports and statements submitted by member administrations and organizations.

Table 3 Study Documents Count by Topical Category

Topical Category	#	Topical Category	#
Astronomy	5	Spacecraft	5
Metrology	2	Time Transfer	6
Navigation and Surveying	2	Horology	1
GNSS	3	Applied engineering	1
Earth science (+ IUGG)	1	**Telecom (ITU-R)**	**17**
		Total	43

As with the book-query data, some categorizations could overlap; the general service announcements could be classified as either GNSS or electronics, for example. The URSI survey result is hard to categorize because of its broad scope; applied engineering was chosen. Table 3 represents the final summary, which includes fewer than half as many categories compared to the book study, and the internally generated ITU-R documents assigned to the *Telecommunication* category makes up almost 40% of the documentation considered by the study.

Figure 5. Word Cloud of Assigned Study-Document Categories (Including ITU-R Documents)

In the interest of comparing the book results with the drafting group documents, a word cloud visualization was also attempted (Figure 5). However, the dominance of internally generated ITU-R documents assigned to the *Telecommunication* category greatly skews the word cloud. To compare to the book study, the word cloud was regenerated and limited to only the external documents contributed to the study group, excluding the *Telecommunication* category (Figure 6).

Figure 6. Word Cloud of Assigned Study-Document Categories (Excluding ITU-R Documents)

The word-cloud views of the study-document categories provide an interesting visual contrast to that of the internet book survey. That the number of study-document categories is much less may suggest that many of the categories derived from the survey of book literature are not really stakeholders in the definition of UTC. This is likely true for some categories such as general-interest literature; however, major technical categories appeared absent from the study-group documentation categories, such as computing, software, databases, and technical reference materials (including almanacs). It must be presumed that computing technology and reference materials will be affected by UTC redefinition in some way, but for this analysis it was unclear how these domains were represented and assessed by the ITU-R study group.

CONCLUSION

Information technology has allowed technical documentation to become easier to store and access, such that it has become much more cost effective to manage and develop documentation over time. However, inaccurate or outdated information can also be easily circulated, widely exposed, and persist inside and outside of cyberspace for a very long time. Because the creation of good technical documentation is labor intensive, the expense is often neglected or deferred. Thus, the risk of technical confusion resulting from a redefinition of UTC is not negligible. At the very least, documentation changes will affect technology domains that are otherwise unaffected by a redefinition of UTC; therefore, this aspect is presumed to be very far reaching.

If UTC is redefined, the identification of affected technical areas will be tricky. This paper proposes that internet search-engine technology could be used to help identify various technical domains that dependent on the definition of UTC. Such investigations could be readily expanded to potentially explore sensitivities to proposed changes within these domains. Although the scale of this study is small and by no means conclusive, it reinforces a perception that technologies related to astronomy and computing might be significantly affected by UTC redefinition. There is also likely to be a significant amount of general reference materials in the form of encyclopedias, dictionaries, and general-purpose almanacs that would be need updating; often these types of resources are managed and edited by those not necessarily familiar with horology. Finally, based on declared study-group documents, it is unclear how thoroughly certain technology domains have been represented and assessed by groups studying the issue of UTC redefinition. Investigation of domains such as computing, software development, programming, databases, and perhaps applied engineering might be especially beneficial.

REFERENCES

[1] Recommendation ITU-R 460-6 (2002), Standard-frequency and time-signal emissions (Question ITU-R 102/7), in ITU-R Recommendations: Time Signals and Frequency Standards Emission, Geneva, International Telecommunications Union, Radiocommunication Bureau.

[2] McCarthy, D.D. (1991), "Astronomical Time." *Proc. IEEE*, Vol. 79, No. 7, pp. 915-920.

[3] Finkleman, D., J.H. Seago, P.K. Seidelmann (2010), "The Debate over UTC and Leap Seconds". Paper AIAA 2010-8391, from the Proceedings of the AIAA/AAS Astrodynamics Specialist Conference, Toronto, Canada, August 2-5, 2010.

[4] Mills, D. (2006), Computer network time synchronization: the network time protocol, CRC Press, Francis & Taylor Inc, Boca Raton. pp. 2, 16, 116, 209-11.

[5] McCarthy, D.D., P.K. Seidelmann (2009), *Time-From Earth Rotation to Atomic Physics*. Wiley-VCH, Chapter 14.

[6] Montenbruck, O., E. Gill, *Satellite Orbits. Models, Methods and Applications*. Springer, 2000, pp. 159, 168-9.

[7] Grenwal, M.S., L.R. Weill, A.P. Andrews (2007), *Global positioning systems, inertial navigation, and integration*. pp. 44-6.

[8] Van Valkenburg, M.E., W. Middleton (2001), *Reference data for engineers: radio, electronics, computer, and communications*. pp. 1-143 – 1-160.

[9] Tenenbaum, D. (2009), *The Why? Files: The Science Behind the News*, Penguin, p. 7.

[10] Steel, D. (2001), *Eclipse: The Celestial Phenomenon That Changed the Course of History*. Joseph Henry Press. p. 219.

[11] Bartholomew, T.R., "The Future of the UTC Timescale (and the possible demise of the Leap Second)–A Brief Progress Report," *Proceedings of the 48th CGSIC Meeting*, Savannah GA, 2008.
(URL http://www.navcen.uscg.gov/pdf/cgsicMeetings/48/Reports/Timing%20Subcommittee/48-LS%2020080916.pdf)

[12] Seago, J.H., and Storz, M.F., "UTC Redefinition and Space and Satellite-Tracking Systems," in: *Proceedings of the ITU-R SRG Colloquium on the UTC Timescale*, IEN Galileo Ferraris, Torino, Italy, 28-29 May 2003.
(URL http://www.ucolick.org/~sla/leapsecs/torino/seago.pdf)

[13] Seago, J.H., and Seidelmann, P.K., "The Times—They Are a Changin'?" (AIAA 2004-4848), *AIAA/AAS Astrodynamics Specialist Conference and Exhibit Proceedings*, 16-19 August 2004, Providence, Rhode Island.

[14] Matsakis, D., et al. (2000), Report of the URSI Commission J Working Group on the Leap Second, July 2, 2000.
(URL http://www.ietf.org/mail-archive/web/ietf/current/msg13828.html)

[15] Nelson, R.A., D.D McCarthy., S. Malys, J. Levine, B. Guinot, H.F. Fliegel, R.L. Beard, T.R. Bartholomew (2001), "The leap second: its history and possible future." *Metrologia*, Vol. 38, pp. 509-529.

DISCUSSION CONCLUDING AAS 11-664

Rob Seaman noted that the number of results that could be seen was only a fraction of those that could not be seen, so there is no reason to assume that those unseen results might not extend into broader categories. John Seago replied that a Google search result declared a certain number of discoveries (several thousands), but as one tried to view those outcomes by displaying them as a list, the search engine would only allow the user to view so many outcomes. Therefore, Seago commented that he had no way to identify the content of outcomes that were not displayable, or even if the report of thousands of un-displayed outcomes was accurate.

Neil deGrasse Tyson commented that clearly the exercise was to get some kind of quantitative understanding of impact. Seago added that the goal of exercise was to possibly discover technical areas that might have some interest in a redefinition in UTC. Tyson said it might be useful to investigate how pockets of research or reporting have to change due to alterations of any other specification that might go on. Standards change regularly and this is why the National Institute of Standards and Technology (NIST) exists. If the experiment was a concern about the impacts of redefining UTC rather than just the awareness of the impact, then it might be that this industry is changing its parameters all the time for other reasons as well. Seago replied that Tyson's point was a very good one, but Seago's effort did not go very far in that direction, because his primary goal was simply an attempt to identify possibly affected domains that might be impacted. Seago noted that while a census of the topical areas was conducted and relative results were reported, as was alluded to by Seaman's comments, it seemed difficult to make good judgments based on relative numbers of "hits" and this caveat was noted in the presentation. As a point of comparison, Tyson replied that it might be interesting to choose another definition from metrology that had been redefined within the last century, like the length of meter, to see which domains might care about that. Seago agreed that would be interesting and that Tyson's point was well taken. Seaman speculated that such a search could emphasize the difference as well, so it is unclear which way the result would go, but agreed that it would be interesting.

Arnold Rots commented that the restriction to English-language books was not surprising considering that the search term was English. Seago agreed, explaining that he is unfamiliar with the technical nomenclatures of other languages and therefore did not feel he could search or categorize results in any other language. Even if Seago could do this, he lacked the time to attempt it; he is not part of an established study effort on this issue and the results are simply informative to demonstrate a proposed approach. With regard to the "Spacecraft" category, Rots asked if there was any difference in emphasis on down-looking versus up-looking spacecraft. Seago said that, while there were apparently interesting aspects to many of the discovered references, he was not able to delve into any details of these references due to the large numbers of references involved. For the purposes of his paper, he was just looking to see if a particular topical domain mentioned "leap seconds" and "UTC" which might indicate some stakeholder domain in the definition of UTC. Even then, the results wouldn't say for sure that a stakeholder domain existed.

Steve Malys asked if each reference was only counted once; Seago replied that he assigned each title to a specific topical category and thereby each was only counted in the tally once. Seago admitted that there is some subjectivity in the decision because some titles are multidisciplinary; also, a few titles may be released as different editions and thereby the separate editions may have contributed to the tallies.

Seaman commented that this exercise would be interesting to extend to other types of technical literature such as journal articles. Seaman also said that just because a book doesn't include the phrase "leap second" doesn't mean that a book's topic doesn't depend on the leap second. If Coordinated Universal Time is redefined to no longer be Universal Time, then the user base interested in a DUT1 correction would suddenly increase. Seago agreed and added that the approach is extensible.

David Simpson noted that the acronym "UTC" is frequently misinterpreted as an abbreviation for *Universal Time Code* among software people; a search on that term might reveal additional discoveries within literature about software.

AAS 11-665

TIME IN THE 10,000-YEAR CLOCK

Danny Hillis,[*] Rob Seaman,[†] Steve Allen[‡] and Jon Giorgini[§]

The Long Now Foundation is building a mechanical clock that is designed to keep time for the next 10,000 years. The clock maintains its long-term accuracy by synchronizing to the Sun. The 10,000-Year Clock keeps track of five different types of time: Pendulum Time, Uncorrected Solar Time, Corrected Solar Time, Displayed Solar Time and Orrery Time. Pendulum Time is generated from the mechanical pendulum and adjusted according to the equation of time to produce Uncorrected Solar Time, which is in turn mechanically corrected by the Sun to create Corrected Solar Time. Displayed Solar Time advances each time the clock is wound, at which point it catches up with Corrected Solar Time. The clock uses Displayed Solar Time to compute various time indicators to be displayed, including the positions of the Sun, and Gregorian calendar date. Orrery Time is a better approximation of Dynamical Time, used to compute positions of the Moon, planets and stars and the phase of the Moon. This paper describes how the clock reckons time over the 10,000-year design lifetime, in particular how it reconciles the approximate Dynamical Time generated by its mechanical pendulum with the unpredictable rotation of the Earth.

INTRODUCTION

The 10,000-Year Clock is being constructed (Figure 1 and Figure 2) by the Long Now Foundation.[**] It is an entirely mechanical clock made of long-lasting materials, such as titanium, ceramics, quartz, sapphire, and high-molybdenum stainless steel. The chamber for the pendulum-driven clock is currently being excavated inside a mountain near the Texas/New-Mexico border in the southwestern United States. Its mechanism will be installed in a 500-foot vertical shaft that has been cut into a mountain to house the clock. The clock is designed to maintain its long-term accuracy by synchronizing to the Sun, through heating of a sealed chamber of air by a shaft of light that shines into the mountain at solar noon. The air pressure change generated by this light drives a piston that adjusts the clock. Diurnal solar heating is also used to wind the weight that powers the pendulum and sun-tracking mechanism.[1,2,3]

---

[*] Co-chair, Long Now Foundation, Fort Mason Center, Building A, San Francisco, California 94123, U.S.A.

[†] Senior Software Systems Engineer, National Optical Astronomy Observatory, 950 N Cherry Ave., Tucson, Arizona 85719, U.S.A.

[‡] Programmer-analyst, UC Observatories & Lick Observatory, University of California, 1156 High Street, Santa Cruz, California 95064, U.S.A.

[§] Senior Analyst, Solar System Dynamics Group, NASA/Jet Propulsion Laboratory, California Institute of Technology, 4800 Oak Grove Drive, Pasadena, California 91109, U.S.A.

[**] http://www.longnow.org.

The mechanical Clock is a digital computer with analog inputs and outputs. The principle inputs are the oscillations of the pendulum, sunlight falling on the solar synchronizer, and the pre-computed correction to solar time as realized in the equation-of-time cam. The outputs include an analog depiction of the sky in the orrery and a digital display of the Gregorian calendar date. These inputs and outputs are discussed below.

FIVE KINDS OF TIME GENERATED BY THE CLOCK

Subtle differences in the definition of time can make a significant difference over 10,000 years. Universal Time will be 2×10^5 seconds behind Terrestrial Time (TT) in 10,000 years, because they tick different seconds. Barycentric Dynamical Time will be 5×10^3 seconds ahead of TT even though both tick *SI* seconds.[*] None of these time bases is more inherently correct than another. They are just different ways of labeling a sequence of instants in time. Over its 10,000 year life the display of the clock, averaged over the course of a year, is designed to remain within 300 seconds of Universal Time (UT).

How does the concept of UT affect the clock? As the word "average" suggests, Universal Time is a mean realization of solar time. The length of the solar day varies continually, seasonally and over the long term. In particular, there is a secular slowing due to lunar tides. By removing the seasonal variations, the implicit usage of UT[†] permits the regular pendulum-driven cadence of the clock to adapt to the long term changes.

The clock ultimately uses but does not explicitly display the solar system barycentric coordinate time of general relativity. This timescale is the independent variable in the equations of planetary motion that emerge from Einstein's space-time field equations and metric tensor. It is therefore a direct expression of our current understanding of the space-time relationship.[4, 5, 6] A defined relationship between coordinate time in the solar system barycentric frame and International Atomic Time (TAI) at a site on Earth (or Earth satellite) can be used to properly relate these timescales.

Current terminology of the International Astronomical Union (IAU) refers to the coordinate time of general relativity as Barycentric Dynamical Time (TDB) and, when properly related to a site on or near the surface of the Earth, Terrestrial Time (TT) can be derived. The distinction between these two dynamical timescales is generally periodic with an amplitude of 0.002 seconds in the course of a year, due to the Earth's elliptical orbit around the Sun. The self-correcting mechanisms of the clock described below mean the distinction is not significant operationally, and so "Terrestrial Time" will be used in the remainder of the paper, and the term "86400-second day" will be used refer to an 86400-second interval of Terrestrial Time.

With these underlying theoretical concepts in mind, there are five different time bases generated within the clock: Pendulum Time, Uncorrected Solar Time, Corrected Solar Time, Displayed Solar Time, and Orrery Time. Each will be described in detail.

---

[*] The *SI* second is defined based on specific transitions of a cesium atom. If this is measured on the rotating Earth, relativistic effects that depend in part on the earth's rotation will influence a cesium clock's rate relative to a clock in an inertial frame. Thus, even the measurement of the *SI* second that is used to define Terrestrial Time will be at least slightly dependent on the unpredictable rate of Earth's rotation, although the effect of variations in the Earth's rotation is not currently within the sensitivity of our measurements.

[†] The proposal to redefine Coordinated Universal Time (UTC) raises the issue of which concept will live longer, UTC or general-purpose "UT" with its meaning of mean solar time. The Clock will likely see many such cultural debates over its long lifetime. In the case of UTC no longer providing actual Universal Time, visitors' wristwatches will wander willy-nilly over the centuries in comparison to the Clock's solar-synchronized display.

Pendulum Time

The generative time base in the clock is called Pendulum Time. It is created by counting the 10-second cycles of the clock's gravity pendulum.[7] Pendulum Time advances every 30 cycles of the pendulum, once every five minutes. All other forms of time in the clock are derived from Pendulum Time and measured to this five-minute resolution. The mechanical pendulum is designed to have a long-term accuracy of better than 800 milliseconds per day, about one tick per year. The choice of a five-minute resolution is governed primarily by the requirement that the clock be robust through sunless periods lasting many decades. This also simplifies the design of the solar synchronizer and provides sufficient precision for the orrery display. The pendulum is driven by stored energy provided by a solar power mechanism or human winding. The clock is designed to operate unattended for as many as 10 millennia between human visits. The power system does not depend on sunny days, but only on diurnal temperature variations.

Uncorrected Solar Time

Uncorrected Solar Time is computed from Pendulum Time by adding a seasonally varying correction for the equation of time. This varies about half an hour or so over the course of a year. The analemma is a two-dimensional graph of the equation of time (Figure 3). The equation of time also varies from year to year, century to century. This variation is predictable within the uncertainty of the Earth's rotation rate, so it is pre-computed over the clock's 10,000-year operating interval and stored in the equation-of-time cam based on values is derived from an extended form of the JPL DE422 solar system solution The function encoded in the cam assumes the predicted slowing of the earth's rotational period at the rate of 1.8 milliseconds per day per century (Figure 4). The cam function also encodes the uncertainty of slowing of the Earth's rotational period, by a mechanism that will be described later.

The equation-of-time cam is driven by Corrected Solar Time, which is derived from Uncorrected Solar Time. Since this in turn is used to generate Corrected Solar Time, this sounds like a circular definition. It is, but because the corrections of Corrected Solar Time create a very small change in the equation of time correction, the "gain" around the loops is very small and the series converges.

Like Pendulum Time, Uncorrected Solar Time advances once every five minutes. Uncorrected Solar Time should be regarded as a purely internal time scale within the Clock as it retains the inevitable drift of the pendulum, corrected in the next step.

Corrected Solar Time

Corrected Solar Time is a realization of local apparent solar time at the site of the Clock. It is intended to stay synchronized over the long-term with the rotation of the Earth. The solar synchronizer adds the required correction automatically whenever the Uncorrected Solar Time deviates from apparent solar time by more than five minutes.

Corrected Solar Time, having ticks straddling solar noon, is computed from Uncorrected Solar Time by adding or subtracting a correction tick if the sun synchronizer detects solar noon while the Uncorrected Solar Time is not within the noon interval. This solar synchronization corrects for both the inaccuracy of the pendulum and the unpredictable component of the Earth's rotation.

Thus, corrected solar time is the start time, plus the total number of pendulum ticks, plus the equation of time correction, plus the sum of the signed correction ticks. The correction is positive if the Sun is detected before the just-before-noon tick, and negative if it is detected after the just-after-noon tick. No correction is made if the Sun is detected in between these two ticks. On any given day a maximum of a single correction tick will be added or subtracted. Thus, if the Clock is ever in a state many minutes or hours from the correct local apparent solar time, it may take multiple sunny days to correct itself.

Because the correction may be negative, Corrected Solar Time as generated is not monotonic; it can go backwards, repeating a short interval of time. To prevent the mechanisms that are driven by Corrected Solar Time from moving backward, a mechanism for storing "borrowed-time" is used to stop the advancing shaft until the time "catches up," creating a monotonic version of Corrected Solar Time that pauses rather than backing up. Since only a single correct tick is added per day, the mechanism will not cause the clock to pause for more than five minutes.

All displays on the clock are designed to maintain accuracy to within a five-minute tick over the entire 10,000-year lifetime of the clock, as long as the clock detects solar synchronization at least once a year. It is possible that an unusual event such as a volcanic eruption could prevent the clock from detecting noon for more than a year. In this case, the clock may temporarily drift away from the correct time, but it will eventually resynchronize when clear skies reemerge as long as it has not drifted more than 12 hours. This allows the clock to successfully recover after more than a century of overcast skies. In that case it will require at least as many sunny days to recover the correct solar time as the number of 5-minute ticks the clock has drifted, 12 days per hour of drift. If the skies are overcast for several centuries, the Clock's calendar display could gain or lose days in its calendar count, but the correct solar time will be restored when the Sun returns.

Displayed Solar Time

Displayed Solar Time is generated from Corrected Solar Time each time the clock is wound by a visitor. The solar power mechanism is not used to advance this part of the mechanism. All of the clock's displays are derived from Displayed Solar Time, or Orrery Time, which is derived from it. Displayed Solar Time pauses between windings. It moves forward each time the clock is wound until it matches the monotonic version of Corrected Solar Time. The displays move forward as Displayed Solar Time moves forward, so the displays move only while the clock is being wound. Depending on how long it has been since the clock has been wound, winding may take a few minutes or many hours. The winding mechanism will stop the winding when Displayed Solar Time matches Corrected Solar Time.

Displayed Solar Time drives the portion of the dial that shows the apparent position of the Sun in the local sky. The displays indicate Displayed Solar Time in increments that straddle solar noon, so the just-before-noon tick, and the just-after-noon tick are displayed as 2.5 minutes before and after solar noon, respectively. Displayed Solar Time also drives a digital calendar display, which displays the exact date according to the Gregorian calendar system.[8]

Orrery Time

Orrery Time is used to compute the astronomical display of the position and phase of the Moon, the tropical year, the sidereal day, orbits of the visible planets, and the precession of the Earth's axis. Orrery Time is intended to approximate the solar system barycentric coordinate time. In that sense it is similar in purpose to Barycentric Dynamical Time (TDB), or the older Ephemeris Time (ET). Orrery Time differs from Corrected Solar Time because the rate of the Earth's rotation is slowing. Historical trends suggest that the average day is currently slowing by about 1.8 milliseconds per day per century. If this trend continues creating a deviation that grows quadratically with time, a Terrestrial Time clock, measuring an 86,400-second day, would differ from Corrected Solar Time by about 3.8 days after 10,000 years. This would not create noticeable error on the displays of the planetary positions, but it would be apparent on the display of the position and phase of the Moon.

Orrery Time is calculated from Displayed Solar Time by subtracting out a correction for the slowing of the Earth's rotation. Orrery Time is automatically corrected for the expected slowing of the Earth's rotation by a cam with a quadratic correction included in the function encoded on the cam. An additional mechanism is also provided for future adjustment to Orrery Time to match the observed slowing, as described below. No attempt is made to subtract the equation of time to produce a daily approximation of Terrestrial Time, since this error is too small to be seen on the display, and the length of the day will average out to the mean solar day over the course of each tropical year.

DISPLAYS

Days

Three types of days are computed for driving the displays: solar days, orrery days and sidereal days. The clock also makes use of the 86400-second day in the design of its cams and gear ratios, but does not represent it explicitly in the displays.

Solar days are demarcated by the motion of the Sun display, which indicates the approximate position of the Sun in the sky (Figure 5 and Figure 6). The Sun travels around a circle divided by a horizon line that adjusts throughout the tropical year so the Sun will show the time of sunrise and sunset. Solar days are also counted on the calendar display, which indicates day of week, calendar month, and day of month. The Sun display and calendar date are both driven directly from Displayed Solar Time.

Orrery Days are the clock's long-term approximation of 86400-second days (TT), used to drive the motions of the planets in the orreries. They are derived from Orrery Time. They differ from the 86400-second days in that they vary in length with the season, but they closely approximate 86400-second days when averaged over the course of a tropical year.

Sidereal days are used to drive the horizon-line indicators that move across the star field display. The horizon-line indicators are similar in form to the "rete" of an astrolabe, although on a classical astrolabe the stars move across the background representing the horizon.[9] In the clock, the horizon indicators move across the background of the stars. The clock sidereal day is produced by adding the sidereal year rate to the motion of the Sun to produce the sidereal days. Since the sidereal year rate is derived from Orrery Time and the motion of the Sun is derived from Displayed Solar Time, the rate of the horizon-line indicator is derived from a combination of both of the time bases.

Years

Four different versions of the year are generated by the clock: the tropical year, the sidereal year, the calendar year and the cam year.

The tropical year, that is the year corresponding to the seasons, is used to compute the horizon dial that adjusts the length of the displayed day for the time of year. The number of 86400-second days in a tropical year is $365.2421896698 - 0.00000615359\,T$, where T is the number of Julian centuries past the year 02000[*].[10] The average length of the tropical year, over the next 10,000 years, will be 365.241882 86400-second days (TT). The tropical period is used to drive the cam that moves the horizon dial. It is calculated from the Orrery Time.

---

[*] The Long Now Foundation uses five-digit dates, the extra zero is to solve the deca-millennium bug which will come into effect in about 8,000 years.

The sidereal year is the time it takes the Earth to move once around the Sun with respect to an inertial coordinate system (formerly the "fixed" stars, now a similar, agreed-upon coordinate system defined by the compact radio sources of the International Celestial Reference Frame (ICRF). The sidereal year is used to display the position of the Earth in the orrery, and its calculation is discussed in the section describing the calculations of the orbits of the planets. It is also used to compute the sidereal day to drive the horizon line indicators in the star display. The sidereal year is derived from Orrery Time.

The calendar year is computed according to the Gregorian calendar system, including the leap-year exceptions for centuries and millennia. There is still some question about whether to incorporate a possible future reform of the Gregorian calendar[11], in which the millennial exception skips every fourth millennium. This would keep the calendar in closer synchronization with the tropical year. The current thought is not to do so, on the grounds that the Gregorian calendar is displayed primarily as a cultural artifact of our time. The calendar year is displayed in a five-digit format.

The equation of time and solar elevation cams are rotated once a "cam year," which is defined to be 365.2222 mean solar days. Since the two-dimensional functions encoded in the equation-of-time cam are actually derived from a one-dimensional time varying function, there is flexibility in the exact ratio of days to cam years. This interval was chosen because the ratio is easily computed by gears and makes a beautifully shaped cam. It is calculated from Corrected Solar Time by gear trains with a ratio of 2958/81 = 365.2222.

Moon

The face of the dial displays the phase of the Moon and the position of the Moon in the sky. The Moon position display is an aperture that moves over the background of a 16 lunar phase display that moves with the Sun display. The aperture normally moves with this background at the same rate as the Sun, but it occasionally moves backwards against it, in discrete steps of 1/16 full rotation, changing its phase with respect to the Sun display. The displayed phase of the Moon is determined by the angle between the Sun display and the Moon display. The Sun display is driven from Displayed Solar Time and the Moon aperture is driven from the sum of the Sun rate and the backwards steps. These backward steps are derived from Orrery Time, so the Moon display is computed by a combination of Orrery Time and Displayed Solar Time.

The phase of the Moon is a very sensitive indicator of the long-term ratio of the month and day lengths, which depends not only on long-term changes in the rotation of the Earth, but also long-term changes in the orbit of the Moon. These are caused primarily by tidal dragging of the Earth's oceans, landmasses and atmosphere. The mean sidereal month over the next 100 centuries is 27.3216719 86400-second days. As explained below, this mean value is derived from an extended form of the JPL DE422 solar system solution. The number of 86400-second days in the mean synodic month (average time from new moon to new moon) is derived by subtracting the mean sidereal period of the Earth's orbit from the mean sidereal period of the Moon.

Stars

The face of the dial shows the position of the stars in the sky, which rotate once per sidereal day. The sidereal day rate is equal to the solar day rate plus the sidereal year rate. In the clock this is generated by the motion of the horizon indicator over the star field, which in turn rotates at the precession rate of the Earth's axis. Both of these are generated from Orrery Time. The horizon should rotate in 0.9972696693 solar days, which is about 23 hours 56 minutes. This is the mean period of the sidereal day over the next 10,000 years, given the predicted slowing.

The stars display also takes the precession of the equinoxes into account. This rotation happens about once every 26,000 years, so there will be less than half a rotation in the lifetime of the clock. In 10,000 years, the bright star Vega in the constellation Lyra will be the pole star. Vega is currently known as one corner of the so-called "Summer Triangle." At the end of the planned 10,000-year design life of the clock, it will be a circumpolar star visible in all seasons of the year.

The clock also generates the precessional rate from Orrery Time. Because the star display is generated by horizon line indicators moving across the star field, and the horizon line indicators are driven in part by Displayed Solar Time, the star display is computed by a combination of Orrery Time and Displayed Solar Time.

Planets

The clock will display the visible planets in two orreries: a heliocentric Copernican orrery (Figure 7 and Figure 8), and a geocentric Ptolemaic orrery. For this purpose the sidereal orbital periods are used, modeling the mean motion of the planets, rather than the true elliptical orbits. The Ptolemaic orrery is driven from these same periods, with a mechanical system that computes the appropriate coordinate transformation. The planetary orbit rates are all derived from Orrery Time.

Table 1. Time Periods displayed by the clock. The "Gear Ratio" column shows the ratio of the gear train connecting the display to the indicated time base, approximating the number of the Predicted Solar Days.

	86,400-Second Days (TT)	Predicted Solar Days	Gear Ratio	Time Base
Solar Day	1.000001042	1	1	Solar
Sidereal Month	27.32167193	27.32164347	27.3216438	Orrery
Synodic Month	29.53060095	29.53057019	29.53057085	Orrery
Sidereal Day	0.997270708	0.997269669	0.997269669	Orrery
Precession	9412982.24	9412972.435	9412882.619	Orrery
Tropical Year	365.242189	365.2418085	365.2415166	Orrery
Sidereal Year	365.2563681	365.2559877	365.2564103	Orrery
Cam Year	365.2226027	365.2222222	365.2222222	Solar
Mercury	87.96925644	87.96916481	87.96914701	Orrery
Venus	224.7007992	224.7005652	224.7037994	Orrery
Earth	365.2563681	365.2559877	365.2564103	Orrery
Mars	686.9798408	686.9791252	686.978544	Orrery
Jupiter	4332.599090	4332.594577	4332.798497	Orrery
Saturn	10759.08080	10759.06959	10755.49679	Orrery

Table 1 shows the sidereal periods of the planets in 86,400-second days, and in predicted mean solar days of 86,400.09 seconds, averaged over the next 10,000 years, based on the projected slowing of the solar day. The sidereal planetary periods, as well as those of the sidereal month, are derived from an extended form of the JPL Development Ephemeris 422 (DE422) solar system solution and the Horizons software and algorithms.* DE422 is a slightly updated version of DE421.[12] This planetary ephemeris is an integration of 2nd-order parameterized post-Newtonian (PPN) n-body equations of motion consistent with general relativity, fit to several centuries of accumulated ground-based measurements and spacecraft tracking data. When these fundamental equations are numerically integrated, many periodicities and secular trends in planetary motion emerge as a result of mutual perturbations such as resonances. Therefore, to properly determine a mean sidereal period, the ephemeris for each object was sampled at intervals between 1 and 50 days (depending on object) over the 10,000-year interval of interest (02011-12011). A single precessing ellipse was then estimated for each planet's dataset in a least-squares sense, iteratively converging on a parameter set that is a least-squares best fit of an ellipse to the computed dataset. The resulting mean sidereal periods above are thus the statistically optimal fit for the full 10,000-year timespan of the clock, relative to the DE422 solution. If a different timespan was considered, slightly different values would result.

DEALING WITH THE UNPREDICTABLE VARIATIONS OF EARTH'S ROTATION

As previously discussed, the Earth's rotation is currently slowing at a rate of about 1.8 milliseconds per day per century.[13] Of course, the trend may not continue, especially if the climate changes. The variation is caused by a variety of effects including tidal drags, shifts in the Earth's crust, changes in ocean levels, and even weather (Figure 9. For example an ice age would put more mass near the poles, making the day shorter. Melting icecaps would make it longer.[14] This creates an uncertainty in the average length of day of about 10 parts per million, an uncertainty of plus or minus 37 solar days over the design lifetime of the clock.†

Since the difference between Orrery Time and Displayed Solar Time is only clearly visible in the Moon display, determining the correct value is not crucial for overall functioning of the clock. For this reason, provision is provided for user adjustment, allowing future observers to tune the Moon display to match their observations of the Moon. The clock will record the history of these observation-based adjustments, making the 10,000-Year Clock a scientific instrument for recording the long-term slowing of the Earth's rotation.

Human observation is a satisfactory method of converting Displayed Solar Time to Orrery Time for display, but it cannot be depended on for the operation of the timekeeping mechanisms of the clock. The displays only matter when they are visited by humans, but the timekeeper is designed to continue to keep track of the correct time without human intervention. There is only one mechanism in the timekeeping of the clock that is sensitive to the unpredictable slowing of the Earth's rotation: Corrected Solar Time is used to drive a light steering prism that guides the noontime light into the mountain. This steering prism is required to steer the light down a 500-foot shaft to the light sensor, adjusting the Sun's varying noontime elevation during the course of the tropical year.

The solar elevation is calculated from a cam that already encodes the predicted slowing of the Earth. Our uncertainty in this prediction translates to a growing uncertainty in the elevation of the noontime Sun. The Sun elevation cam encodes this uncertainty by seeking for the Sun in different positions of the uncertainty window on successive days, scanning the entire uncertainty window back and forth so that it will always be within the acceptable accuracy for at least one day during each scan. This works because the clock does not require a synchronization event every day. The

---

* Giorgini, J.D., NASA/JPL Horizons On-Line Ephemeris System, 2011 (http://ssd.jpl.nasa.gov/?horizons).
† Busch, M.W., "Climate Change and the Clock." (http://blog.longnow.org/category/clock-of-the-long-now/page/2/.)

scans are performed twelve times every tropical year. Since the uncertainty scanning grows over time, the window of uncertainty over which the scans are performed goes accordingly. Thus, the scanning deviates very little from the predicted value in the early centuries, but the perturbations become more noticeable as time progresses.

Since the equation of time is also sensitive to deviations in the phase of the tropical year, the function encoded in the equation-of-time cam can also encode the same time scanning as the Sun elevation cam. Thus, when the Sun is detected, Uncorrected Solar Time will be adjusted with the appropriate equation of time value to create corrected solar time.

CONCLUSION

Human societies have always organized their activities around the rising and setting of the Sun. Civilization required agriculture. Agriculture required sunlight. Much of human culture is organized around a diurnal or annual cadence. Systems whose primary duty cycles are much shorter or much longer than a day will still have some superposed diurnal signature resulting from human-mediated interactions, such as maintenance and administrative activities. In this case, both the solar-powered and regulated mechanism of the Clock, as well as its concept of human-mediated operations to update the orrery and calendar displays, will exhibit strong diurnal peaks of activity. The Sun matters to humans, even to their devices in the depths of space and on the surfaces of other planets. It has done so throughout history and the Clock makes the statement that it will continue to matter thousands of years into the future.

Yet the Clock is more than a sundial. It also represents the apparent positions of other astronomical objects: the Moon, the planets and the stars. Unlike a sundial, the representation of apparent solar time becomes an engineering choice, not the result of directly measuring the hour angle or azimuth of the Sun from its shadow. To implement a living simulation of the solar system in a shaft drilled 500 feet into a mountain requires precisely the same dynamic conversation between a carefully tuned physical oscillator and the changing syncopations of the natural world In the case of the 10,000-Year Clock, the former is its pendulum and the latter the orrery. In the case of civil timekeeping, it is an ensemble of climate controlled atomic chronometers versus the rotation of the Earth as represented by Universal Time.

As the centuries pass the occasional resynchronization of the Clock on some sunny day will correct for the slight residual drift of the high-precision pendulum. But it will also accommodate the quirky residuals in the Earth's rotation remaining after compensating for the pre-computed equation of time. These residuals are precisely the result of variations in the mean solar length-of-day as shown in Figure 9 Like all conscientious timekeepers the clock eventually just reconciles the steady dynamical swing of its pendulum with the wavering rotation of the Earth.

ACKNOWLEDGMENTS

The authors thank William Folkner (JPL) for extending the DE422 planetary ephemeris integration to cover the clock's entire interval of operation and Robert Jacobson (JPL), whose natural satellite software was adapted to compute the mean sidereal periods of the planets and Moon over this timespan, and Michael Busch for his help in making the astronomical calculations required for the design of the clock.

APPENDIX: FIGURES

Figure 1. A drive gear to be used in the 10,000-year clock.

Figure 2. Tunnel that is currently being excavated to reach the underground site of the clock

Figure 3. Analemma with the Temple of Apollo, photographed by Anthony Ayiomamitis by making multiple exposures at 09:00:00 UT+2, on multiple days between January 7 and December 20, 2003.[*]

---

[*] http://www.perseus.gr/, *A Vibrant Universe in Vivid Colour: Astrophotography by Anthony Ayiomamitis* (used with permission).

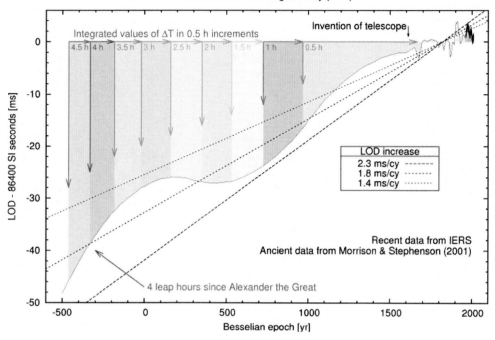

Figure 4. Historical variation of the length of the day.

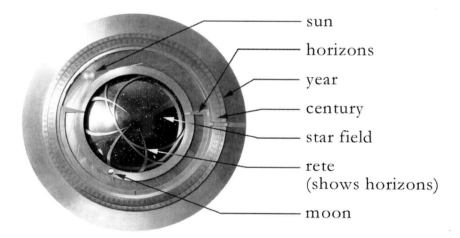

Figure 5. The face of the prototype 10,000-Year Clock, showing the positions of the Sun, Moon, and stars. The actual 10,000-Year Clock will have a similar design, except that the calendar year and century will be on separate displays.

Figure 6. Prototype 10,000-Year Clock, currently on display in the London Science Museum.

Figure 7. Prototype of the 10,000-Year Clock's Copernican orrery, currently on display at the Long Now Foundation's museum in San Francisco, CA

Figure 8. Close up of the prototype of the 10,000-Year Clock's Copernican orrery, showing the stone spheres that represent the visible planets.

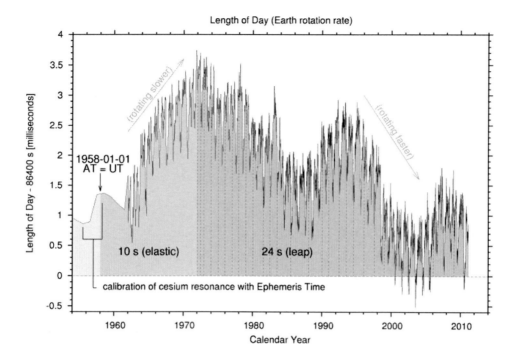

Figure 9. Recent variations in Length of Day from IERS data. The number of seconds that must be added into the atomic time scale (AT) to track Universal Time (UT) is proportional to the area under the curve.

REFERENCES

[1] Hillis, W.D., "The Millennium Clock," Wired, Scenarios: The Future of the Future, special edition, 1995, (http://www.wired.com/wired/scenarios/clock.html).

[2] Brand, S., *The Clock of the Long Now: Time and Responsibility*. Basic Books, 1999, ISBN-10: 0-465-00780-5.

[3] Hillis, D., "Introduction to the Long Now Clock Project." *Horological Science Newsletter*, Feb. 2000 p. 2

[4] Brans, C., and Dicke, R.H., "Mach's Principle and a Relativistic Theory of Gravitation", Phys. Rev, vol. 124, pp. 925-935, 1961.

[5] Estabrook, F.B., "Post-Newtonian n-Body Equations of the Brans-Dicke Theory," Astrophys. J., Vol. 158 pp. 81-83, 1969.

[6] Shahid-Saless, B., and Ashby, N., "Relativistic Effects in Local Inertial Frames Including Parameterized-Post-Newtonian Effects," Phys. Rev. D, 38, pp. 1645-1657, 1988.

[7] Hillis, D., "A Simple Method for Constructing Compensated Pendulums." to be published in *British Horological Journal*.

[8] Dershowitz, N. and Reingold, E.M., *Calendrical Calculations*. Cambridge University Press, 2008, ISBN 978-0-521-88540-9.

[9] Chaucer, G., *A Treatise on the Astrolabe*, 1391.

[10] *Explanatory Supplement to the Astronomical Almanac*, University Science Books, Edited by Seidelmann, K.P., U.S. Naval Observatory, Washington, D.C., 1992.

[11] Gregorius P. P. XIII, "Inter gravissimas", in Clavius, C., *Romani Calendarii A Gregorio XIII P. M. Restituti*, Rome, 1603.

[12] Folkner, W.M., Williams, J.G., Boggs, D.H., "The Planetary and Lunar Ephemeris DE 421." JPL Memorandum IOM 343R-08-003, Jet Propulsion Laboratory, March 2008.

[13] Morrison, L.V., Stephenson, F.R., "Historical eclipses and the variability of the Earth's rotation." Journal of Geodynamics, Volume 32, Issues 1-2, August-September 2001, Pages 247-265, ISSN 0264-3707, 10.1016/S0264-3707(01)00024-2.

[14] Nakada, M. & Okuno, J. (2003), "Perturbations of the Earth's rotation and their implications for the present-day mass balance of both polar ice caps." Geophysical Journal International, 152: 124-138 doi: 10.1046/j.1365-246X.2003.01831.x.

DISCUSSION CONCLUDING AAS 11-665

Neil deGrasse Tyson jested that the Long Now should put some signage on the 10,000 Year Clock so that a post-apocalyptic Earth will not think that the world will end when the clock stops working. Rob Seaman replied that they had discussed the long-term signage designed for nuclear waste dumps. Tyson noted that an entire religion might develop around the discovery of such a device.

Tyson said that he considers time-keeping devices not just within the context of telling correct time, but within their capacity of measuring intervals of time. An elaborate mechanism for chiming local noon is not needed; the shadow of a stick in the ground suffices for making that determination. The 10,000 Year Clock wouldn't actually predict the phase of the moon, but rather it displays it; however, the phase of the moon can be determined just by looking up. Tyson therefore wondered about the motivation for building this clock. Seaman reiterated from his talk that was he was not in a position to speak authoritatively on the motivations of the project, but in his opinion the Clock was a philosophical statement, particularly on long-term engineering. Tyson surmised that the project was therefore more of an exercise in engineering rather than an exercise in horology. Steve Allen noted that James Lick left his fortune to build a pyramid on Market Street, but that is not what happened.

Steve Malys asked if the project was entirely privately funded or if tourism would contribute to its financing (and also asked in jest whether the property taxes had been paid for the next 10,000 years). Seaman said he thought it was all privately funded by the leadership of the Long Now and encouraged people to visit the web pages for more information.[*] He added that this is not the only project of the Long Now. There is a Rosetta-stone project related to language preservation. They host a series of seminars on long-term thinking. Seaman said it possible to join the Long Now organization; he is a member and members receive a stainless-steel membership card. The fundamental philosophy is that there are so few areas today that think long term. Seaman mentioned New York City, from where Tyson grew up, noting that it is constantly reinventing itself. The Long Now attempts to be something entirely different than that. Tyson supposed that the Earl of Orrery was a wealthy chap who just decided to make the orrery as an engineering curiosity, so perhaps this project is similar: "rich people wanting to do something really cool with gears."

Frank Reed said that he had a long connection with some of the contributors to the project as customers of his software. He said that they are the wealthy "Silicon Valley intelligentsia" and these types of projects are basically public art. Seaman noted that talking to Danny Hillis was not like to talking to most people he knows; astronomy is full of clever guys and Hillis is a *really* clever guy.

---

[*] http://www.longnow.org/

Paul Gabor noted that Clavius predicted the Gregorian calendar only to the year 5000 because the calendar would likely need to be reformed before that time. Unfortunately the calendar does not have a built-in mechanism for reform, such as dropping out one the leap day, but it will clearly be in need of reform within the next one- to two-thousand years. Seaman noted that there has been some debate as to whether to include another correction every 4000 years which is not yet part of the official calendar. The thinking may be to not include that aspect within the design of the clock in order to give future generations some historical insight regarding the origins of clock, which makes for an interesting problem of symbolism. As an aside, Seaman noted that the colloquium was organized such that the earlier presentations should evoke broad cosmic concepts related to timekeeping to lay the foundation for more technical presentations later.

Mark Storz asked if the Clock displayed the occurrences of solar eclipses. Seaman replied that he had inquired explicitly about eclipses and believed the response was negative, but agreed that would be an interesting aspect to such a device. Seaman said the project would likely welcome expert feedback on their design.

Ken Seidelmann noted that the design included a star map that accounted for precession, and asked if it would account for proper motion. Seaman replied that he had suggested this possibility to Hillis, who seemed intrigued with the notion of possibly including the path of Barnard's Star over the lifespan of the clock on the face of the planisphere, guessing that it might travel almost 30 degrees. Seidelmann said that over that time period, many stars will have moved significantly. Seaman agreed it was an interesting question, particularly because the planisphere will be depicting the brighter stars which tend to be closer and thereby are more prone to exhibit proper motion.

Seaman said that one aspect of the project was to invoke an interesting conversation, which the clock certainly does; Gabor concisely characterized the Clock as "a conversation piece".

George Kaplan noted that 10,000 years is over one-third of the precession cycle and the obliquity of the ecliptic would be expected to change over that amount of time, which effects the equation of time. Allen said that the final model for the gears will be based on the calculations of Jon Giorgini of JPL who is running integrations of the JPL Development Ephemeris 422 (DE422) for a full 10,000 years. Tyson said that each gear is in some sense like an epicycle; each cycle within the system has to be represented by another gear. Seaman added that the final clock intends to have two orreries: one Copernican and one Ptolemaic.

Storz asked if the Clock will attempt to synchronize itself with the Moon like it does with the Sun, because it would seem that lunar motion would be hard to predict in the very long term. Seaman noted that the lunar model seemed to be one of the greatest challenges but there is no known way to synchronize to the Moon. He added that the solar synchronization of a purely mechanical device is already challenging.

Wolfgang Dick asked if the Clock is expected to run for 10,000 years without any maintenance. Seaman said that the philosophy is to not build a device to which no one pays attention, so there is a notion of human interaction in the design. If there is no sunlight, then the clock will need to be wound, for example. But the Clock is being designed to minimize wear such that it will not require substantial maintenance to operate. If the clock required some tweaking in the future, then that would not be considered a "design failure." Also there is some idea of making the mechanism extensible such that features could be added in the distant future as desired, such as different chimes. Reed asked if any security measures would be in place; Seaman said that issue gets back to Tyson's issue of signage.

Malys asked if there was an engineering issue driving the massive size of the Clock, such as the length of the weight drop, or was that an artistic decision. Seaman replied that big things obviously tend to survive longer. Tyson added that a 10-second pendulum would require a certain size. Seaman added that the "tick rate" is thirty times longer than the pendulum rate, but admitted that he was not sure how the final dimensions were arrived at. Seidelmann noted that John Harrison learned that more accuracy could be obtained from a smaller mechanism.

MIDDAY ROUND-TABLE DISCUSSION OF OCTOBER 5, 2011

MIDDAY ROUND-TABLE DISCUSSION OF OCTOBER 5, 2011

> The context for decoupling civil timekeeping from Earth rotation includes the long history of international regulation of time signals in radio broadcasts. This discussion filled in the background for many of the legal and technical aspects of the national processes which contribute to the decisions made by the ITU-R.

John Seago recalled that Steve Malys had asked about, and Dennis McCarthy had commented on, how governments might be arriving at their positions. Seago therefore asked Malys if his organization was aware of a US DoD-wide survey conducted by the US Naval Observatory in 2008 and whether his organization participated as a DoD entity.[*] Malys said that his organization participated in the survey and, in preparation for this colloquium, became more aware of an existing DoD memorandum on leap seconds.[†] However, Malys' questions have been mainly with the ITU-R process. He was particularly curious as to how the voting delegation operated, how many votes each nation had within the Radiocommunications Assembly, and the required percentage to approve the Recommendation. Ken Seidelmann said that he had been told that a 70% supermajority was needed, but it was unclear to him whether this was 70% of voting delegates within the assembly or 70% of the member administrations. George Kaplan asked who gets to vote and how many votes are there. Paul Gabor suggested that voting is by nation and Seidelmann added that there are approximately 190 nations. Kaplan asked if that meant every nation gets one vote equally, to which several attendees seemed to respond affirmatively.

Malys wanted confirmation that a vote was scheduled to take place in January 2012, to which several attendees responded affirmatively. Steve Allen added that the agenda of the 2012 Radiocommuncation Assembly has not been published publicly, but all the preparations appeared to have taken place for that vote to happen. Malys was surprised that the vote was happening so quickly, as coordination thus far appeared inadequate. Specifically, there seemed to be a number of varied opinions on this topic, some of which were being discussed in this colloquium, yet many other viewpoints were not being represented at the colloquium. Malys thought that the figures from Paper AAS 11-664 indicated many stakeholder communities, and it may be an outstanding issue as to whether those communities have had any voice in the process.

Seidelmann said that ITU-R Study Group 7 had three votes against the proposed Recommendation to redefine UTC (with about eight to ten votes in favor), but it was ruled that the Recommendation should advance to the Radiocommunication Assembly because the three dissenting votes were not addressing a technical issue relevant to telecommunications. Seidelmann said this may be evidence that other communities or issues were not considered, including national legal issues. Allen suggested this may have been the case when leap seconds were first established.

---

[*] http://tycho.usno.navy.mil/leap_second_poll.html
[†] http://tycho.usno.navy.mil/Discontinuance_of_Leap_Second_Adjustments.pdf

Specifically, the International Radio Consultative Committee (CCIR)[*] first decided to have leap seconds, leaving various other communities to sort out the consequences of that decision, and make technical suggestions back to the CCIR. Those suggestions became the implementation details appended to Recommendation 460.[1] As evidence, Allen noted that the first version of Recommendation 460 simply said that there would be leap seconds and implementation details would follow.[2] Seidelmann added that was a different era, back when the distribution of time signals was the primary issue and there was a pressing need to transmit both atomic time interval and astronomical time of day; also, time services needed to coordinate time signals over long distances by radio.

McCarthy offered that the "process of the ITU is byzantine at best." The recommendation process starts with the creation of an acceptable Study Question. Once the "question is out there, they expect technical responses" from whoever wants to respond. These responses may be papers, publications, opinions, *etc*. As McCarthy understands it, this information is "stuck away in a file drawer somewhere in Geneva until somebody looks at it." They try to establish a drop-dead date after which the responses will be considered, and then the collected information is handed over to the appropriate Study Group. The Study Group in turn passes the information to a Working Party which reports back to the Study Group. Each country can have its own component national Working Party; McCarthy is a member of US Working Party 7A. International Working Parties assembled from national Working Parties are thereby assigned these study questions. Some questions have language which grows so old "that they eventually just fall off the table." Others questions are deemed important enough to push through to the International Working Party. At this stage a Recommendation can either be advanced up to the Study Group or dropped by the Working Party. Thus, the national Working Parties funnel into an international Working Party, and each international Working Party reports to a Study Group which advances Recommendations to the ITU.

Regarding the vote in January, McCarthy said there is a huge bureaucracy involved with the ITU headquarters. This bureaucracy decides whether issues and concerns are technical versus non-technical, not so much whether the issues are relevant to telecommunications. For example, if we'd like to keep GMT because we like the name *Greenwich*, that is not considered a technical argument. Consequently, if that type of argument is put before the ITU it will not be accepted as technically relevant. Eventually, the Recommendation is voted upon at the Radiocommunication Assembly where the final decision is made. That vote is at a very high political level with delegates sent from departments of State and foreign ministries.

Seago commented that he was not sure of the degree to which Study Question 236/7 mentioned *technicality*, but the Study Question explicitly mentioned *legality* when it noted "…considering that UTC is legal basis of timekeeping… what are the requirements…?" Seago was unsure how McCarthy's description and GMT example correlated with the language within the Study Question. McCarthy replied that "technicality is the basis for accepting someone's negative or positive vote." McCarthy added that it also comes down to what the chairs of the Working Parties and Study Groups decide; there are decisions made at that level and there are decisions made at ITU headquarters, and "some of these may just be tough calls." David Terrett suspected that interpretation of the word "technical" might be rather broad, and perhaps a legal argument could very much be interpreted as a "technical" argument.

Malys asked if the ITU was involved in the original decision to introduce leap seconds back in 1972. Seidelmann said at that point it was the CCIR, but that organization is now under the

---

[*] The CCIR is the predecessor of the ITU-R.

ITU-R. Malys asked if anyone had any visibility into the decision process within the United States, as it was Malys' impression that various departments of the executive branch (Defense Department, Commerce Department, *etc.*) had contributed to the position decided by the Department of State. Allen replied that the US State Department has a number of committees which receive documents and those committee meeting notices are all part of the public record. Tracing down the committee structures is not an easy task from the outside, but information may have been available about those meetings by those who knew about them via notices in the Federal Register. Seidelmann understood that only the US Department of Defense and NASA issued position statements favoring the proposed revision, and that the Departments of Transportation and Commerce offered "no position" on the proposed revisions.[*] McCarthy responded that the group within the US State Department under Cecily Holiday makes the decision. David Simpson asked if the US State Department had made its decision yet, and McCarthy replied that the US decided it would support a redefinition of UTC.

McCarthy added it must be understood that within the US State Department "this has got to be one of the things that they almost don't even care about." This would not be true within the US only, but also within other nations. A much bigger ITU-R issue is spectrum allocation. Seidelmann added that this issue doesn't have its own lobbying group representing an activity where lot of money is being spent. Seago wondered if the US State Department was treating the issue as a telecommunication issue primarily, where the advisory committees primarily represented telecommunication interests and broader considerations outside telecommunications were lacking. Malys suggested that if some graphical information similar to the contents of Paper AAS 11-664 were presented to the responsible person at the US DoS, then perhaps that person could gain a greater appreciation of the fact that there is more at stake than just telecommunications. Allen noted that the colloquium proceedings would be made available, but that doesn't mean that decision-makers would necessarily care about them. Seidelmann said that the US State Department seemed to pay attention to this issue after an article appeared on the front page of the *Wall Street Journal*.[3] Malys said that the colloquium attendees had gathered because they knew very well that the decision would certainly impact various communities, and we should therefore try to make the responsible person(s) at our departments of State aware that there is more than telecommunications at stake.

Seago noted that the US Department of Defense and NASA apparently commissioned data calls and surveys before issuing their position statements supporting (or at least not opposing) the proposed revision. Yet, it was unclear how agencies and departments handled the collected responses from these surveys and data calls. The USNO survey responses were not publicly reported and it was not clear how the information received from survey responses affected the DoD position statement. Seago clarified that the DoD statement essentially consisted of language drafted by the USNO before its survey; the final DoD statement simply changed the date of leap-second cessation one year later from the original USNO language.[†] Seago also reported hearing anecdotal complaints from NASA employees who responded to an agency-wide data call but never received any acknowledgement or feedback from NASA headquarters, and were therefore surprised to learn that NASA had issued its statement supporting UTC redefinition. Seago acknowledged that the issue was an international one and it was even more unclear to him how

---

[*] According to US Code Title 15, Chapter 6, Subchapter IX, § 260, the Secretary of Transportation is responsible for time-zones (§ 260) and the Secretary of Commerce is responsible for Coordinated Universal Time.
[†] The US DoS still recommends adoption of a draft revision to Recommendation 460-6 which calls for leap-second cessation one year earlier than the US DoD requested.

other nations are arriving at their positions. Allen reminded the attendees that we would be hearing the results of at least one international survey later in the day (Paper AAS 11-668).

REFERENCES

[1] CCIR, "Detailed instructions by Study Group 7 for the implementation of Recommendation 460 concerning the improved coordinated universal time (UTC) system, valid from 1 January 1972" in *XIIth Plenary Assembly CCIR,* (New Dehli, India, 1970) III, p. 258 a-d (ITU, Geneva, Switzerland, 1970).

[2] CCIR, "Standard-frequency and time-signal emissions." (Recommendation 460), in *XIIth Plenary Assembly CCIR*, (New Dehli, India, 1970), III, p. 227 (ITU, Geneva. Switzerland. 1970).

[3] Winstein, K.J., "Why the U.S. Wants To End the Link Between Time and Sun," *Wall Street Journal*, 29 July, 2005, p. 1 (URL http://www.post-gazette.com/pg/05210/545823.stm)

Session 3:
EARTH ORIENTATION

USING UTC TO DETERMINE THE EARTH'S ROTATION ANGLE

Dennis D. McCarthy[*]

The Earth's rotation angle is a critical component of the suite of five Earth orientation parameters used to transform between terrestrial and celestial reference systems. This angle is defined mathematically using an adopted conventional relationship between UT1 and the mathematical quantity known as "Earth Rotation Angle" (ERA). For practical purposes, then, UT1–UTC provides a convenient means to obtain UT1, knowing UTC, and thus the ERA. Because the Earth's rotational speed is variable, it is not practical to model UT1 as a function of time with the accuracy needed for many applications. Consequently astronomical and geodetic institutions from around the world share observations of the Earth's rotation angle and these data are then used to provide users the latest observations of UT1–UTC as well as predicted estimates with accuracy that depends on the prediction interval. This process can provide users with daily updates of UT1–UTC with accuracy of the order of tens of microseconds and predictions with accuracy better than 1 millisecond up to ten days in advance. The International Earth Rotation and Reference Systems Service (IERS) was established in 1987 by the International Astronomical Union and the International Union of Geodesy and Geophysics to provide this information operationally. In addition to the services routinely providing UT1 with sub-millisecond accuracy, UTC is currently adjusted to keep |UT1–UTC| < 0.9 seconds, and this definition provides a means to access UT1 automatically with accuracy of the order of one second. Should UTC be defined without the restriction keeping |UT1–UTC| < 0.9 seconds, the low accuracy estimate of UT1 (± 1 second) would no longer be assured. However the existing national and international services can be expected to provide the current products as they do now via paper bulletin and electronic means. It is assumed that the accuracy of those products will always reflect the state of the art. In the future, high-speed transfer of high-quality observational astronomical, meteorological, oceanic and geophysical data promise to decrease the latency of the observations and provide UT1–UTC at sub-daily intervals with increasingly improving accuracy. In addition to the current means of distribution, increasing access to electronic communication services has the potential to provide near real-time, state of the art UT1–UTC to users when and wherever it is needed. If there were sufficient demand, we might even envision a UT1–UTC application being made available for future hand-held devices.

---

[*] U. S. Naval Observatory, 3450 Massachusetts Avenue, N.W., Washington DC 20392, U.S.A.

INTRODUCTORY BACKGROUND

It is important to distinguish between reference systems and reference frames when discussing the use of Earth orientation parameters. Reference systems, either terrestrial or celestial, have an origin, specified directions of three fundamental dimensional axes, and a set of conventional models, procedures, and constants used in the actual realization of the system. A reference frame, on the other hand, is the realization of that system through a list of coordinates, either angular or Cartesian.

Celestial reference systems generally have their origins at the barycenter of the solar system, and their polar axes (z-axes) related in some way to the rotational axis of the Earth. The second axis (x-axis) then lies in the equatorial plane perpendicular to the z-axis and is directed toward a fiducial point in that plane. The third axis is chosen to complete a right-handed orthogonal system. In astronomical applications the International Celestial Reference System (ICRS) is the idealized barycentric coordinate system to which celestial positions are referred. It is kinematically non-rotating with respect to distant extragalactic objects. It was aligned close to previous astronomical reference systems for continuity. Its orientation is independent of epoch, ecliptic or equator and is realized by a list of adopted coordinates of extragalactic sources. The Geocentric Celestial Reference System (GCRS) is a system of geocentric space-time coordinates defined such that the transformation between BCRS and GCRS spatial coordinates contains no rotation component, so that GCRS is kinematically non-rotating with respect to BCRS. The spatial orientation of the GCRS is derived from that of the BCRS. The International Celestial Reference Frame (ICRF), then, is a set of extragalactic objects whose adopted positions and uncertainties realize the ICRS axes.[1] It is also the name of the radio catalog listing the directions to defining sources. Successive revisions of the ICRF are intended to minimize rotation from its original orientation. Angular coordinates of optical stars, consistent with that frame, are provided by the Hipparcos Catalogue.[2]

Terrestrial reference systems generally have their origins at the center of mass of the Earth with their polar axes related to the direction of an axis fixed with respect to the Earth's crust. The origin of longitudes in the equatorial plane provides the second direction. Again, a third axis is chosen to complete a right-handed orthogonal system. The Geocentric Terrestrial Reference System (GTRS) is a system of geocentric space-time coordinates co-rotating with the Earth. The International Terrestrial Reference System (ITRS) is a specific GTRS for which the co-rotation condition is defined as no residual rotation with regard to the Earth's surface, and the geocenter is understood as the center of mass of the whole Earth system, including oceans and atmosphere. It was aligned close to the mean equator of 1900 and the Greenwich meridian, for continuity with previous terrestrial reference systems. The International Terrestrial Reference Frame (ITRF), is a realization of ITRS by a set of instantaneous coordinates (and velocities) of reference points distributed on the topographic surface of the Earth (mainly space geodetic stations and related markers).[3] Its initial orientation of the ITRF is aligned closely to previous terrestrial systems for continuity.

The Celestial Intermediate Reference System (CIRS) is a geocentric reference system related to the GCRS by a time-dependent rotation taking into account precession-nutation. It is defined by the intermediate equator of the Celestial Intermediate Pole (CIP) and the Celestial Intermediate Origin (CIO) on a specific date. The CIP is a geocentric equatorial pole defined as being the intermediate pole in the transformation from the GCRS to the ITRS, separating nutation from polar motion. Its GCRS orientation results from the part of precession-nutation with periods greater than 2 days, the retrograde diurnal part of polar motion (including the free core nutation, FCN) and a reference frame bias. Its ITRS orientation is comprised of the part of polar motion which is outside the retrograde diurnal band in the ITRS and the motion in the ITRS corresponding to nutational motions with periods less than 2 days. The motion of the CIP is realized by the IAU precession-nutation plus small time-dependent corrections called "celestial pole offsets." The CIO is the origin for right ascension on the intermediate equator in the CIRS. It is the non-rotating origin in the GCRS originally set close to the GCRS meridian and throughout 1900-2100 stays within 0.1 arc seconds of this alignment. The CIO was located on the CIP equator of J2000.0 at a direction 2.012 milli-arcseconds (mas) from the ICRS prime meridian at right ascension 0h 0m 0s.000 134 16 in the ICRS. As the true equator moves in space, the path of the CIO in space is such that the point has no instantaneous east-west velocity along the true equator. In contrast, the equinox defined by the intersection of the equator and the plane of the ecliptic has instantaneous velocity along the equator.

The Terrestrial Intermediate Reference System (TIRS) is a geocentric reference system defined by the intermediate equator of the CIP and the Terrestrial Intermediate Origin (TIO). It is related to the ITRS by polar motion and the TIO locator. It is related to the Celestial Intermediate Reference System by the Earth Rotation Angle (ERA) around the CIP that realizes the common z-axis of the two systems. The TIO is the origin of longitude in the ITRS. It is the non-rotating origin in the ITRS that was originally set at the ITRF origin of longitude and throughout 1900-2100 stays within 0.000 1″ of the ITRF zero meridian.

The terrestrial system rotates in the celestial system and its orientation in that system is affected by precession, nutation, polar motion and variations in the Earth's rotational speed. The fact that the Earth is not strictly a rigid body means that non-rigid body effects need to be considered in models of the Earth's rotational motions, and because the Earth's core experiences a free wobble with respect to the mantle, existing geophysical models of nutation may not account for all of the observed motions. Further, motions caused by redistribution of mass in the Earth, its oceans and atmosphere, along with relatively high-frequency variations in global meteorology and hydrology may also need to be taken into account.

With the introduction of a new reference system in the 1992–2004 period, the CIO replaced the moving vernal equinox; the TIO replaced the Greenwich Meridian; and the Earth Rotation Angle (ERA) replaced the Greenwich Sidereal Time. The alternative system based on the equinox, mean and true positions, and the Greenwich Mean Sidereal Time is still supported and when properly applied can provide equivalent accuracies.[4]

EARTH ORIENTATION PARAMETERS

The transformation between celestial and terrestrial frames is specified by five angles called Earth orientation parameters. The rigorous details are outlined in the publications of the International Earth Rotation and Reference Systems Service (IERS), specifically in the *IERS Conventions (2010)* and its updates, which are available electronically at http://tai.bipm.org/iers/conv2010/conv2010.html.[5] Three would be sufficient, but five angles are used in order to describe the physical processes involved and to make the transformations easier to apply. Two angles are used to model the changing direction of the CIP due to the precession and nutation of the Earth. These phenomena are driven by the gravitational attraction of the solar system bodies, principally the Sun and the Moon, on the non-spherical Earth. Precession refers to the aperiodic portion of the motion and nutation refers to the periodic portion. Both motions depend on the positions of the solar system bodies and the internal structure of the Earth, but they can be modeled mathematically with reasonable accuracy.

Two more angles are used to describe the motion of the CIP with respect to the Earth's crust. This phenomenon called "polar motion" is driven by geophysical and meteorological variations within the Earth and its atmosphere. Polar motion is difficult to model because the forces driving the motion are difficult to predict. As a result these angles must be observed astronomically and made available to users operationally.

The last of the five angles characterizes the rotation angle of the Earth and is described by the time difference UT1–UTC. UT1 is a measure of the Earth's rotation angle expressed in time units and treated conventionally as an astronomical time scale defined by the rotation of the Earth with respect to the Sun. It is expressed in time units rather than in degrees, minutes, and seconds of arc because of its historical use in providing a standard international time scale. In practice, UT1 was defined until 1 January 2003 by means of a conventional formula (Aoki, et al., 1982).[6] It is now defined as being linearly proportional to the ERA, and the transformation between the ITRS and GCRS is specified using the ERA. The ERA is the angle measured along the intermediate equator of the CIP between the TIO and the CIO, positively in the retrograde direction and increasing linearly for an ideal, uniformly rotating Earth. It is related to UT1 by a conventionally adopted expression in which ERA is a linear function of UT1.

$$ERA(T_U) = \theta(T_U) = 2\pi(0.779\ 057\ 273\ 264\ 0 + 1.002\ 737\ 811\ 911\ 354\ 48 T_U), \tag{1}$$

where T_U = (Julian UT1 date - 2451545.0), and UT1 = UTC + (UT1–UTC). Its time derivative is the Earth's angular velocity (Figure 1).

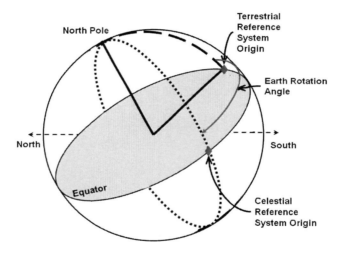

Figure 1. The Earth Rotation Angle.

UT1 is determined operationally by astronomical observations (currently from VLBI observations of the diurnal motions of distant radio sources and other observations), and was originally defined as a kind of time determined by the rotation of the Earth. It is obtained in practice as a time difference from the international time standard, Coordinated Universal Time (UTC), by using the quantity UT1–UTC. The use of the Earth as a timekeeping device became problematic, when it became apparent that its rotational speed was variable and the need for a uniform time scale began to grow. Principal variations in the rotation speed of the Earth include a constant deceleration due to tidal deceleration and de-glaciation, decadal variations due to changes in the internal distribution of the Earth's mass, largely seasonal, meteorologically driven variations and tidally driven periodic variations. As with polar motion, UT1–UTC is difficult to model and predict, and must be observed astronomically and reported to users routinely.

Greenwich Sidereal Time (GST) is an angle that is the sum of the ERA and the angular distance between the CIO and a conventional equinox along the moving equator. This distance is called the Equation of Origins (EO) which is the CIO based right ascension of the equinox along the moving equator and corresponds to the accumulated precession and nutation in right ascension from the epoch of reference to the current date. This relationship can be written as

$$GST = ERA(UT1) - EO, \qquad (2)$$

where EO is given by

$$EO = -0.01450600'' - 4612.15653400''t - 1.391581700''t^2 + 0.0000004400''t^3 \\ - \Delta\psi \cos\varepsilon_A - P, \qquad (3)$$

and $t =$ (Terrestrial Time (TT) − 2451545.0 TT) / 36525, $\Delta\psi \cos\varepsilon_A$ is the classical equation of the equinoxes, and P represents a series of periodic terms given in Table 5.2e of the *IERS Conventions (2010)*.[5]

RELATING THE ITRS TO THE GCRS

The transformation to be used to relate the ITRS to the GCRS at the date t of the observation can be written as:

$$[\mathbf{GCRS}] = \mathbf{Q}(t)\,\mathbf{R}(t)\,\mathbf{W}(t)\,[\mathbf{ITRS}], \qquad (4)$$

where $\mathbf{Q}(t)$, $\mathbf{R}(t)$ and $\mathbf{W}(t)$ are the transformation matrices arising from the motion of the celestial pole in the celestial reference system, from the rotation of the Earth around the axis associated with the pole, and from polar motion respectively. The details are provided in the *IERS Conventions (2010)*.[5] UT1 is only involved in $\mathbf{R}(t)$. The CIO based transformation matrix arising from the rotation of the Earth around the axis of the CIP can be expressed as $\mathbf{R}(t) = R_3(-ERA)$ or

$$\mathbf{R}(t) = \begin{bmatrix} \cos(ERA) & -\sin(ERA) & 0 \\ \sin(ERA) & \cos(ERA) & 0 \\ 0 & 0 & 1 \end{bmatrix} \qquad (5)$$

The equinox based transformation matrix $R(t)$ for Earth rotation transforms from the TIRS to the true equinox and equator of date system using Apparent Greenwich Sidereal Time (*GST*), i.e. the angle between the equinox and the TIO, to represent the Earth's angle of rotation, instead of the ERA, is $\mathbf{R}(t) = R_3(-GST)$ or

$$\mathbf{R}(t) = \begin{bmatrix} \cos(GST) & -\sin(GST) & 0 \\ \sin(GST) & \cos(GST) & 0 \\ 0 & 0 & 1 \end{bmatrix} \qquad (6)$$

It is then apparent that two kinds of time are involved in forming $\mathbf{R}(t)$, i.e. TT and UT1. Both are provided by their relationships to the international time standard, Coordinated Universal Time (UTC). The TT epoch can be determined in relation to a UTC epoch by using

$$TT = \text{International Atomic Time (TAI)} + 32.184s = UTC + [(TAI - UTC) + 32.184s] \qquad (7)$$

and the values of TAI-UTC given in Table 1. UT1 can be determined by using

$$UT1 = UTC + (UT1 - UTC) \qquad (8)$$

and values of UT1–UTC provided by a variety of sources. In cases where the application may permit, some users also may just neglect UT1–UTC by effectively setting it to zero knowing that, through the use of leap seconds, |UT1–UTC| < 0.9s. However for users who require accuracy better than 1 second of time (equivalent to 15 seconds of arc or 464 meters at the Earth's equator) non-zero values of UT1–UTC must be used.

Table 1. Values of TAI-UTC.

FROM	TO	TAI-UTC
1961 Jan. 1	1961 Aug. 1	1.4228180s + (MJD-37300) x 0.001296s
Aug. 1	1962 Jan. 1	1.3728180s + (MJD-37300) x 0.001296s
1962 Jan. 1	1963 Nov. 1	1.8458580s + (MJD-37665) x 0.0011232s
1963 Nov. 1	1964 Jan. 1	1.9458580s + (MJD-37665) x 0.0011232s
1964 Jan. 1	April 1	3.241300s + (MJD-38761) x 0.001296s
April 1	Sept. 1	3.341300s + (MJD-38761) x 0.001296s
Sept. 1	1965 Jan. 1	3.441300s + (MJD-38761) x 0.001296s
1965 Jan. 1	March 1	3.541300s + (MJD-38761) x 0.001296s
March 1	Jul. 1	3.641300s + (MJD-38761) x 0.001296s
Jul. 1	Sept. 1	3.741300s + (MJD-38761) x 0.001296s
Sept. 1	1966 Jan. 1	3.841300s + (MJD-38761) x 0.001296s
1966 Jan. 1	1968 Feb. 1	4.3131700s + (MJD-39126) x 0.002592s
1968 Feb. 1	1972 Jan. 1	4.2131700s + (MJD-39126) x 0.002592s
1972 Jan. 1	Jul. 1	10s
Jul. 1	1973 Jan. 1	11s
1973 Jan. 1	1974 Jan. 1	12s
1974 Jan. 1	1975 Jan. 1	13s
1975 Jan. 1	1976 Jan. 1	14s
1976 Jan. 1	1977 Jan. 1	15s
1977 Jan. 1	1978 Jan. 1	16s
1978 Jan. 1	1979 Jan. 1	17s
1979 Jan. 1	1980 Jan. 1	18s
1980 Jan. 1	1981 Jul. 1	19s
1981 Jul. 1	1982 Jul. 1	20s
1982 Jul. 1	1983 Jul. 1	21s
1983 Jul. 1	1985 Jul. 1	22s
1985 Jul. 1	1988 Jan. 1	23s
1988 Jan. 1	1990 Jan. 1	24s
1990 Jan. 1	1991 Jan. 1	25s
1991 Jan. 1	1992 Jul. 1	26s
1992 Jul. 1	1993 Jul 1	27s
1993 Jul. 1	1994 Jul. 1	28s
1994 Jul. 1	1996 Jan. 1	29s
1996 Jan. 1	1997 Jul. 1	30s
1997 Jul. 1	1999 Jan. 1	31s
1999 Jan. 1	2006 Jan. 1	32s
2006 Jan. 1	2009 Jan. 1	33s
2009 Jan. 1		34s

SOURCES OF UT1

As pointed out earlier, the Earth's variable rate of rotation makes it necessary to determine UT1–UTC observationally using astronomical and geodetic techniques. However, for low-accuracy applications some might choose to ignore the difference between UT1 and UTC, and effectively set UT1–UTC = 0 for all time, assuming that the current definition of UTC will ensure that |UT1–UTC| < 0.9s (13.5 seconds of arc).

However, many users can benefit from the use of the more accurate data that have been provided by international service agencies. Figure 2 shows these observational values as a function of time. It demonstrates the effect of the definition of UTC in place since 1970 whereby UTC is adjusted through the use of leap seconds.

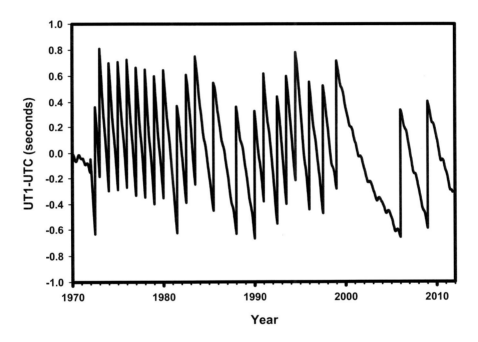

Figure 2. Observed values of UT1–UTC.

The IERS provides a variety of sources for UT1 that can be obtained quite readily. It was established as the International Earth Rotation Service in 1987 by the International Astronomical Union and International Union of Geodesy and Geophysics. In 2003 it was renamed to the International Earth Rotation and Reference Systems Service. The IERS serves astronomical, geodetic and geophysical communities by providing:

- International Celestial Reference System (ICRS) and its realization, the International Celestial Reference Frame (ICRF),
- International Terrestrial Reference System (ITRS) and its realization, the International Terrestrial Reference Frame (ITRF),
- Earth orientation parameters required to transform between the ICRF and the ITRF and for research,

- Geophysical data to interpret time/space variations in the ICRF, ITRF or earth orientation parameters, and model such variations, and
- Standards, constants and models (*i.e.*, conventions) encouraging international adherence.

In partial fulfillment of its mission, then, the IERS provides UT1–UTC in a variety of forms. All are supplied in electronic and paper formats. Definitive observed values with errors less than 10 µs (0.000 15 seconds of arc) are made available about one month after the observations were made. Rapid Service values with accuracy between 20 and 30 µs are also provided with a latency of a few days. In addition, predicted values are supplied for short terms (< 1 year in advance) and for long term (up to twenty years in advance). Obviously the errors of those predictions increase with the length of the forecast, and Figure 3 shows the maximum error that can be expected in IERS products in comparison with the maximum error achieved by assuming that UT1–UTC can be ignored. These predictions are based on the most recent astronomical and meteorological data as well as statistical models.

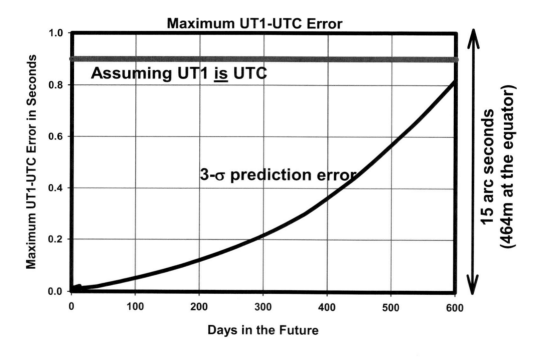

Figure 3. Maximum error in IERS estimates of UT1–UTC.

The effects of daily and sub-daily variations in the Earth's rotation are not included in the UT1–UTC values provided by the IERS. These very small additional components should be added to the UT1–UTC values derived after interpolating the daily estimates for the epoch of interest to account for the effects of ocean tides and libration. The procedures to do this are provided in the IERS Conventions 2010.

In addition to the products provided directly by the IERS Product Centers it is expected that UT1–UTC values will be part of the navigational message broadcast by the GPS III satellites. That would make accurate values of UT1–UTC easily accessible to users with the appropriate receivers. If there were sufficient user demand, we might even envision a UT1–UTC application being made available for future hand-held devices.

DISCUSSION

UTC was defined in 1970 so as to ensure that |UT1–UTC| < 0.9s, and consequently some applications have made use of this definition to neglect the effect of non-zero values of UT1–UTC. When the definition was originally formulated in 1970, it was designed principally to accommodate the needs of celestial navigators.[7] Since that time celestial navigation has been largely surpassed by electronic navigation and the requirement for celestial navigation is no longer so critical.

Knowledge of the Earth Rotation Angle, however, remains an important requirement for many uses in geodesy, astronomy and particularly in various space applications. In these fields the need for improved accuracy continues to grow, and it is likely that it will continue to grow in the future. The fifteen arcsecond accuracy provided by assuming that UTC can be used as a proxy for UT1 is unlikely to meet those future requirements. To meet those needs it is important to plan for ways not only to improve observations, but also for ways to provide this information to users.

UT1–UTC estimates are now available routinely at the level of accuracy of ±0.000 15 seconds of arc as opposed to the ±0.015 seconds of arc available in 1970. At that time UT1–UTC estimates were available via paper bulletins one month after the astronomical observations. Now they are available electronically in near real-time. These facts would suggest that those applications that require knowledge of the relationship between celestial and terrestrial reference systems be re-investigated to determine if they can benefit from the accuracy and availability of UT1–UTC estimates provided routinely by the IERS. Significant improvements may be possible if actual observations were used as opposed to using UTC as a proxy for UT1.

CONCLUSION

Relating celestial and terrestrial reference systems requires knowledge of the Earth's rotation angle in space. This information is proved by a mathematical expression for the Earth Rotation Angle that requires numerical values of the angle UT1. Those data can be obtained by combining estimates of UT1–UTC with Coordinated Universal Time (UTC). The difficulty in predicting the Earth's variable rate of rotation makes it necessary for UT1–UTC to be determined observationally and for those values, as well as forecast values, to be disseminated in a convenient way for access by users.

Currently low-accuracy (±15″) values of UT1 can be obtained by ignoring the difference between UT1 and UTC. However, high-accuracy (±0.000 15″) values are also routinely available electronically in near real time at no cost to the user. It is likely that the demand for high-accuracy will grow in the future, and international and national agencies are prepared to meet that need. Applications requiring UT1 data might benefit from investigating potential improvements to operations made possible by taking advantage of high-accuracy estimates of UT1–UTC instead of just using UTC.

REFERENCES

[1] Ma, C., and Feissel, M., eds., 1997, *Definition and Realization of the International Celestial Reference System by VLBI Astrometry of Extragalactic Objects*, International Earth Rotation Service Tech. Note 23, Observatoire de Paris, Paris

[2] Perryman, M.A.C., Lindegren, L., Kovalevsky, J., Høg, E., Bastian, U., Bernacca, P.L., Creze, M., Donati, F., Grenon, M., Grewing, M., van Leeuwen, F., van der Marel, H., Mignard, F., Murray, C.A., Le Poole, R.S., Schrijver, H., Turon, C., Arenou, F., Froeschle, M., Petersen, C.S., 1997, "The Hipparcos Catalogue," *Astron. Astrophys.*, **323**, L49-L52.

[3] Boucher, C. Altamimi, Z., Sillard, P., Feissel-Vernier, Martine, 2004, *The ITRF200*, International Earth Rotation and Reference Systems Service (IERS). IERS Technical Note, No. 31, Frankfurt am Main, Germany: Verlag des Bundesamtes für Kartographie und Geodäsie.

[4] Kaplan, G. H., 1981, *The IAU Resolutions on Astronomical Constants, Time Scale and the Fundamental Reference Frame*, U. S. Naval Observatory Circular No. 163.

[5] *IERS Conventions (2010)*, 2010, (Edited by Gérard Petit and Brian Luzum), International Earth Rotation Service Tech. Note 36, Verlag des Bundesamts für Kartographie und Geodäsie, Frankfurt am Main.

[6] Aoki, S., Guinot, B., Kaplan, G. H., Kinoshita, H., McCarthy, D. D., and Seidelmann, P. K., 1982, "The New Definition of Universal Time," *Astron. Astrophys.*, Vol. 105, pp.359-361.

[7] Nelson, R. A., McCarthy, D. D., Malys, S., Levine, J., Guinot, B., Fliegel, H .F, Beard, R. L., Bartholomew, T. R., 2001, "The leap second: its history and possible future," *Metrologia*, Vol. **38**, 509-529.

DISCUSSION CONCLUDING AAS 11-666

Ken Seidelmann noted that there will still be a requirement for predicted UT1 for space missions, where UT1 must be uploaded onboard spacecraft in advance. Dennis McCarthy replied that those sorts of things will probably still be there, but McCarthy thought that the process could be streamlined so that a good percentage of the user community could take advantage of the accuracy available in real time. McCarthy said the problem with predictions is that they degrade with time and will always degrade with time, and that predictions on the order of a week to a month will "do as well as you can predict the weather." Seidelmann said that some spacecraft missions do not maintain continuous communication but contact may be available daily. McCarthy said that this issue exists for systems where there are concerns that space communications might be disrupted for an extended period. Rob Seaman said communication lapses are also an issue for ground-based systems, because modern telescopes tend to operate untended in remote locations. Arnold Rots asked if this is only an issue for Earth-looking spacecraft, rather than sky-looking spacecraft. Seidelmann replied that most Earth-orbiting spacecraft have ground access only periodically and operational spacecraft may need reasonably accurate knowledge of UT1 for antenna pointing far in advance to maintain space-based communications. Mark Storz added that if field of view is narrowly constrained then a sufficiently accurate prediction of UT1 may be required in advance.

Seaman said that he was able to follow McCarthy's presentation until the mention of software where "you waved your hands and said that software will just naturally take this into account." Seaman said that it is people like himself, Allen, Rots, and others within the community of astronomical software-development that will have to create this software. Seaman said he was also lost at the comment "You don't care if people sell these time signals." The missing link when it gets to the IERS appeared to be the network time protocol (NTP) which connects all these computers and keeps clocks running accurately until they disconnect from the Internet. Seaman said that the math seemed correct but the infrastructure needed to be closed; the people that need to be involved in that discussion would be NTP folks who wouldn't necessarily accept responsibility for transmitting UTC and UT1 *both*. McCarthy thought "they would build it;" Seaman agreed that it will be built but only if someone is told they must build it. McCarthy said it was an opportunity to get people to use the "full-blown accuracy" of UT1-UTC. Allen quipped that perhaps he and Seaman could go into business selling an NTP service hacked to provide UT1.

Seidelmann noted that UT1 distribution is an issue independent of leap seconds. However, Terrett offered that if leap seconds were dropped, then more people would need this type of service, so the matter may not be independent. Seidelmann added that the redefinition of UTC forces everyone to accommodate a higher level of accuracy (perhaps whether it is needed or not). Terrett said now is the perfect opportunity to attempt some kind of service, while some attention is being paid to the issue. Seaman said if UTC is redefined, something like this *certainly* must happen.

George Kaplan asked if UT1-UTC is broadcast as part of the GPS navigation messages; McCarthy replied "No, not now." Kaplan asked if there was a space for it. McCarthy said that there is a prediction formula for UT1-UTC in the navigation message now. Malys clarified that this prediction is put into the so-called *five-line elements* used in the daily processing of the GPS ground control segment but it is not put into the broadcast messages. McCarthy concurred. Malys added that he would address ground operations in his presentation. McCarthy said that, per his understanding, GPS III is expected to have UT1-UTC as a broadcast element. Allen asked if its value would be constrained; McCarthy thought that the absolute value might be constrained to 99 seconds. Storz said his recollection of the ICD was that it would be limited to 64 (2^6) seconds. McCarthy added that he knew it was a double-digit value. Storz said the magnitude of that value would likely get us into the next century if UTC is redefined.

AAS 11-667

THE IERS, THE LEAP SECOND, AND THE PUBLIC

Wolfgang R. Dick[*]

Bulletin C with announcements of leap seconds is the most popular of IERS products. A large part of requests from the public received by the IERS Central Bureau concerns the leap second. Although other IERS products may be of more importance, leap-second announcements produce a maximum of attention with a minimum of efforts. IERS has plans for an UT1 time service in case that UTC would be redefined. However, with respect to the public relations of the IERS, a possible abolishment of the leap seconds has to be compensated by other outreach activities which will attract similar public attention.

INTRODUCTION

This paper will discuss the consequences of a redefinition of UTC without leap seconds with respect to the public outreach of the IERS. It reflects mainly the experience and thoughts of the author as a staff member of the IERS Central Bureau, who is responsible also for public relations, and is not an official statement of the IERS. The paper does not intend to argue in favor or against a redefinition of UTC.

The *public* meant here is mainly a technically and scientifically interested audience, like high-school and university students, science teachers, engineers, astronomers, geo-scientists, physicists, etc. This public includes also specialists, who need a more precise time scale than in everyday life, but who are otherwise not involved in Earth rotation matters.

THE INTERNATIONAL EARTH ROTATION AND REFERENCE SYSTEMS SERVICE

The International Earth Rotation and Reference Systems Service (IERS)[1] was established by the International Astronomical Union and the International Union of Geodesy and Geophysics in 1987. Its primary objectives are to serve the astronomical, geodetic and geophysical communities by providing the following:

- the International Celestial Reference System (ICRS) and its realization, the International Celestial Reference Frame (ICRF),

- the International Terrestrial Reference System (ITRS) and its realization, the International Terrestrial Reference Frame (ITRF),

- Earth Orientation Parameters (EOP) required to study Earth orientation variations and to transform between the ICRF and the ITRF,

---

[*] Dr, IERS Central Bureau, Bundesamt für Kartographie und Geodäsie, Richard-Strauss-Allee 11, 60598 Frankfurt am Main, Germany.

- geophysical data to interpret time/space variations in the ICRF, ITRF or Earth Orientation Parameters, and model such variations,

- standards, constants and models (*i.e.* conventions) encouraging international adherence.

In 2003 the IERS, formerly known as the International Earth Rotation Service, got its current name to reflect the equal importance of reference systems besides the Earth Orientation Parameters in its tasks. The IERS is a non-governmental organization, based on voluntary contributions by many institutions around the world, providing personal and financial resources. The IERS does not have its own budget and thus also does not have any staff besides that in the host organizations. This is of certain importance here because it influences how the IERS addresses its users and the public.

The work of the IERS is currently being done by 6 Product Centers (one of them consisting of 4 special bureaus, another supported by 3 ITRS Combination Centers), 4 external Technique Centers (IGS, ILRS, IVS, and IDS), and 4 Working Groups. Its work is coordinated by a Directing Board, an Analysis Coordinator, and the Central Bureau. These components run more than 15 independent web sites and several ftp servers.

The central IERS web site,* maintained by the Central Bureau, provides more detailed information on IERS and gives access to all other IERS web sites. It includes the IERS Data and Information System, providing more general information related to Earth rotation and reference systems, collecting all IERS products from the individual Product Centers, and storing and offering them in different formats together with metadata.[2] Among these products are Bulletins C and D with leap-second announcements and UT1 data.

The IERS web sites include also some information for the public, although their main target groups are scientific and technical users of IERS products. Currently, there are no press releases by the IERS itself, but only by the host organizations of the IERS components.

IERS BULLETINS C AND D

The decision to introduce a leap second in UTC is the responsibility of the IERS, specifically of the IERS Earth Orientation Center, hosted by Paris Observatory, which issues announcements of leap seconds as IERS Bulletin C. This Bulletin is mailed every six months, either to announce a time step in UTC, or to confirm that there will be no time step at the next possible date.[†]

The decision on the introduction of a leap second is a byproduct of the IERS's work on the determination of Universal Time (UT1) and its prediction for the next months. Hundreds of people all over the world are involved in measurements and calculations of the Earth Orientation Parameters, but once these have been derived for the past and predicted into the future, the estimation of whether a leap second will be necessary or not is an easy task. Probably the most time-consuming task regarding Bulletin C is to maintain its mailing list.

The IERS Earth Orientation Center issues also Bulletin D containing announcements of the value of DUT1 = UT1−UTC to be transmitted with time signals with a precision of 0.1s, which is also being distributed by e-mail and is available for download.

---

*http://www.iers.org/
†See http://hpiers.obspm.fr/eoppc/bul/bulc/BULLETINC.GUIDE for more details.

IERS BULLETIN C, THE LEAP SECOND, AND THE PUBLIC

IERS Bulletin C seems to be the most popular of IERS products: there are approximately 1600 subscribers to it, about twice as much as the other IERS bulletins. Although this figure has to be treated with caution because there are no download statistics for other IERS products like reference frames and continuous Earth Orientation Parameter (EOP) series, it shows a tendency. The increased popularity of Bulletin C with the leap-second announcements is feasible because precise time is needed in more applications than reference systems or EOPs. An overview of the fields of activity of the users of Bulletins C and D was gained by the surveys done by the Earth Orientation Center in 2002 and 2011.[3,4]

Also a large part of requests from the public received by the IERS Central Bureau concerns the leap second. There was a climax around the end of the year 2008, when the last leap second was introduced. It is noted that the inquiries did not come immediately after the publication of the corresponding Bulletin C, but rather before or after the end of the year 2008, when many public media published reports about the forthcoming leap second introduction. Also a significantly larger amount of new subscriptions to Bulletin C were observed at that time.

The requests received by the IERS Central Bureau came from journalists, who asked for interviews or more information for writing articles, and from the readers of these articles, drawing attention to errors in them or asking questions like the following ones:

- "Why would the leap second have an impact on the accuracy of GPS satellites?"

- "I understand 24 seconds have been added since 1972 but I see other web sites that state only 2.2 seconds need to be added in 100,000 years. Will you be adding about 1 second every 1 1/2 years in the future as you have in the past?"

- "Is there a similar 'slowing' of the earth in orbit around the sun each year? I realize there are leap years but was there a time when a year was shorter and will this too require an adjustment in the future?"

- "When will the next leap second occur?"

Very often these questions are not easy to answer, but in any case they show the attention of the public.

In December 2010 the Bundesamt für Kartographie und Geodäsie (BKG) in Frankfurt am Main, Germany, which is the host organization of the IERS Central Bureau, issued a press release stating that there would be no leap second at the end of 2010, noting that the leap second was overdue compared with past years when they were introduced more often, giving a geophysical explanation for this, and drawing attention to the work of the IERS. A science writer of a large German news agency wrote a well researched article on the basis of this press release. The article was reprinted in many national and local German newspapers between Christmas and New Year's Eve, and several inquiries from readers were received after this.

Generally, press releases and articles about leap seconds and their background seem to gain rather good attention. There are several reasons for this: civil time concerns many more people than other data related to Earth rotation and reference systems. Leap seconds are an integrated effect of the deceleration of Earth rotation, for which mainly the Moon is responsible – a very interesting astronomical, geophysical, and physical phenomenon which can be used nicely in science popularization. The introduction of a leap second is a single event, and it occur mostly at New Year's Eve (at least during the last years), i.e., at a rather distinguished moment. At that time of the year other news is rare, such that mass media readily include news about a forthcoming leap second. Partially, discussions about the future of UTC also caused public interest.

The situation described above applies to the last decade, and to future years, when the introduction of a leap second is perceived as a rather rare event. This will eventually change in the future as Earth's rotation slows down and leap seconds will be needed more routinely like in the 1970s (Figure 1). Therefore, one might expect less attention from the public in this situation.

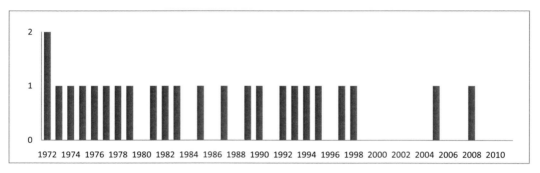

Figure 1. Leap seconds per year between 1972 and 2011

CONCLUSION: THE IERS WITHOUT LEAP SECONDS

Leap seconds are a byproduct of other IERS activities; thus, even without leap seconds the IERS will have an important and continuing role with its other products.

The IERS has plans for a UT1 time service should UTC be redefined to no longer be an approximation to UT1, e.g., through the use of the Virtual Observatory[5] or by other means.[6] This will be a task in addition to the existing ones and will probably demand more efforts than the current Bulletins C and D.

However, such a UT1 service is of interest mainly for those who really need it in practice, and not so much for a broader audience. As a permanent service, it would not be useful for public outreach. Although other IERS products may be of more scientific importance, the current leap-second announcements, especially in combination with press releases, produce maximum attention with minimal effort.

Thus, an abolishment of leap seconds (but equally the expected higher frequency of leap seconds in the future) would result in an loss of publicity for the IERS and geo-sciences. This would have to be compensated by other activity attracting similar public attention – a not so easy task for the IERS doing mainly routine service work. However, IERS might still increase and improve its public outreach activities, e.g., by the following:

- include more information for non-specialists in the web sites;

- publicize more strongly the scientific research relating to global climate change and sea level rise, which are popular topics in the media;

- issue IERS press releases;

- define a policy for public relations and public outreach, which currently does not exist.

ACKNOWLEDGMENTS

Galina Dick, Daniel Gambis, Chopo Ma, Dennis D. McCarthy, Bernd Richter, and John Seago read draft versions of this paper and made valuable suggestions, for which I would like to thank them.

REFERENCES

[1] W. R. Dick and B. Richter, "The International Earth Rotation and Reference Systems Service (IERS)," *Organizations and Strategies in Astronomy, Vol. 5* (A. Heck, ed.), Vol. 310 of *Astrophysics and Space Science Library*, pp. 159–168, Dordrecht, Boston, London: Kluwer, 2004.

[2] B. Richter and W. Schwegmann, "IERS Data and Information System," *Observation of the Earth System from Space* (J. Flury, ed.), pp. 321–332, Berlin, Heidelberg, New York: Springer, 2006.

[3] D. Gambis, P. Baudouin, C. Bizouard, M. Bougeard, T. Carlucci, N. Essaifi, G. Francou, and D. Jean-Alexis, "Earth Orientation Centre," *IERS Annual Report 2001* (W. R. Dick and B. Richter, eds.), pp. 36–46, Frankfurt am Main: Verlag des Bundesamts für Kartographie und Geodäsie, 2002.

[4] D. Gambis, "Results of the 2011 IERS Questionnaire Concerning a Possible Redefinition of UTC," *This volume*, No. AAS 11-668, 2011.

[5] F. Deleflie, C. Barache, J. Berthier, and D. Gambis, "Dissemination of DUT1 Through the Use of Virtual Observatory," *This volume*, No. AAS 11-680, 2011.

[6] D. D. McCarthy, "Using UTC to Determine the Earth's Rotation Angle," *This volume*, No. AAS 11-666, 2011.

DISCUSSION CONCLUDING AAS 11-667

Mark Storz asked if any studies had been performed to see if global mass transport (ice melt, erosion, *etc.*) might change the Earth's moment of inertia enough to explain the decreasing trend in the number of leap seconds. Wolfgang Dick replied that his discussions with Peter Brosche suggested that the effect seems mainly due to core-mantle coupling inside the Earth. Specifically the core and the mantle may rotate slightly differently, with the mantle experiencing more variation due to tidal effects caused by gravitational interaction with the Moon and Sun. The coupling between the mantle and the core has a regulating effect on the mantle such that whenever the mantle slows down, contact with the core will spin the mantle back up.

McCarthy said that rotational deceleration due to tides is computable, but the theory is different than what is observed. The difference between theory and observation has been attributed to an acceleration caused by rising sea levels and deglaciation; the Earth speeds up as it becomes rounder. Storz asked if mass transport had been analyzed sufficiently to support the theory, and McCarthy replied that "the theory matches the observations fairly closely." Steve Allen noted that, among the fifteen web sites of the IERS, some point to research in geophysical fluids where specialist studies can be reviewed.

Neil deGrasse Tyson concurred with Dick's observations that there is a public appetite for leap seconds. Public interest is "very high; they love it, they love talking about it, they love trying to understand it. The explanation involves the Moon, things they've heard about, words they've used before." Based on his life experience, Tyson "agreed emphatically" that the topic makes for "quite an entrée" into the domain of solar-system dynamics. Dick added that he uses leap seconds to explain the work of the BKG to visiting students. Particularly, it is the only effect that an ordinary person can observe themselves, because a precise watch will reveal the one-second difference in the time of day the next morning after a leap second.

Steve Malys asked how observed trends in Earth rotation might be related to the fact that the *SI* second was based on some average duration of the mean solar second from the 19th century. McCarthy said that if we were to define the *SI* second relative to Earth rotation now, we wouldn't have the issue of leap seconds but the issue would instead exist a century from now. This is because there is a deceleration that accumulates as a parabola. Seago added that McCarthy's statements were accurate but only over very long time periods. He said that if the trend were truly parabolic we would have experienced more and more leap seconds over time but fewer and fewer have been needed since their introduction in 1972. Over the scale of human lifetimes so-called *decadal fluctuations* tend to dominate the underlying long-term trend. As evidence, Seago noted that length of day was not increasing in the plots presented by Dick. Dick added that if the second had been redefined in 1972 then the origin of such plots would shift to that epoch.

George Kaplan affirmed the thrust behind Malys question: we have adopted a conventional rotation rate of the Earth based on 19th century observations, but there is "a lot of slop" in how the definition of today's *SI* second came about. The *ephemeris second* was defined in terms of the tropical year of 1900 as defined by Newcomb's theory of the solar system. Markowitz calibrated the atomic second to the ephemeris second that was uncertain to about one part in 10^9. Kaplan concluded that the conventional value that we put in our formulae for Earth rotation may be somewhat arbitrary.

AAS 11-668

RESULTS FROM THE 2011 IERS EARTH ORIENTATION CENTER SURVEY ABOUT A POSSIBLE UTC REDEFINITION

Daniel Gambis, Gérard Francou and Teddy Carlucci[*]

The Earth Orientation Product Center is responsible for the prediction and announcement of the leap second (Bulletin C) and the announcement of the value of DUT1 truncated at 0.1 s for transmission with time signals. A first survey made in 2002 show that 89% of IERS users were satisfied by the current determination of UTC, including leap seconds introductions. With the increasing number of users belonging to the various communities, it was felt necessary to take a new survey to find out the strength of opinion for maintaining or changing the present system before the proposal of redefining UTC is discussed at the ITU-R meeting which will be held in Geneva in January 2012.

INTRODUCTION

The legal time scale UTC (Coordinated Universal Time) is derived from TAI (Temps Atomique International) by the insertion of leap seconds in order to maintain UTC within ±0.9 s of the time scale based on the Earth's rotation UT1, *i.e.* |UT1-UTC| < 0.9 s. This system was introduced in 1972. Several years ago, some communities particularly involved in telecommunications and navigation systems proposed a revision of the UTC definition, aiming to eliminate leap seconds in order to have a continuous time scale. This has been a topic of discussions for nearly 20 years.

The Earth Orientation Product Center of the International Earth Rotation and Reference Systems Service (IERS) is responsible for the prediction and announcement of the leap second (Bulletin C) and the announcement of the value of DUT1 truncated at 0.1 s for transmission with time signals.

The first survey taken in 2002 showed that a large majority of IERS users were satisfied by the current determination of UTC, including leap seconds introductions.[1] With the increasing number of users belonging to the various communities, it was felt necessary to take a new survey to find out the strength of opinion for maintaining or changing the present system before the proposal of redefining UTC be discussed at the ITU-R meeting, which will be held in Geneva in January 2012.

---

[*] Earth Orientation Center of the IERS, Observatoire de Paris, 61 av. de l'observatoire, 75014 Paris, France.

QUESTIONNAIRE TO SURVEY OPINIONS CONCERNING A POSSIBLE REDEFINITION OF UTC

The Survey Language

Universal Time, the conventional measure of Earth rotation is the traditional basis for civil timekeeping. Clocks worldwide are synchronized via Coordinated Universal Time (UTC), an atomic time scale recommended by the Radiocommunications Sector of the International Telecommunications Union (ITU-R) and calculated by the Bureau International des Poids et Mesures (BIPM) on the basis of atomic clock data from around the world.

UTC is computed from TAI by the introduction of leap seconds such that UTC is maintained within 1 second of UT1. Since 1972, these leap seconds have been added on December 31 or June 30, at the rate of about one every 18 months. Since 1 January 2009, 0:00 UTC, UTC-TAI= -34 s.

After years of discussions within the scientific community, a proposal to fundamentally redefine UTC will come to a conclusive vote in January 2012 at the ITU-R in Geneva. If this proposal is approved, it would be effective five years later. It would halt the intercalary adjustments known as leap seconds that maintain UTC as a form of Universal Time. Then, UTC would not keep pace with Earth rotation and the value of DUT1 would become unconstrained. Therefore UTC would no longer be directly useful for various technical applications which rely on it being less than 1 second from UT1. Such applications would require a separate access to UT1, such as through the publication of DUT1 by other means.

The objective of the survey is to find out the strength of opinion for maintaining or changing the present system.

Two references:

1. Nelson, R.A., McCarthy, D.D., Malys, S., Levine, J., Guinot, B., Fliegel, H.F., Beard, R.L., and Bartholomew, T.R. "The leap second: its history and possible future." *Metrologia*, Vol. 38, 2001, pp. 509-529 http://www.cl.cam.ac.uk/~mgk25/time/metrologia-leapsecond.pdf

2. Finkleman, D., Seago, J.H., and Seidelmann, P.K. "The Debate over UTC and Leap Seconds." Proceedings of the AIAA/AAS Astrodynamics Specialist Conference, Toronto, Canada, 2010. http://www.agi.com/downloads/resources/user-resources/downloads/whitepapers/DebateOverUTCandLeapSeconds.pdf

The Survey Options

1. I am satisfied with the current definition of UTC which includes leap second adjustments.

2. I prefer that UTC be redefined as a uniformly increasing atomic timescale without leap seconds and constantly offset from TAI. Consequently, UTC would increasingly diverge from the Earth's rotation.

3. I have another preference.

4. I have no opinion or preference.

5. Comments.

The Survey Results

The following figures give respectively the global results and statistics concerning the domains of activities, as well as the number of answer per country.

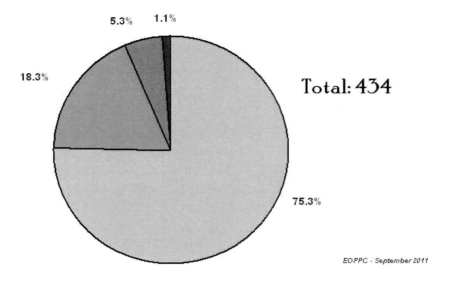

Figure 1. Global results.

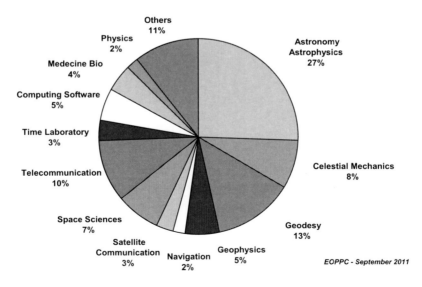

Figure 2. Fields of activities.

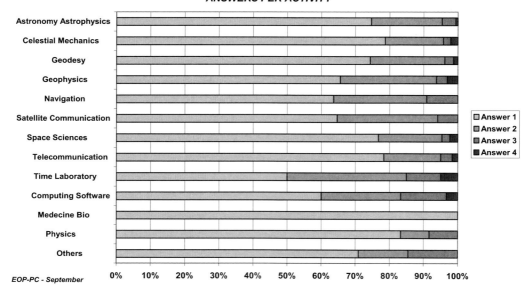

Figure 3. Percentage of answers per field of activity.

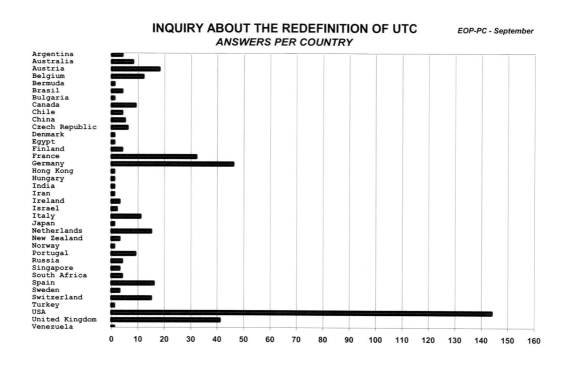

Figure 4. Statistics concerning the number of answers per country.

COMMENTS TO THE QUESTIONNAIRE

We have selected hereafter the most pertinent and representative answers. These are listed without any hierarchical order.

General Arguments Favoring the Status-Quo (75%)

- The present system, working well, is a good compromise between Earth rotation and atomic time.
- Arguments to change are not sufficient compared to the advantages of a coordinated UTC time scale linked to the Earth rotation.
- Any changes in these areas will likely cause substantial confusion and disasters (principle of security).
- In particular, risk of confusion and problems in the case of the increase of the tolerance UT1-UTC
- A majority of UTC users are not aware of the difference between UT1 and UTC. If the new definition is adopted, they should. When the difference DUT1 increases, 30s, 10 min, 1 hour, a lot of problems will arise.
- There is too much software with the assumption of UTC being coordinated with the Earth rotation. The costs of change would be important. Unforeseen problems could happen.
- No reason to maintain 3 separate time scales (GPS, TAI, and UTC) differing by a constant offset.
- In many countries legal time is based on solar time.
- No strong argument to change. The current system works. "If it ain't broke, don't fix it!"
- Few problems were reported after the 2009 leap second introduction.
- In a few decades, who will remember the origin of the procedure?
- There is no strong justification to adopt a time scale no longer related to the rotation of the Earth.
- More time should be needed to evaluate the consequences of such a change (UNESCO statement).

General Arguments Favoring a Change (19%)

- Ambiguity of date at the occurrence of a positive leap second which is potentially dangerous.
- Separating the two concepts (angle for UT1, time for UTC) would be an improvement for high-accuracy applications.
- Leap-seconds were a good idea in 1972 when people just had a few inaccurate analog clocks, but now so much equipment has a clock, it is a nightmare to correct it all.
- Ignoring leap seconds will not be a significant problem for civil purposes.
- Analyzing the performance of the time servers during the 2008/2009 leap second showed a worrying percentage of (otherwise well configured and well maintained) systems being a second out of sync with everyone else for hours and in some cases even days!

- Having a time scale that is discontinuous causes a lot of problems with writing and maintaining software for processing non-ground-based astronomical missions.
- There is no technical reason for keeping the existing system other than TRADITION.
- Designing, operating and testing time service equipment for leap seconds require tremendous efforts.
- Most databases can't deal with leap seconds and interval calculations can't. For this reason we need to unwind leap seconds.
- The handling of leap seconds adds a considerable complexity for equipment manufacturers and for operators in order to prepare and pre-program for the insertion/removal of a leap second.
- The leap seconds represent a nuisance for the modern applications requiring time synchronization

Other Proposals

- It would be useful for leap seconds to be scheduled further in advance.
- With the ubiquitous use of NTP, I believe there is now an opportunity to separate civil time from the high-precision time/frequency dissemination services.
- Time correction applied on a deterministic date, and more rarely on 01 January 00:00 every 10 years. Or even better, to apply them each 29 February.
- A better representation can preserve the existing and traditional meaning of UTC as civil time while also alleviating the problems faced by software systems.

SUMMARY

The statistics mostly reflect the statements of communities of time scales users. As of 10 September 2011, there were 443 responses to the questionnaire; 9 of these were discarded having been considered wacky.

Over the 434 remaining answers, about 75% favor the *status quo*, i.e., no change in the current definition including leap seconds. 19% favor switching to the new UTC definition, *i.e.*, continuous time scale no leap second. 5% favor another solution, mostly requiring the prediction of the leap second with a longer schedule in advance. 1% has no opinion.

Let us note that globally and except for the time community where 50% are for the *status quo*, the percentages of users favoring and opposed to the *status quo* is similar, with a majority favoring the *status quo* whatever the domain of activity.

Answers and comments to the questionnaire are fully available at the following web site: http://hpiers.obspm.fr/eop-pc/questionnaire/result.php

REFERENCES

[1] Gambis, D., Bizouard, C., Francou, G., and Carlucci, T., "Leap Second Results of the Survey made in Spring 2002 by the IERS," in: Proceedings of the ITU-R SRG Colloquium on the UTC Timescale, IEN Galileo Ferraris, Torino, Italy, 28-29 May 2003. (URL http://www.ucolick.org/~sla/leapsecs/torino/gambis_leap.pdf)

EPILOGUE TO AAS 11-668: SURVEY RESULTS

The editors have included responses to the 2011 IERS Earth Orientation Center survey as an epilogue to Gambis et al. (2011).[*] Each available recorded response includes:

i. The date of response, followed by the contributor's declared name.

ii. The contributor's declared professional affiliation and nationality.

iii. The contributor's declared domain of activity.

iv. The contributor's preference indicted by the following number:

1. Satisfaction with the status quo (with leap seconds).
2. Preference that UTC be redefined as a uniformly increasing atomic timescale without leap seconds.
3. Another preference.
4. No preferential opinion.

v. Optional commentary (up to 1600 characters).

To enhance the readability of the responses, some typographical errors were corrected. Majuscule (upper case) or minuscule (lower-case) typing for names, locations, *etc.*, was also changed to enhance readability. To conserve space, salutations, closings, signature blocks, and personal information (such as business URLs, email, and postal addresses), were omitted. Some country names were also abbreviated to conserve space. The sequencing of a few responses was changed for more efficient pagination; the included date reveals original sequencing.

As of October 26, 2011, 447 responses were available, which includes responses unavailable when Gambis *et al.* (2011) was submitted. An updated tabulation of the response percentages is provided in Table 1 of this epilogue, but these are not significantly different than before. A separate letter was received by Royal Institute of Navigation (RIN), which is also appended.

Table 1. Tally of Available Responses (as of October 26, 2011)

Response #1 (*status quo*)	341	(76.3%)
Response #2 (decouple UTC and Earth rotation)	80	(17.9%)
Response #3 (another preference)	21	(4.7%)
Response #4 (no opinion)	5	(1.1%)

---

[*] As available via http://hpiers.obspm.fr/eop-pc/questionnaire/result.php on October 26, 2011.

08 July 2011; Woltz, Lawrence
NASA, USA
Satellite precipitation measurement
Response = 1

08 July 2011; van Schellen, Remco
Omroep Zeeland, the Netherlands
Media
Response = 1

08 July 2011; Savoie, Denis
SYRTE-Observatory of Paris, France
Astronomy-Astrophysics
Response = 1
The redefinition of UTC will cause serious difficulties for fans of sundials. The conversion of solar time in standard time will be even more complicated to explain to the public and students! The calculation of analemma (directly indicating Universal Time by integrating the equation of time and the longitude) with will become problematic. Sundials are nothing in front of the lobby GPS; but their role in teaching astronomy is very important

08 July 2011; Marmet, Louis
NRC Canada, Canada
Time-laboratory
Response = 1
The time scale TAI is already implemented for applications where a leap second would be a problem. I consider that a change of the definition of UTC will reduce the credibility of our institution (time standards community) in the eye of the public. Seriously affected will be the users who use the position of the sun. A redefinition of UTC will change the calendar date of these events and have serious social impacts once it becomes known to the public. There is not enough room here to bring more arguments...

08 July 2011; Olsson, Sten
Lockheed Martin Corporation, USA
Air Traffic Control
Response = 1

08 July 2011; Tang, Jingshi
Astronomy department, Nanjing University, China
Astronomy-Astrophysics Celestial-mechanics Geodesy
Response = 1

08 July 2011; Cooper Jr., Peter
None, USA
Hobbyist that finds time interesting
Response = 1

08 July 2011; Gambis, Perceval
Thales, France
Air Traffic Control
Response = 2
In the field of Civil Air Navigation, UTC time is the reference. We don't really care about TAI. Leap seconds have always been an issue and sometimes a 1 second jump in the clocks can cause majors problems in our complex air traffic control systems and subsystems. Getting rid of the leap seconds is preferable.

08 July 2011; Gonzalez, Hervé
Airbus Operations SAS, France
Aeronautics
Response = 1

08 July 2011; Williams, David
Fidelity Bank, USA
Telecommunication Information Technology
Response = 1
Civil time reckoning has always been tied with the Earth's rotation. Leap seconds are small enough that few people are inconvenienced and frequent enough that there are tested procedures on how to deal with them. Leap minutes or leap hours would be very disruptive. To drop the relationship with Earth's rotation is to not deal with the issue and to kick the can down the road for someone else to deal with at a later time.

08 July 2011; Enfinger, Eugene Bryan
Enfinger & Assoc., LLC, USA
Astronomy-Astrophysics Celestial-mechanics Geodesy
Response = 1
LEAVE THE CURRENT SYSTEM AS IS!!!!!!

08 July 2011; Finch, Tony
Univ. of Cambridge Computing Service, England
Telecommunication
Response = 2
If we are to continue leap seconds, they will be much more easy to handle if they are announced several years in advance, such that leap second tables can be distributed as part of a computer system's software.

08 July 2011; Smith, Eric
Total Spectrum Software, Canada
Computer software
Response = 3
I am generally satisfied with the current definition of UTC which includes leap seconds. However, I think it would be useful for leap seconds to be scheduled further in advance (for example several years, rather than 6 months). This would allow makers of computer systems to more readily schedule and prepare for leap seconds. This advantage would, I think, outweigh the difficulties in keeping DUT1 within 1 second over such a long period -- particularly if the alternative is to allow DUT1 to grow without bound!

08 July 2011; Hall, Shannon
JHU/APL, USA
Astronomy-Astrophysics Geophysics
Response = 1

08 July 2011; Seaman, Rob
National Optical Astronomy Observatory, USA
Astronomy-Astrophysics
Response = 1
I would support lengthening the forecast interval immediately to whatever value permits remaining within the 0.9s DUT1 limit - note that no change would be required to TF-460 to do this. The state of the art of EOP forecasts has improved dramatically since 1972 and we should benefit from that. I might additionally consider supporting the relaxation of the 0.9s limit to later permit lengthening the forecast interval further. Care should be taken with planning for any change to UTC. Due diligence has not been met by the current ITU-R process.

08 July 2011; Vince, Peter
BBC Television, UK
Broadcast radio and television
Response = 2
UTC currently gives an accurate indication of the Earth's orientation - something most people have no interest in, particularly to the degree of accuracy achieved. Most people assume noon (12:00) to be when the sun is at its highest, but with the analemma effect, and especially daylight savings time, that is certainly not true.
Modern broadcasting and communication equipment needs an accurately synchronised reference, so variable frequencies tracking the Earth's rotation is not an option. With 24-hour broadcasting of mainly pre-recorded programmes, it is essential for professional continuity that the duration of the programme is known, and changing the clock time during the transmission of a programme negates this accuracy.
Leap-seconds were a good idea in 1972 when people just had a few inaccurate analogue clocks, but now so much equipment has a clock, it is a nightmare to correct it all. There is also a cost penalty to do this, for the time and effort of the staff involved, and the confusion, if not danger, of them not being corrected and synchronised.
I believe daylight-saving time should also be abolished, but that is another argument. At least let us take this opportunity to simplify time-keeping for the majority. There will be a cost penalty for the astronomers, but that is nothing to the cost currently incurred by everyone else. They already have to compensate for sidereal time and polar wobble - a slightly larger DUT-1 should be a very minor change.
But please ensure DUT-1 *IS* available to everyone, including modifying the LF radio time-signal data formats.

08 July 2011; Müller, Ulrich
Institut für Kernphysik, Univ. Mainz, Germany
Astronomy-Astrophysics
Response = 1

08 July 2011; Theodosiou, Georgios
None, France
jobless
Response = 3
Redefinition, or better, new name, for example: Universal Civil Time (UCT) with its own unit, UCT second, defined as 1/86400 of mean solar day, under condition |UT1-UCT| < 1 sec.
Due to earth slowing and variation in LOD some slight increase (10-20 nanosec) every 19 years (metonian cycle, largest periodic element in LOD) will be needful in UCT sec. It's possible this increase be allocated each year or even each day. Atomic time and its second will remain time scale for scientific and technical purposes, GPS etc., and UT1 and its second for astronomical purposes.

08 July 2011; Sokolov, Michael
Citizen of the Universe, Republic of New Poseidia
moral and political philosophy
Response = 1
For thousands of years the effective definition of a day has been the mean solar day. Hours, minutes and seconds are merely subdivisions of the millennia-old concept of the day. In other words, for thousands and thousands of years the definition of "day" and the time of day has been given by Mother Nature, i.e., by the Sun in the sky. What the outrageous ITU proposal is effectively asking us to do is to give up our trust in Mother Nature in the matters of time of day and to vest our trust instead in the racks of strange equipment operated by a bunch of guys in lab coats.
Universal Time means mean solar time. Anyone who attempts to redefine UTC as something that isn't Universal Time should be arrested and prosecuted for treason against nature / crimes against humanity.

---

08 July 2011; Deniel, Laurent
Thales, France
Telecommunication
Response = 2

---

08 July 2011; Griesbach, Jacob
Analytical Graphics, Inc., USA
Astronomy-Astrophysics
Response = 1
I believe it's important that UTC retain its celestial meaning. The rather infrequent use of leap seconds is only a light burden to maintain this synchronicity.

---

08 July 2011; Hujsak, Richard
Analytic Graphics, Inc, USA
Celestial-mechanics orbit determination and prediction
Response = 1
There are too many software systems with the current definitions embedded. The costs of changing that many systems for a frivolous change in definition is too great to be worthwhile. The danger is some systems would convert to the new definition, while others would not. And that mismatch can have expensive consequences. In the words of a famous procrastinator "If it ain't broke, don't fix it."

---

08 July 2011; Viceré, Andrea
Università di Urbino, Italy
Astronomy-Astrophysics Gravitational Waves
Response = 2
In my field, we rely on GPS counts as a uniformly increasing timescale. UTC would replace it very well and serve as a reference solution.

---

08 July 2011; Gupta, Sanjeev
DCS1, Singapore
Telecommunication
Response = 1
I prefer that UTC closely follow a smoothed UT1. I would accept an increase in the allowed value of DUT1, if it would help in long-range predictions of leap seconds.

---

08 July 2011; Candey, Robert
NASA, USA
Space-sciences
Response = 2

---

08 July 2011; West, Michael
Geophys. Inst., Univ. Alaska Fairbanks, USA
Geophysics
Response = 2

---

08 July 2011; Meagher, Kevin
University of Maryland, USA
Astronomy-Astrophysics
Response = 1
I think that is important for astronomy to keep UT and UTC as close together as possible, As far as I know, all of the problems associated with leap seconds are due do substandard software. These problems could be mitigated by standard software libraries to handle leap seconds.

---

08 July 2011; Dewar, Duncan
none, Scotland
Astronomy-Astrophysics Telecommunication
Response = 1

---

08 July 2011; Clark, Richard
National Solar Observatory, USA
Astronomy-Astrophysics
Response = 1
There is already TAI and the GPS timescale. Why do we need STILL ANOTHER constant uniform timescale?
Keep UTC as it is.

08 July 2011; Finkleman, David
CSSI and ISO TC20/SC14, USA
Astronomy-Astrophysics Space-sciences
Response = 1
I appreciate that we are referenced, but this matter is scientific and concrete, not abstract or a matter of opinion. I ask why opinion is important and whether the results of this survey will be cited to support or claim any collective consensus. Also, you might cite our American Scientist Magazine article instead of the AIAA paper. The former is more widely accessible at no cost and captures the issues more concisely.

08 July 2011; Kenworthy, Matthew
Leiden Observatory, The Netherlands
Astronomy-Astrophysics
Response = 1

08 July 2011; Wyatt, Wiliam
Smithsonian Astrophysical Observatory, USA
Astronomy-Astrophysics
Response = 1
The worst case would be for UTC to be redefined as proposed, i.e. without a name change to distinguish it from earlier UTC.

08 July 2011; Hochschild, Peter
Google, USA
Large Scale Distributed Computing
Response = 2

08 July 2011; Laney, C. David
Brigham Young University, USA
Astronomy-Astrophysics
Response = 1

08 July 2011; Buinoud, Maxime
French Navy, France
Studies
Response = 1

08 July 2011; Scott, Mike
Vercet LLC, USA
Design of Geophysical Recording Systems
Response = 2
I would like a system that makes corrections no more than once every 10 years, and gives a minimum of 1 year's notice of any changes, thank you for giving me the opportunity of having my say

09 July 2011; Townsend, Gregg
University of Arizona (retired), USA
Telecommunication Computer Science
Response = 2

09 July 2011; Osvaldo, Osvaldo Fernández
Ex professor of The Patagonia University, Argentina
Astronomy-Astrophysics Geodesy
Response = 1
I enjoy making astronomical measures of latitude and longitude by theodolite and chronograph. I need DUT1.

09 July 2011; Pfyffer, Gregor
Royal Observatory of Belgium, Belgium
Geodesy Geophysics
Response = 1

09 July 2011; Hansen, Ask
The NTP Pool project, USA
Time synchronization; computer systems
Response = 2
I operate a system providing time services for tens of millions of computers via about 2000 volunteered NTP servers. Analyzing the performance of the time servers during the 2008/2009 leap second showed a worrying percentage of (otherwise well configured and well maintained) systems being a second out of sync with everyone else for hours and in some cases even days! For computer system operations at both small and late scale the leap second comes at a great cost. For less time critical systems it "just" means that any logs or any other timed information around the leap second are unusable or at best suspect. For time critical systems to cost is shutting down the system around the leap second or if that isn't possible then great and difficult engineering around it.

09 July 2011; Barnes, Howard
Georgi Dobrovolski Solar Observatory, New Zealand
Astronomy-Astrophysics
Response = 1
Perhaps, a "leap minute" once a century might do. That would be better than this silly idea of a "leap hour".

09 July 2011; Wilkinson, James
Google, Australia
Internet
Response = 1

09 July 2011; Withers, Laurence
Güralp Systems Ltd, UK
Geophysics
Response = 2

For seismology, which my company generally focuses on, and other areas of geophysics, we must use a global time reference so that data from geographically distant measurement stations can be correlated. Furthermore, this time reference must be constant (i.e. the definition of one second must not vary), as otherwise any frequency-based calculations would be inaccurate. Unfortunately, seismologists universally use UTC and not TAI for this time source. As a software engineer dealing with acquisition, transmission and processing systems I know from my own experience and from observing other software in the field that leap seconds are an area of huge complexity, often doubling the amount of code required for any timestamp-related task. Furthermore, each leap second occurrence leads to a raft of system failures across all manufacturers. Changing the definition of UTC to be TAI with a constant offset would greatly simplify the task of writing and maintaining software and remove an area of significant concern.

09 July 2011; Trueblood, Mark
National Optical Astronomy Observatory, USA
Astronomy-Astrophysics
Response = 1

Can you imagine the havoc created by this proposal to make UTC an atomic time? The civil time of day MUST be tied to the Earth's rotation. This rotation is gradually slowing due to tidal friction with the Moon. Therefore, we need to continue to introduce leap seconds into the time to keep our clocks in synch with where the Sun is in the sky. Over a period of centuries, the proposed change to an atomic time would make us rise at odd hours of the day, and make it impossible to point telescopes accurately. This proposal is sheer nonsense.

09 July 2011; McCartney, Craig
On-Site Training International, USA
Telecommunication Time-laboratory
Response = 1

09 July 2011; Kulda, Tomas
Charity, Czech Republic
computer programmer
Response = 2

10 July 2011; Deines, Steve
Donatech Corporation, Inc., USA
Navigation
Response = 1

Until the timekeeping community understands Universal Time, it is best to keep the status quo with UTC. Tidal friction is a torque that causes a quadratic deceleration of Earth's orientation. The simplest model of tidal friction is a constant deceleration that will cause the angular velocity (Earth's inertial spin) to decrease linearly and its angular displacement to lag in a quadratic curve. The current definition of UT1 comes from the formula from Capitaine et al, and that formula converts Earth orientation angle into UT1. Since the formula was first published in 1986, the epoch associated with the data is circa 1980, maybe 1981. The last two leap seconds roughly fit a quadratic curve after that epoch, which remarkably fits the tidal friction curve. Christodoulidis et al (1988) obtained $-5.98\pm.22\text{E}-22$ rad/sec2 from analyzing 17 artificial satellites. From the fossil record, I obtained a value of $-5.99\pm1.77\text{E}-22$ rad/sec2 was obtained from 17 studies involving 44 fossils. It is a straightforward statistical test to show that the last two leap seconds do not come from the same population of leap seconds between 1972 and 1998. The Dec 2005 leap second was 11.96 st. dev. off and the Dec 2008 leap second was 3.96 s.d. off. Both fail the 99.9% acceptance test. Knowing that result, the divergence between TAI and UT1 between 1958 and 1998 is nearly linear. If you review the processing of that time, (See Markowitz 1968 in Telescopes) the operational epoch was advanced every day when PZT data were taken and processed. Moving the epoch would hide tidal friction under the noise in the measurements. Now, VLBI gets Earth orientation data very precisely. The operational epoch is now frozen (embedded in Capitaine et al formula), and tidal friction is now revealed in divergence between UT1 and TAI. There is no uniform divergence anymore. In a few decades, tidal friction will invalidate the Capitaine et al formula that obtains UT1, because tidal friction was never incorporated into the derivation. I firmly believe that the timekeeping community should postpone the vote until it thoroughly reviews the effects of tidal friction, which is not incorporated into the current prediction models for the divergence between UT and TAI, and definitely not in the Capitaine et al formula that defines UT1.

10 July 2011; Bertou, Xavier
Centro Atómico Bariloche, Argentina
Astronomy-Astrophysics
Response = 2

10 July 2011; Goodwin, Julien
-, Australia
Telecommunication
Response = 1

10 July 2011; Lynch-Aird, Nicolas
Independent, UK
Telecommunication
Response = 1
Leap seconds should be retained as the ongoing mechanism for maintaining UTC close to UT1. Allowing UTC to drift away from UT1 will necessitate larger corrections to be made at some unspecified point in the future which will be far more disruptive than the current system of introducing leap seconds. It would be of greater benefit to extend the time code transmission standards in such a way as to enable automated systems to be able to detect an upcoming leap second in advance of the event. Assuming that the transmission of DUT1 is retained then this only requires one extra bit of information - the sign of the leap second can be determined from the sign of the preceding value of DUT1. It would be beneficial also to include additional flags in the transmitted data to indicate an upcoming change in the transmitted value of DUT1. Finally the time code standard should be made freely and publicly available. In this way equipment manufacturers will be more readily able to develop systems that can respond in a wholly automated manner to changes in DUT1 and the introduction of leap seconds.

10 July 2011; Scott-Thoennes, Yitzchak
Shiftboard, USA
Telecommunication
Response = 1

10 July 2011; Kapounek, Petr
Comerce Sphere, Czech Republic
Telecommunication
Response = 1

10 July 2011; Verhaege, Christophe
Laboratoire de Météorilige Physique, France
Astronomy-Astrophysics
Response = 1

11 July 2011; Lim, Peter
Nil, Singapore
Nil
Response = 2

11 July 2011; Siddiqui, Hassan
ESA/ESAC, Spain
Astronomy-Astrophysics
Response = 2
Having a time scale that is discontinuous causes a lot of problems with writing and maintaining software for processing non-ground-based astronomical missions, and in particular the link to the spacecraft and the ground segment. It would help immensely if UTC is redefined such that it represents terrestrial time in as simple a way as possible.
Of course, a counter-argument for my request is to simply use TAI right now instead of UTC - if in the future the difference would be a constant. This is in fact my preference. However, a lot of ground based activities are heavily intertwined to UTC, it is best to work on making that system simpler.

11 July 2011; Herrero, Javier
HV Sistemas S.L., Spain
Space-sciences
Response = 2

11 July 2011; Meyer, François
Observatoire de Besançon, France
Time-laboratory
Response = 1
It seems that addressing the main engineering concerns generated by leap seconds, could be greatly simplified by enhancing the accessibility and timespan of the leap second table : such a table, would list not only past leap seconds but also scheduled leap seconds for the next 10 years (instead of the 6 month notice that is in use today), and should be made widely available. Involving only minimal changes, this would be a good, conservative compromise, both preserving the UT feature of UTC (which should not be thrown away lightly in my opinion) and smoothing its engineering drawbacks, at least for the next centuries as long as the average frequency of leap seconds remains below one per month.

11 July 2011; Ochsenbein, Francois
CDS, Obs. Strasbourg, France
Astronomy-Astrophysics
Response = 2
Some adjustment would however be necessary in the future to avoid a too large difference (>15min? >1hr?) with Earth rotation but there will be plenty of time to converge on a consensus :-)

---

11 July 2011; Vallado, David
Center For Space Standards And Innovation, USA
Celestial-mechanics
Response = 1
There are a great number of systems that include processing for leap seconds. Adding the leap seconds maintains uniformity between the actual earth rotation and time systems. I see no need to change that.

---

11 July 2011; Bernstein, Gary
University of Pennsylvania, USA
Astronomy-Astrophysics
Response = 1

---

11 July 2011; Francis, Gribbin
Isaac Newton Group, Spain
Astronomy-Astrophysics
Response = 1
At the Isaac Newton Group we use UTC (from atomic clock) as input to the Telescope Control System. We receive notifications of leap seconds via the IERS bulletin. Our systems can be programmed such that the leap second is introduced automatically: this involves setting hardware switches on the clocks to specify when the leap second is to be introduced and updating control system parameter to say when it is to be expected. Since the clock is autonomous we have also omitted programming the leap second (e.g. on 31-Dec-2011) and introduced manually it later (when it's not a holiday). So we can cope fine with leap seconds. Omitting leap seconds will create some more work (although with our independent clocks we could avoid this). Overall I think the argument is about whether UTC should be related to the sun. Since all of us still live on the planet (earth) it make sense to continue the current regime.

---

11 July 2011; Swaters, Robert
NOAO, USA
Astronomy-Astrophysics
Response = 1

---

11 July 2011; Main, Andrew
no affiliation, UK
software engineering
Response = 3
I have no strong opinion on whether the main broadcast time scale, or the basis of civil time, should continue to track UT1. However, I have opinions about other aspects of the process. Any time scale that does not closely track UT1 would not be a form of UT, and should not have a UT-related name. Specifically, the time scale resulting from initially synchronising with present UTC and then not applying leap seconds would not be a form of UT, and so should not be called "UTC". The name "International Time" with initialism "TI" has been proposed for such a time scale, and I would find that entirely satisfactory. It should, of course, be clearly defined whether proleptic TI matches UTC over UTC's prior period of operation or remains a constant offset from TAI; I think the latter is more manageable. Although many users would prefer a leap-second-less time scale, and many more can at least accept one, it is not feasible to force such a time scale on all present users of UTC. Some would continue to desire a time scale behaving like the present form of UTC, with leap seconds. If such a scale is not readily available then it will be necessary to invent one, but local reinvention repeated by many users would cause a proliferation of badly-managed not-quite-compatible time scales. Thus it would remain useful for IERS to issue canonical leap second decisions for such users, defining a standard time scale that would continue to behave as the current UTC does. This time scale should probably be named "UTC". The decision about the time scale that is used in broadcast dissemination of time should be divorced from other questions about UTC. The broadcast time scale may sensibly be UTC as presently defined, TAI plus an offset (TI as described above), or plain TAI. Whichever is chosen as the primary broadcast time scale, broadcasts should where possible carry the parameters needed to convert between UTC-with-leap-seconds and TI/TAI. Where those parameters are readily available, the exact choice of primary time scale becomes much less significant.

11 July 2011; Murray, Stephen
Johns Hopkins University, USA
Astronomy-Astrophysics
Response = 1
I do not think a change in the current definition of UTC is good for astronomy and celestial navigation activities. There is a great deal invested in the current definition and the software that uses it and any change would likely lead to errors for many years as the transition would need to propagate across many systems and users.

11 July 2011; Flanders, Tony
Sky & Telescope, USA
Astronomy-Astrophysics
Response = 1
Presumably, civil time would continue to be tied to UTC. This would cause sunrise and sunset times to become unpredictable, which seems like a very bad thing in the long run. Julius Caesar tried adopting a simple, uniform time scale in his eponymous calendar; it turned out to be a short-sighted solution. Until the day when we all live in underground enclosures, as foreseen by many science-fiction writers, let's not allow the convenience of a few technologists to take precedence over the Sun!

11 July 2011; Paget, James
The Aerospace Corporation, USA
Astronomy-Astrophysics Celestial-mechanics Space-sciences
Response = 1
Please be sure to make UT1 or UT1C available (such as by radio signals) if you decide to allow UTC to drift more than 1 second from UT1. Please consider renaming UTC if leap seconds are no longer included.

11 July 2011; Martin-Mur, Tomas
JPL, USA
Celestial-mechanics
Response = 2

11 July 2011; Greve, Tora
Tycho Brahe Observatory, Sweden
Astronomy-Astrophysics
Response = 1

11 July 2011; Lee, Steven
AAO, Australia
Astronomy-Astrophysics
Response = 1

11 July 2011; Dicker, Simon
Upenn, USA
Astronomy-Astrophysics
Response = 2
5 years may be too soon to switch.

12 July 2011; Laidler, Victoria
Space Telescope Science Institute, USA
Astronomy-Astrophysics
Response = 1
I am primarily a developer and maintainer of astronomical software.
Although it is somewhat annoying to have to update some software to account for the latest leap second, it would be far more annoying to have to use "a separate access to UT1, such as through the publication of DUT1 by other means" and implement support for both kinds of time.
From my perspective, the current system works. It is not broken. Let's not fix it.

12 July 2011; Fulco, Charles
Port Chester Middle School, USA
Astronomy-Astrophysics Space-sciences
Response = 1

12 July 2011; Kamp, Poul-Henning
The FreeBSD Project, Denmark
Telecommunication Operating System Design & Implementation
Response = 3
The main operational problem with leap seconds is the very short warning. 6-10 months is not nearly enough for operating systems to propagate this information to all installed copies. If leap seconds were announced 10-20 years in advance, tables could be distributed with operating systems and their updates, and computer systems consequently could be trusted to always have up to date tables when leap seconds strikes. If this is not a possible compromise, leap seconds should be abolished.

12 July 2011; Tricarico, Pasquale
Planetary Science Institute, USA
Astronomy-Astrophysics Celestial-mechanics Space-sciences
Response = 1
I think that UTC should stay as it is. If you want to create another timescale, like UTC but without leap seconds, go ahead, just call it something else than UTC. How difficult is that? That said, scientific arguments should prevail over surveys and votes. If there is a strong scientific argument for changing UTC, so be it. But it seems to me that this is not the case, and as you state, a UTC without leap second would be of lesser value than the current UTC definition, so really I don't see the point of it.

12 July 2011; Chenal, Jonathan
Institut Geographique National, France
Geodesy
Response = 1
Continuous timescales still exists, as TAI. In my opinion, it is important to have a basis for legal times (UTC) which follows solar times (UT1). If UTC is a source of problems because of its discontinuities, UTC should simply disappear et be replaced by TAI. UTC is useful precisely because of its discontinuities. A temporary solution could be to create a new timescale, continuous, in parallel to UTC, which would stay the basis of legal timescales. This new continuous timescale would be used for tests only, and could have a permanent entire offset with TAI, which could be the actual value of TAI-UTC. But my preference is to keep the actual definition of UTC, which includes leap second.

12 July 2011; Tang, Jingshi
Astronomy department, Nanjing University, China
Astronomy-Astrophysics Celestial-mechanics Geodesy
Response = 1

12 July 2011; Schittel, Christoph
Plusnet GmbH & Co. KG, Germany
Telecommunication
Response = 1

12 July 2011; Schrama, Ernst
TU Delft, The Netherlands
Geodesy Space-sciences
Response = 1
Please do not change standards, we agreed once upon a time on a definition, textbooks spend text on this problem, etc., so why change that.

12 July 2011; Thivillon, Alain
N/A, France
Telecommunication
Response = 1

12 July 2011; Maisonobe, Luc
CS Communication & Systèmes, France
Celestial-mechanics
Response = 1
Leap seconds are already well understood and well implemented in many space systems. Systems that handle several time scales (say TAI and UTC) either already support leap seconds introduction in real time or have a constant TAI-UTC offset in a configuration file and need a restart a few days after the leap. Systems that handle only one time scale simply don't see anything and run seamlessly when leap seconds occur. So for ALL these systems, regardless of their implementation, leap seconds are clearly not a problem. However, ALL these systems are based on assumption DUT1 remains small (less than 0.9s in the current setup). A few high precision systems track this value from IERS files, almost all systems do not track it and consider it to be 0. Removing the leap second would mean ALL systems should track DUT1 as the simplifying assumption would not hold anymore. This would imply modifying data handling, importing external data in operational systems that did not import anything beforehand, modifying ALL software layers to propagate this DUT1 down to the lower layers for frames transforms, revalidating EVERY space flight dynamics in the world. So for all systems except the very few high precision and costly ones that have already done this work, removing the leap second would in fact induce a lot of difficult work. There are plenty of fixed time scales already available (TAI, GPS, Galileo ...) and only one time scale that is a convenient compromise between purely geometric TU1 and regular physics TAI, let's keep it.

12 July 2011; Defraigne, Pascale
Royal Observatory Of Belgium, Belgium
Time-laboratory
Response = 2

12 July 2011; Street, Jim
N/A, USA
Telecommunication
Response = 2

12 July 2011; Poggi, Jerome
-, France
Government
Response = 1

12 July 2011; Saers, Paul
private, Sweden
Computing industry
Response = 1

12 July 2011; Aerts, Wim
ROB, Belgium
Telecommunication Time-laboratory
Response = 3
Why not introducing leap minutes instead of leap seconds?

12 July 2011; Helk, Frank
-, Germany
process computing
Response = 1
If there's a need for another time reference - like the proposed "UTC without leap seconds" or otherwise - it should be defined as a new entity and be distributed separately. Redefining a widely used standard would only lead to problems ... if anybody needs the new reference, he should use it on a "new service" base.

12 July 2011; Widdas, Brian
n/a, UK
Telecommunication
Response = 1

12 July 2011; Brouw, WN
Groningen University, Netherlands
Astronomy-Astrophysics Celestial-mechanics
Response = 1

12 July 2011, 11h34; West, Richard
University of Leicester, UK
Astronomy-Astrophysics
Response = 2

12 July 2011, 11h34; Clarke, Peter
Newcastle University, UK
Geodesy Geophysics
Response = 1

12 July 2011, 11h51; Mueller, Juergen
Institute of Geodesy, Univ. of Hannover, Germany
Celestial-mechanics Geodesy
Response = 1

12 July 2011, 12h41; Nothnagel, Axel
IGG, University of Bonn, Germany
Geodesy
Response = 1

12 July 2011, 13h01; Pardo, Jeff
SES, USA
Celestial-mechanics
Response = 1

12 July 2011, 13h57; Ewell, Douglas
Individual, USA
Software development
Response = 1

12 July 2011, 14h15; Wallace, Patrick
RAL Space, UK
Astronomy-Astrophysics observatory automation
Response = 3
Leap seconds are a nuisance, and surprisingly difficult to deal with reliably in software. However, there are unknown numbers of applications in existence which, explicitly or implicitly, rely on the distributed time to be close to UT1. So the choice is between continuing to distribute an approximation to UT1 or accepting that problems will occur. With the ubiquitous use of NTP, I believe there is now an opportunity to separate civil time from the high-precision time/frequency dissemination services. This would be done by providing UT1-based NTP servers, for dissemination of ordinary time-of-day and expressly intended for applications not requiring accuracies of better than 0.1s. We could call it GMT, which many countries still refer to in their laws. (The fact the US law was changed not long ago to say UTC is regrettable but should not be allowed to influence the debate.) The existing time/frequency dissemination services would by default distribute leap-less UTC. As the difference between this and UT1 grows, developers of computer applications would become used to the idea that they had to make a choice - which they do now, in principle.

12 July 2011, 14h16; Jubier, Xavier
None, France
Astronomy-Astrophysics
Response = 1
I have nothing against a change. However, I would point out that each definition has its merit. So why change if it doesn't really bring anything or simplify computations. As for predictions of eclipses in future years (especially within the next 100 years) the proposal to cease inserting leap seconds (that is, keeping UTC fixed with respect to TT) has significant merit – it will allow accurate UTC predictions to be issued many years before the event. Nevertheless it doesn't change anything since the difference between UT1 and UTC would henceforth be unconstrained! At the same time, the current prediction methodology contains two 'unknowns' for future predictions: the conversion from TT to UTC, and the rotational position of the Earth (UT1-UTC). Of these, the effect of the uncertainty in prediction times resulting from the conversion from TT to UTC is an order of magnitude greater than the effects of the rotation of the Earth over the same time period. However if leap seconds are discontinued, the two uncertainties are reduced to just one – the rotational orientation of the Earth. And of the two uncertainties, this is the one that has the lesser impact on actual prediction times.

12 July 2011, 14h16; Ray, Jim
U.S. National Geodetic Survey, USA
Geodesy
Response = 1

12 July 2011, 14h18; Goerres, Barbara
Instiutut for Geodesy, University Bonn, Germany
Geodesy
Response = 1

12 July 2011, 14h21; Willmott, Paul
AMSAT-BDA, Bermuda
Astrodynamics
Response = 1
We have no wish to reprogram thousands of lines of complex astrodynamics code. Given that we will have to maintain the original UTC definition for our historical data, it will all be very confusing.

12 July 2011, 14h28; Schoene, Tilo
GFZ Potsdam, Germany
Geodesy
Response = 1

12 July 2011, 14h29; Grant, Mike
Plymouth Marine Laboratory, UK
Remote Sensing (satellite and airborne)
Response = 2

12 July 2011, 14h45; Newhal, X X (Skip)
Jet Propulsion Laboratory (Retired), USA
Astronomy-Astrophysics Celestial-mechanics Planetary Ephemerides
Response = 1

12 July 2011, 14h53; Beyerle, Georg
GeoForschungsZentrum Potsdam, Germany
Geodesy GNSS Remote Sensing
Response = 2

12 July 2011, 14h57; Lennon, Christopher
MIT Lincoln Laboratory, USA
Radar Systems
Response = 1
GPS time exists as a free running clock alternative to UTC. One needs to maintain the list of leap seconds to go back an forth, but that is only a minor pain in the neck.
I have a mild preference that UTC maintain its connection with the rotation of the earth.

12 July 2011, 15h02; Kidger, Mark
ESA, Spain
Astronomy-Astrophysics
Response = 1
The current system has worked for years, why change it? Keep the day linked to the rotation of the Earth.

12 July 2011, 15h05; Svalgaard, Leif
Stanford University, USA
Space-sciences
Response = 2

12 July 2011, 15h08; Fani, Paolo
-, Italy
Astronomy (amateur)
Response = 1

12 July 2011, 15h19; Visser, Pieter
Delft University of Technology, Netherlands
Celestial-mechanics Geodesy Geophysics Space-sciences
Response = 1

12 July 2011, 15h42; Seidelmann, P. Kenneth
University of Virginia, USA
Astronomy-Astrophysics Celestial-mechanics Space-sciences
Response = 1
I think the definition of UTC should be considered by much wider scientific and administrative organizations than the ITU. The full impact of the change and its implications need to be considered.

12 July 2011, 15h44; Hestroffer, Daniel
IMCCE, France
Astronomy-Astrophysics Celestial-mechanics
Response = 4

12 July 2011, 15h47; Mueller, Ivan
The Ohio State University, USA
Geodesy Geophysics
Response = 1

12 July 2011, 16h11; Dr. Federspiel, Martin
Planetarium Freiburg, Germany
Astronomy-Astrophysics
Response = 1

12 July 2011, 16h20; Hase, Hayo
Bundesamt für Kartographie und Geodäsie, Germany
Astronomy-Astrophysics Geodesy
Response = 1
Based on my personal experience: The existence of the leap second convinces common people to understand the need of geodetic VLBI and justify its expensive operation. If the product "leap second" becomes officially superfluous, the current VLBI programmes are put in danger. The importance of VLBI is not only based in the "leap second". But it is the easiest argument to communicate to politicians and administrators of financial resources. Both cited articles mention "VLBI" only once and do not focus on the global VLBI infrastructure which is still contributing to the "leap second" determinations. The number of arguments for pro and contra shows the need for both timescales: - the atomic time scale - the earth rotation time scale.

12 July 2011, 17h19; Martin, Thomas
Van Martin Systems, Inc., USA
Geodesy Space-sciences Precision satellite orbit determination
Response = 1
UTC serves a very useful purpose. For those for whom UTC leap seconds present a problem, we already have TAI and GPS time which are uniformly increasing atomic time scales. With the advent of GNSS, anyone anywhere in the world has access to GPS time at very little cost. GLONASS, and I believe some SBAS systems, provide UTC. Additional systems coming on-line, including QZSS and Galileo, essentially also provide GPS time with extremely small offsets. Conversion algorithms between UTC and TAI or GPS border on trivial and are readily available. Those of us who perform precision calculations will continue to require time transformations, even if the leap seconds are eliminated going forward. If you want to hide leap seconds from public view, simply coordinate public clocks to TAI or GPS time!

12 July 2011, 17h28; Johnson, Thomas
National Geospatial-Intelligence Agency, USA
Astronomy-Astrophysics Geodesy Geophysics
Response = 1
While one may think that knowledge and understanding of our universe is a true goal of science, in reality, it is not. The ultimate goal of science and its pursuit of greater knowledge is for the improvement of society. While predicting and keeping UTC aligned with the earth's rotation is not an easy task, it has benefits to society and therefore, should be maintained. For example, there are many users of UTC from around the world that have built their systems on the assumption of UTC being co-ordinated with the earth rotation. The decoupling of these systems would result in a great deal of work and financial expense to correct, all of which is unnecessary. Furthermore, this is one indirect benefit to having leap seconds. Every time a leap second is inserted, the public media has to reach out to the scientific community to educate its consumer on the physics behind the need for this adjustment. Therefore, the general population gets a science lesson reminding them of the importance of astronomy and geophysics in their daily lives.

12 July 2011, 17h42; Stefan, Krista
Royal Astronomical Society of Canada, Canada
Astronomy-Astrophysics
Response = 1

12 July 2011, 17h42; Capitaine, Nicole
Bureau des longitudes& Paris Observatory, France
Astronomy-Astrophysics Celestial-mechanics Geodesy
Response = 2
- Separating the two concepts (angle for UT1,time for UTC) would be an improvement for high-accuracy applications.
- UT1 is defined by a conventional linear relation to ERA and benefits from the accuracy of that angle; for its best scientific use, that angle varying with time must be referred to a uniform time scale.
- The definition of UT1 is such that it is kept approximately (but not strictly) in phase with the mean solar time; so it in fact differs from the mean solar time +12 h and the difference is increasing with time.
- The definition of UTC based on leap seconds was designed to provide sufficient approximation to UT1 to celestial navigation; this is obsolete. For scientific applications, the use of the best uniform time scale is required (without leap seconds).
- If leap seconds are removed, the gap between UTC and UT1 will reach 3 min in 2100, 30 min in 2700, differences that are below those between legal time and solar time (+12h) that we tolerate.
- Scientific applications requiring prediction of UT1-UTC, such as precise astronomical ephemerides, can be established based on an IERS UT1-UTC prediction, leading to an accuracy at least as good as access to the UT1 derived from UTC with leap seconds.
- The responsibility of the IERS will be increased with the new interesting charge of providing predictions of the difference between UT1 and UTC, or UT1(UTC), in order to provide access to UT1 in real time. These values can easily be disseminated by positioning systems, such as GPS, which would give access to UT1 in real time to wide categories of users.

---

12 July 2011, 17h48; Mabie, Justin
NOAA, USA
Space-sciences
Response = 1
No such change in the definition of UTC should be considered. Instead, the community should propose a new timescale that is referenced off of UTC. There are several reasons for this.
1) It seems there is no proposed mechanism for the transition, between conversion of data to the new timescale or to update algorithms, instruments, and models that depend on the current definition of UTC.
2) Any change in the definition of UTC would require a secondary timescale so that conversions can be made from the old definition to the new definition. Any such timescale would in of itself serve the same purpose as the proposed change and is therefore unnecessary.
3) This proposal would create unnecessary confusion that could be avoided with an alternate such as the option of creating a new timescale.
4) Any errors or failure to adequately track conversions from the old to the new definitions could seriously affect research, particularly with regard to dynamics on timescales of one second or less. It should be noted that international metadata standards, although robust, do not even adequately address the needs of modern datasets, and are not used sufficiently to provide the capabilities for which they are intended. Certainly, application of a redefined timescale would add serious problems to metadata tracking. In addition to this, historical datasets that have not yet been fully modernized, do not abide by any metadata standard, and risk having their proper time stamps corrupted during the modernization process.

---

12 July 2011, 18h09; Wilson, Keith
Jet Propulsion Laboratory, USA
Geodesy
Response = 1
I am also concerned about the change in delivery of the dEps, dPsi,EOP parameters. These seem to lag their dX and dY counterparts by 1 month. Is there a way to convert these dX and dY parameters to dEps and dPsi?

---

12 July 2011, 18h14; Byun, Sung
Jet Propulsion Laboratory, USA
Celestial-mechanics Geodesy Geophysics
Response = 3
If there is no leap second (keeping up with Earth rotation) what is the point of having UTC time scale? It doesn't have much meaning other than some offset from TAI. But I do understand that inserting UTC will become more frequent in the future and will become quite a nuisance. I am wondering there has been enough discussion regarding introducing 'leap minute' instead of leap second.

12 July 2011, 19h12; Horan, Karen
NOAA, USA
Space-sciences
Response = 4

12 July 2011, 19h24; Haywood, Gerald
Jubilee Office Supplies, England
Business.
Response = 1
It isn't broken. Please don't fix it.

12 July 2011, 19h31; Melnick, Jorge
ESO, Chile
Astronomy-Astrophysics
Response = 1
Human time is a measure of the position of the Sun on the sky, which is determined by the rotation of the Earth and the portion of the Earth on its orbit around the Sun. Decoupling human time from the rotation of the Earth would take humanity one step further on the path to virtual existence. This is probably inevitable, but should be delayed as much as possible. Our organisms are still ruled by night/day cycles.

12 July 2011, 20h22; Francou, Gerard
Observatoire De Paris - SYRTE, France
Astronomy-Astrophysics Celestial-mechanics Geodesy
Response = 1

12 July 2011, 20h23; Bolotin, Sergei
NVI, Inc./NASA GSFC, USA
Astronomy-Astrophysics Geodesy
Response = 1
I believe that redefining of UTC time scale is unwise move. If one-second leap time adjustment is too complicated for civilian time keeping they can invent an appropriate time scale or use one of already existing continuous time scales, e.g. TAI.

12 July 2011, 20h36; Podesta, Ricardo
Observatorio Felix Aguilar (OAFA), San Juan, Argentina
Astronomy-Astrophysics Geodesy
Response = 1

12 July 2011, 21h05; Boyson, Andrew
Home, UK
Time enthusiast
Response = 1
Would like to see NTP provide TAI. Internal PC clocks and file timestamps in TAI. PCs could easily adjust the displayed time from infrequently downloaded leap seconds and daylight savings information. Telescopes could map TAI against predicted earth rotation to provide an accurate position. in a few thousand years we would need to redefine the earths angular second as some fraction of the TAI second in order to not exceed more than about 10 leap seconds per year.

12 July 2011, 21h16; Abarca del Rio, Rodrigo
DGEO, Chile
Geodesy Geophysics Space-sciences
Response = 2

13 July 2011, 00h32; young, larry
jet propulsion lab, USA
Space-sciences
Response = 2

13 July 2011, 02h15; Wildermann, Eugen
Universidad del Zulia, Venezuela
Geodesy
Response = 1
The current close connection with earth rotation seemed to me a great advantage of UTC, so eliminating this purpose UTC afterwards mainly will be a simple TAI offset. I don't see much sense of this at my current workspace (I'm interested at UTC TAI difference mainly for tide calculations for precise gravity observation would be influenced).

13 July 2011, 02h45; Carter, Bill
University of Houston, USA
Geodesy
Response = 1

13 July 2011, 03h02; Hu, Songjie
Aerospace Flight Dynamics Lab, China
Celestial-mechanics Space-sciences
Response = 1

13 July 2011, 03h09; McGlaun, Daniel
none, USA
Astronomy-Astrophysics
Response = 1
I am an advanced amateur total eclipse chaser, involved also with calculating local circumstances. I see no tangible benefit to modifying the current definition of UTC; in fact, I see that for the purposes of maintaining the ability to perform historical calculations, the community would have to maintain two different sets of time measurement.

13 July 2011, 04h43; Senne, Joseph
Univ. of Missouri Science & Technology, USA
Geodesy
Response = 1

13 July 2011, 07h55; Roberto, Roldan
European Satellite Services Provider, Spain
GNSS
Response = 2

13 July 2011, 09h59; De Greef, Didier
ESSP, Spain
Navigation
Response = 1

13 July 2011, 10h03; Loyer, Sylvain
CLS, France
Geodesy
Response = 1
Unless with a large consensual opinion to change something, it is better to keep the conventional usages as they are, since "they are JUST conventions".

13 July 2011, 11h24; DENIS, Carlo
IAGO Liège, Belgium
Geodesy Geophysics
Response = 1

13 July 2011, 11h30; Piriz, Ricardo
GMV, Spain
Celestial-mechanics Geodesy Geophysics
Response = 1
Our department is involved in software development for GNSS orbit determination, timing, and positioning. In general we are quite satisfied with the current definition of UTC, leap seconds do not pose a problem for us. In any case, for detailed UT1 information we need to access specific IERS files, and this would not change if the UTC definition changes, so there is no impact. So the only benefit for us of a new UTC definition would be a constant offset between GPS Time and UTC, this means that we would have one interface less, we would not need to update GPSt-UTC when leap seconds happen (normally a text file in our system). On the other hand, having UTC tied to UT1 (current definition) is very nice for approximate calculations and simple software tools. For example, if you are using Two Line Elements (TLEs) to calculate approximate satellite orbits, it is quite useful to know that if you interpolate the model using UTC (current) instead of UT1 the resulting accuracy will be within the noise of the TLEs. There is also the issue of backward compatibility, if the UTC definition changes, there might be some side effects in our software that could make it fail, we would have to review the current code carefully. From a "philosophical" point of view, I feel more comfortable knowing that UTC, the time on my watch, is also linked to the Earth rotation and not only to atomic clocks, I believe the current definition is a good compromise between the "two worlds" and that is why it was invented.

13 July 2011, 15h42; Woodburn, James
AGI, USA
Celestial-mechanics
Response = 1
The proposed redefinition of UTC would cause a gradual degradation of many satellite systems which assume alignment between UT1 and civil time. While these systems could be updated, at considerable cost, I am concerned that many operators or users of these systems may not even be aware of the assumption and therefore would not recognize the need for change. As the degradation would be very gradual, system performance would slowly suffer but perhaps not come to a breaking point until the redefinition of civil time was a distant memory.

13 July 2011, 17h10; Guinot, Bernard
Obs. de Paris, Bureau des longitudes, France
Astronomy-Astrophysics Geodesy Space-sciences Time-laboratory Former sailor (navigation officer)
Response = 2
Present definition of UTC causes an ambiguity of date at the occurrence of a positive leap second which is potentially dangerous. It favors the existence of several time scales differing by an integer number of seconds. The present system was devised in 1972 in order to provide directly by radio time signals the needed accuracy of UT1 for celestial navigation (+/- 1 second). This need (which may persist for safety reasons) can be fulfilled by expressing hour angles in printed nautical ephemerides as a function of a continuous UTC (based on a prediction which can be made at the second level over 3 years). For a better precision, UT1 is easily available by internet in real time at the level of a few milliseconds. I recall that a continuous UTC will diverge from UT1 by one or two minutes in 2100 and will reach half an hour toward 2500-2600. Presently the offset of legal time with respect to solar time may exceed two hours in some countries...

13 July 2011, 16h18; Dries, Jan
VITO, Belgium
Earth Observation
Response = 1

13 July 2011, 16h35; Chandler, John
Smithsonian Astrophysical Observatory, USA
Astronomy-Astrophysics
Response = 1

13 July 2011, 17h59; McCann, Stephen
Private, UK
Telecommunication
Response = 1
The rotation of the Earth is not constant. A system that allows dynamic updates must be maintained. Although a change to the definition of UTC will not be noticeable to the vast majority of people who use accurate time keeping, it is not irrelevant. Initially inconvenient 'leap seconds' will be no longer required, but after several decades errors will accrue and at some point in the future a correction will be required. Who will care, as by then we'll all have passed on anyway.

13 July 2011, 19h06; Levine, Judah
National Institute of Standards and Tech, USA
Telecommunication Time-laboratory
Response = 2
It would also be possible to change the leap second to a leap hour when the dut1 correction was greater than about 2000 s. This method would limit the divergence of UT1 from UTC while minimizing the disruptions that occur when a leap second is realized.

13 July 2011, 19h18; McCarthy, Dennis
U. S. Naval Observatory, USA
Astronomy-Astrophysics
Response = 2
This questionnaire is poorly written and appears to be one of a series of such questionnaires. It makes the IERS look bad when it keeps sending out such poorly written and repetitive questionnaires. The second reference listed above is not in a peer-reviewed journal and should not even be listed as a reference. This questionnaire serves no useful purpose for either side of this issue.

13 July 2011, 19h35; Oltrogge, Daniel
1Earth Research, LLC, USA
Celestial-mechanics Space-sciences
Response = 1
I feel that the current system has worked sufficiently well and there is intrinsic value in having UTC tied closely with Earth rotation. I understand the desire to eliminate leap seconds and the discontinuities that they cause. But I feel that we'd have to do extensive surveys, research and careful evaluation before we could determine the extent (likely quite a large impact) of software modifications, financial impacts and programmer/developer time required to eliminate the leap second (thereby potentially causing existing applications to 'break').

13 July 2011, 19h45; ACTIS, Eloy
Observatorio m Astronómico Félix Aguilar, Argentina
Astronomy-Astrophysics Geodesy Satellite Laser Ranging
Response = 1
Current definition of UTC works very good and I don't feel necessity of changing it.

13 July 2011, 19h58; Kouprianov, Vladimir
Pulkovo Observatory, Russia
Astronomy-Astrophysics Space-sciences
Response = 1

13 July 2011, 21h23; Connors, John
Private Individual, USA
Astronomy-Astrophysics GPS commercial user, Astro & Celest Nav
Response = 1
After thoroughly reading the provided references, the cost of changing the UTC leap second system / time standard (ref #2, V The Debate, D Costs of Changing) is not justified or technically warranted. The assumptions favoring the change are weak and favor academia and the scientific community. The commercial realm and public "users" are not well represented in these papers and government (tax payers), including commercial end users, are expected or assumed to absorb the cost (of redesign). Removing the leap second or making UTC "more" dynamically linked to the earth's rotation is an expensive step backwards.

13 July 2011, 23h36; Simpson, David
NASA Goddard Space Flight Center, USA
Astronomy-Astrophysics Celestial-mechanics
Space-sciences
Response = 1
Keep UTC defined as it currently is. The most compelling reason for eliminating leap seconds seems to be that they will have to be introduced with increasing frequency in the future; however, that should not become an important concern for several centuries. Meanwhile, eliminating leap seconds now would leave us with four time atomic scales that differ from TAI only by a fixed offset (TAI, TT, GPS, and UTC), while providing no atomic-based time scale that maintains synchronization with UT1 (an important consideration for civil timekeeping). Dropping leap seconds from the definition of UTC now would be, at best, premature.

14 July 2011, 01h09; Shawhan, Peter
University of Maryland, USA
Astronomy-Astrophysics
Response = 2

14 July 2011, 04h25; Bizouard, Marie-Anne
Laboratoire de l'Accelerateur Lineaire, France
Astronomy-Astrophysics
Response = 1

14 July 2011, 07h54; Cannon, Kipp
Canadian Institute for Theoretical Astr., Canada
Astronomy-Astrophysics
Response = 1
The motivation for removing the leap seconds from UTC is a mystery. If UTC is not the correct time scale for an application, then there are many more to choose from. UTC is just one of a dozen or more time scales that are in regular use: TAI, UT0, UT1, UT1R, UT2, UT2R, UTC, GMST, GPS time, Julian day number, Unix time, ... In particular, several of them are atomic time scales free of leap seconds. For example, TAI, the count of GPS seconds, Unix time, and so on. Anyone who wishes to use a leap-second-free atomic time scale for their application is already free to use one of these. The conversions between these time scales and UTC are simple and well-documented.

14 July 2011, 10h08; McIver, Jessica
University of Massachusetts Amherst, USA
Gravitational wave physics
Response = 1

14 July 2011, 11h52; Skinner, Laurence
-, England
Host an NTP Pool time server
Response = 2

14 July 2011, 13h26; Loh, Jürgen
Alpermann+Velte e.e. GmbH, Germany
Telecommunication
Response = 2
We're a manufacturer of Timecode systems for radio and television broadcasting stations. SMPTE/EBU Timecode is often used to synchronize the equipment to civil time.

14 July 2011, 20h28; Standish, E Myles
Caltech/JPL - Retired, USA
Astronomy-Astrophysics
Response = 1
UTC now approximates the earth's rotation (within 0.9 seconds). There is a lot of software throughout astronomy and navigation which subtly makes use of this fact. To change it would cause many unforeseen problems.

14 July 2011, 20h40; Scott, Stephen
Caltech/Owens Valley Radio Observatory, USA
Astronomy-Astrophysics
Response = 2

14 July 2011, 22h18; Poutanen, Markku
Finnish Geodetic Institute, Finland
Geodesy
Response = 1
It is a much deeper principle than a technical or practical question about the leap second. Quitting the leap second we accept that UTC is no more fixed to the rotation of the Earth and our concept of time is not related to the variation of day and night. But we cannot quit the fact that half of the Earth is illuminated by the Sun, half is in darkness, and due to the rotation of the Earth we see the regular variation of day and night. If we accept the concept that this has no meaning in our life, we can quit the connection of the UTC to the rotation of the Earth. We can as well quit then the time zones, length of 24h day or incompatible length of the year with leap days every fourth year. All these are as well technically possible. But if we want follow day and night variation, then within decades we'll need a leap minute or within millennia a leap hour... Are these any better than the leap seconds?

15 July 2011, 01h15; Boriani, Azelio
SSBT spa, Italy
Digital TV broadcast equipment
Response = 1

15 July 2011, 07h41; Spencer, Mark
Aligned Solutions, Canada
Telecommunication Information technology
Response = 1

15 July 2011, 09h33; Orlati, Andrea
INAF-IRA, Italy
Astronomy-Astrophysics
Response = 1

15 July 2011, 09h51; Young, Iain
n/a, UK
Telecommunication Time-laboratory
Response = 1
TAI and GPS timescales are already available should folks need or want a timescale without leap seconds. It seems to make little sense to me to add a third. Maybe we should consider having a different name for a UTC based timescale w/o leap seconds. But changing the current standard is most likely to just cause confusion, especially amongst the general public

15 July 2011, 09h58; Fenn, David
Of Materials, UK
Astronomy-Astrophysics Telecommunication Materials engineering
Response = 1
If it ain't broke, don't fix it!

15 July 2011, 10h15; Plant, Hannah
Physics, England
Physics
Response = 1

15 July 2011, 11h10; Maccaferri, Giuseppe
Institute of Radioastronomy, Italy
Astronomy-Astrophysics Geodesy Space-sciences Time-laboratory
Response = 1

15 July 2011, 11h12; Verkindt, Didier
LAPP, CNRS, France
Astronomy-Astrophysics gravitational waves detection
Response = 2
I prefer to put the operation of leap seconds addition when getting the UT1 (or local time) and to have a universal UTC date which is earth independent.

15 July 2011, 14h47; Combrinck, Ludwig
HartRAO, South Africa
Geodesy
Response = 1
Changes required to existing software, precompiled library binaries etc. (some of which may not have original source code, so that they cannot be modified) will create chaos. The result will be unworkable and un-fixable software. Who will foot the bill for this? Who will do this work? It is easy enough to maintain UTC, so I say do not fix that which is not broken.

15 July 2011, 15h15; Hildebrand, Andreas
ALC NetworX GmbH, Germany
Professional Broadcast
Response = 3
Speed up Earth rotation accordingly...
If this does not work out, leave UTC as it is - there are many reasons why it has been defined the way it is. If it would be changed to TAI + offset, you could use TAI at first instance.

15 July 2011, 17h04; Hohenkerk, Catherine
HM Nautical Almanac Office (UKHO), UK
Astronomy-Astrophysics
Response = 1
It is useful to have |UTC-UT1|<0.9s. Navigation almanacs, produced in advance, may be inaccurate as the prediction of UT1-UTC at the time of production that is needed to determine GHA may not be good enough. Textbooks etc. will become invalid. The fact that sunrise/sunset times repeat over a 4-year cycle will no longer be necessarily true. Science & Technology ought to be able produce a solution without having to drop leap seconds.

16 July 2011, 03h06; Pogorelc, Scott
USG contractor, USA
Satellite Navigation / OD
Response = 2

16 July 2011, 17h12; Fesler, Jason
Yahoo, Inc, USA
Internet related industry
Response = 2
UTC affects *every computer* on this planet. And every OS implements coping with it differently. The last leap second event caused a global outage for me - across 50,000+ machines, and affecting 100M+ customers - due to a bug in the way leap second was handled. We now have to test every kernel version we operate (300+ kernels across 300,000+ machines) to simulate leap second. Even without a hard lock up, leap seconds across devices that don't handle leap second correctly (not in the kernel, or ntpd not receiving the notice 24 hours in advance) cause the machines to have to skew to make up for it. This means that around the event, I can't correlate events between machines. Not until everything is back within tolerance.

17 July 2011, 17h22; Schoedel, John
Myself, USA
Telecommunication
Response = 1

17 July 2011, 20h25; Noel, Jean-Louis
Education, Belgium
Telecommunication
Response = 1

18 July 2011, 05h39; GULYAEV, SERGEI
Auckland University of Technology, New Zealand
Astronomy-Astrophysics Geodesy
Response = 1

18 July 2011, 10h28; Kutoðlu, Þenol Hakan
Zonguldak Karaelmas University, Turkey
Geodesy
Response = 2

18 July 2011, 10h50; Yule, Andy
u-blox, UK
Telecommunication
Response = 1
If it ain't broke, don't try to fix it!

18 July 2011, 11h17; Planesas, Pere
Observatorio Astronomico Nacional, Spain
Astronomy-Astrophysics
Response = 3
A leap minute could be introduced preferably at the end of June 30th whenever the UT1-UTC difference is predicted to reach 60 s. The announcement would have to be made several years ahead so by the time it is applied the difference would be strictly larger than 55.0 seconds (goal > 60 s) and smaller than 65.5 s. A new DUT1 would need to be defined, and its resolution likely increased down to 1 ms to fulfill high precision applications, Main advantages:
- Keep UTC close to the mean solar time.
- Keep UTC's name and legal status.
- Fewer changes per century (TAI-UTC difference constant for decades),
- Able to cope with the UT1-TAI quadratic growth.
- DUT1 would become more widely used for those who really need it (astronomers, navigators) resulting in higher precision calculations, being a better representation of the astronomical time UT1.
Moreover:
- No need to allow for negative corrections.
- The first leap minute would take place in several decades, allowing for all clocks, time-aware devices, software and time dissemination standards to be able to cope with the extra ("60") minute.
- Might lead to the unification of the time systems by forcing them to follow a unique (new) standard.
- A change in June 30th is less disruptive than on New Year's Eve.
- The new DUT1 will give more visibility to those who determine it. The new DUT1 could be disseminated in a 20-bit word:
- 1 sign bit
- 16 bits to cover time from 0 to 65535 ms.
- 1 measured/predicted bit
- 1 checksum bit
- 1 spare bit
A second 20-bit word could contain the DJM (as an integer), up to the year 4595.

18 July 2011, 13h43; Possenti, Andrea
INAF-Osservatorio Astronomico Cagliari, Italy
Astronomy-Astrophysics
Response = 2

18 July 2011, 16h15; Cooper, Stanley
Johns Hopkins Univ. Applied Physics Lab., U.S.A.
Space mission timekeeping systems
Response = 2

18 July 2011, 17h04; Colomer, Francisco
National Astronomical Observatory, Spain
Astronomy-Astrophysics Geodesy
Response = 1
Alternatively, the concept would remain for DUT1 but change only when added up to a "leap minute".

18 July 2011, 19h02; Talty, Richard
Science Horizons, Inc., USA
Geophysics
Response = 1

It seems to me that there are two classes of users of UTC, and neither should have problems with leap seconds:
1) Users who do not need 1-second accuracy. Typical human scheduling (stores, classes, trains, etc.) operates with more than 1 second of slop. A leap second can be considered another type of unexpected time error and absorbed into that budget.
2) Users who need synchronization to 1 second or better. Stock trading, electricity grid, possibly traffic lights, etc.
Type 2 applications are, in practice, automatically synchronized to some UTC source. From this source, they can both measure their own time error, and obtain warning of upcoming leap seconds. While it is theoretically possible to track UTC to within 1 second on 1-year timescales using a rubidium oscillator, it is far cheaper and more common to use a GPS receiver which provides ample warning of leap seconds.
Some UTC broadcasts provide little (DCF77) or no (MSF) leap-second warning, but that seems like a simpler technical problem to solve. As internet connectivity is more and more widely used, it gets easier to disseminate leap second information.
The great benefit of leap seconds over less frequent larger corrections to maintain 12h00 at roughly mid-day is that they can be ignored by a large number of time users, and that they are (barely) frequent enough to allow software to be tested. People arguing for fewer, larger time scale jumps are just throwing the problem over the wall to some future legislators who will have to redefine local time to UTC offsets. which will invariably not be dome in a coordinated way, leading to a mess similar to the introduction of the Gregorian calendar. (If hopefully without the Protestant Reformation.)
Given free choice, I would suggest more frequent smaller time corrections, but 1-second leap seconds are deeply entrenched and not worth changing now.
If you want a time scale without leap seconds, use TAI. Or GPS time. UTC, like all historical universal times, should remain basically coupled to the position of the sun.

19 July 2011, 03h05; Seago, John
Analytical Graphics, Inc., USA
Celestial-mechanics Space-sciences
Response = 1

19 July 2011, 10h24; Brumfitt, Jon
ESAC, Spain
Astronomy-Astrophysics Space-sciences
Response = 2

19 July 2011, 13h15; Miguel J., Sevilla
Facultad de Matematicas. UCM, Spain
Geodesy
Response = 1

19 July 2011, 17h01; Thornton, Tim
Smartcom Software, UK
Astronomy-Astrophysics Geodesy
Response = 1

Although dropping leap seconds would no doubt be convenient for some members of the scientific community, for everyone else it is necessary for "official" time to be in synch with "natural" time. Also, it appears that the implications and means of management of time without leap seconds has not yet been fully thought out.

19 July 2011, 18h37; Santos, Marcelo
University of New Brunswick, Canada
Geodesy
Response = 1

19 July 2011, 19h02; Hrudkova, Marie
Isaac Newton Group of Telescopes, Spain
Astronomy-Astrophysics
Response = 1

19 July 2011, 19h07; Ghigo, Frank
National Radio Astronomy Observatory, USA
Astronomy-Astrophysics
Response = 1

20 July 2011, 02h28; Robinson, Rob
International Occultation Timing Assn, USA
Astronomy-Astrophysics Celestial-mechanics Space-sciences
Response = 1

20 July 2011, 02h47; Breit, Derek
IOTA, USA
Astronomy-Astrophysics
Response = 1

20 July 2011, 09h30; Quiles, Alfredo
ESA, Netherlands
.
Response = 1

20 July 2011, 12h11; Woan, Graham
University of Glasgow, UK
Astronomy-Astrophysics
Response = 2
My personal experience is that leap seconds create more problems than they solve, especially when implemented by non-experts, and that time differences between events should be easily computable. One of the technicians at the Lords Bridge Observatory in Cambridge had tape measure with 2 feet missing in the middle. *He* had no problem using it, but I don't think it was a popular tape measure.

20 July 2011, 14h49; Gonzalez, Francisco
ESA, The Netherlands
Geodesy
Response = 2

20 July 2011, 15h04; Zebhauser, Benedikt
Hexagon Technology Center, Switzerland
Geodesy Surveying
Response = 1
Why having another timescale without leap seconds parallel to TAI? That makes no sense. TAI can already accessed precise enough for the most applications in real-time e.g. from GPS time with a constant offset of 19 sec. The introduction of leap seconds into UTC in 1972 was made for practical reasons that are still valid today. Many applications would have to acquire current corrections from services. Why complicating? In case of unnecessarily changing the definition of UTC one would have to re-introduce a further time-scale with the current UTC definition including leap-seconds.

20 July 2011, 17h35; Buie, Marc
Southwest Research Institute, USA
Astronomy-Astrophysics Celestial-mechanics Space-sciences
Response = 1
Decoupling UTC from the Earth's rotation is sheer madness. We already have a dynamical time definition and that serves for the computational needs of a uniformly increasing time scale. UTC with its coupling to local time needs to stay as it is. From my point of view there is no advantage to changing the present system. The disadvantages are many, including the modification of ALL data acquisition and data reduction software for astronomical and spacecraft observations. The fact that this rewrite leads to no benefit argues strongly against the change.

20 July 2011, 18h58; Graham, Francis
Kent State University, USA
Astronomy-Astrophysics Space-sciences
Response = 1

21 July 2011, 00h34; Ray, Paul
Naval Research Laboratory, USA
Astronomy-Astrophysics
Response = 2

21 July 2011, 13h08; Wood, Derek
Open University, Scotland
Marine
Response = 1
None

22 July 2011, 12h24; Doom, Claude
Hogeschool-Universiteit Brussel, Belgium
Astronomy-Astrophysics
Response = 2

22 July 2011, 19h31; Karimbi, Mahesh
Faculdade Ciência e Tecnologia / UNL, Portugal
physical sciences
Response = 1
Everyone agrees that the occurrence of astronomical events are not exactly periodic, so also, the relative movement of the Earth, Sun, Moon and some considered Stars. Since the known history and in known civilisations, the measurement of time depended on the astronomical events. Further, it is still practised at almost all the fields ranging from the civil life, cultural observations, military practices, sea navigations, to judge the future astronomical events etc., except, in the laboratory scientific experiments. Therefore, practice of having the 'leap' time magnitudes, such as, year, month etc., and recently, seconds have been in course. Further, in the regions where, 'day light saving phenomenon' is observed, the adjustment of the time is again a mandatory. The only difference in the adjustment process of the time in the all the cases, except 'leap second', is that they are predefined and predetermined. As with the latest technology, the job of predicting and publishing the introduction of leap second is already done by IERS twice a year, it can be well implemented for majority of the purposes except the laboratory scientific experiments.

22 July 2011, 23h42; Wheatley, Peter
University of Warwick, UK
Astronomy-Astrophysics
Response = 2

23 July 2011, 01h40; Douglas, White
NA, Australia
Astronomy-Astrophysics
Response = 1

24 July 2011, 03h46; Anonymous, Anonymous
None, USA
none
Response = 1
If you want a timescale with a constant offset from TAI, why not just use TAI? UTC is still important for keeping track of Earth's actual rotation (corrected for accuracy). The purpose of timekeeping is to keep a stable relationship with the cycles of the day and year. This proposed redefinition would end this link and leave UTC completely arbitrary. This will have a disastrous impact on professions such as astronomy, which require a timescale that corresponds with planetary cycles. Why rob it from them? If you hate leap seconds, use TAI, while people who need UTC can use it for themselves. The system, as it currently is, works.

24 July 2011, 07h05; Banhatti, Dilip G.
Madurai Kamaraj University, India
Astronomy-Astrophysics Physics, Science Numeracy / Outreach
Response = 3
Comments meant to generate a middle path between the two alternatives.
In early 1950s, Megh Nad Saha headed a team of scientists charged with developing a suitable calendar. The team came up with a solar calendar displaced from our usual Jauary-to-December one, but otherwise marking time at the same rate, along with leap years & so forth, so as to serve festive Indian culture better. This was adopted legally by Government of India, and is, in principle, legal from then on, even to date. However, in practice, everyone in India uses what the world does, perhaps mainly for commercial reasons. In fact, very few people are aware that another "Indian" calendar IS (also) legal!
For pulsar timing (especially), and pretty much all other time variable / cyclic astro phenomena, Julian Day (JD) is used. Actual observations have time markers of the observatory making them. Any astro calculation then must convert to JD using standard conversion which includes any jumps (like leap seconds).
(# Paul A M Dirac used laser lunar ranging data in a most imaginative way. Was his use of these data in the way he did possible only due to some subtle issue of timekeeping?)
My preference: Retain both the alternatives for the different purposes where they are needed, with the overheads for the needed change(s) minimized / optimized in each of the two domains.
I guess astrodynamicists essentially use the same standard data on timekeeping that astronomers (at least currently) use to a lesser extent. Eventually, we may have timekeeping tied to solar system barycentre.

25 July 2011, 00h12; Manchester, Richard
CSIRO Astronomy and Space Science, Australia
Astronomy-Astrophysics
Response = 2

25 July 2011, 07h05; Powers, Patrick
Formerly Logica PLC, UK
Information Technology
Response = 1
We are in tune with the sun. We have an important circadian rhythm tuned to the sun. This is the essence of time as we experience it. High noon when the sun is at its apex is midday, halfway between sunrise and sunset, not 13:23:34 DST (Digital Daylight Standard Mean Civil Clock Time). We cannot get past our innate sense of solar time, the rhythm and duration of solar cycles. It is part of our being. This is what sundials show, true solar time. Granted the change to atomic time is a minor adjustment from solar astronomical time, at this time. But the difference is cumulative. The difference will accumulate through the centuries. In the future we would be getting up in the morning at 12:00 or whatever abstract number is defined by vibrating Cesium atoms. The odd leap second can adjust for the slower rotation of the earth. Is this better than the riots when an abrupt shift like the Gregorian correction is required? Computers are easier to reprogram than people. It will simply not be possible for all humanity to be aware of a time measure that is out of synchronism with the sun and this will generate years of requests for a return to the present status. Let us not even go there. "Cogito ergo sum". Thinking people rule, technology serves.

25 July 2011, 09h42; Foschini, Luigi
INAF Osservatorio Astronomico di Brera, Italy
Astronomy-Astrophysics Space-sciences
Response = 1

25 July 2011, 13h23; McEnery, Julie
NASA/GSFC, USA
Astronomy-Astrophysics
Response = 2

25 July 2011, 15h16; Roth, Martin
AIP, Germany
Astronomy-Astrophysics
Response = 1

25 July 2011, 17h54; Noël, Fernando
National Astronomical Observatory, Chile
Astronomy-Astrophysics
Response = 2

25 July 2011, 19h50; Grandi, Steven
National Optical Astronomy Observatory, USA
Astronomy-Astrophysics
Response = 1

26 July 2011, 21h40; Rundle, Nicholas
Rockwell Collins, USA
Telecommunication
Response = 2
In satellite communications, time needs to be very accurate between terrestrial terminals and the payload on the space vehicle. The time source is entered from UTC. Since most computer systems have no notion of a leap second, they must be added to the UTC time in order to create the actual time used by the communications network. This creates enormous complexity especially when leap seconds are added. Since the communications systems are all computer controlled, the notion of time in relation to the Earth's rotation is not important as it is to the human population. Please redefine UTC to be uniformly increasing without leap seconds.

27 July 2011, 03h14; Brown, Kyle
Unassigned, USA
Web Applications Programmer
Response = 4

27 July 2011, 10h39; Ron, Cyril
Astronomical Institute of Acad. Sciences, Czech Republic
Astronomy-Astrophysics Celestial-mechanics Geodesy Geophysics
Response = 1

27 July 2011, 16h04; Kaplan, George
(Contractor to U.S. Naval Observatory), USA
Astronomy-Astrophysics
Response = 1
Applications that require a continuous time scale (without leap seconds) should use TAI.
In previous discussion of this issue, the necessity of the proposed change has not been clearly articulated, while the consequences of the proposed change, for large numbers of software systems, have been discounted. A full assessment of the number of software systems that assume that UT1=UTC has not been carried out. In fact, such an assessment would be difficult to carry out as a limited exercise because the UT1=UTC assumption is often implicit. Leap seconds are a well-defined international standard that, although inconvenient, are within the capabilities of current technology, just as they were within the capabilities of the technology of 1972. "Inconvenience" is not a justification for so fundamental a change. Furthermore, the ITU is not the correct international entity to change the definition of the worldwide system of civil time. This is more than just a change to a radio signal; it involves the very definition of what we mean by civil time and potentially affects every person within the developed or developing world.

27 July 2011, 20h02; Pinto, Heitor David
NASA, USA
Global navigation satellite systems
Response = 1
The problems created by leap seconds are rare and manageable, and therefore do not justify the adoption of a time scale no longer related to the rotation of the Earth.

28 July 2011, 08h11; Mohasseb, Mohamed
Arab Acdemy for Science and Technology, Egypt
Geodesy
Response = 1

28 July 2011, 10h53; Vondrak, Jan
Astron. Inst.., Acad. Sci. Czech Rep., Czech Republic
Astronomy-Astrophysics Celestial-mechanics Geodesy
Response = 2

28 July 2011, 22h29; Himwich, William
NASA GSFC, USA
Astronomy-Astrophysics Geodesy
Response = 1

It is more than that we are satisfied with the current definition of UTC. We depend on it. It is built implicitly into many systems that we use and support world-wide for radio astronomy and geodesy. It will be a significant perturbation on these systems, many extremely difficult to modify, if the definition changes. If there is no way to stop eliminating leap seconds, the proposal to have a "leap hour" is unrealistic and appears to just be an attempt to make time coordination someone (who hasn't been born yet) else's problem. This option also has serious undesirable effect. A more realistic option with less undesirable effects would be a "leap minute", but that would also defer difficult issues irresponsibly. The fundamental problem is that most (if not all) computer operating systems as they exist now do not properly recognize leap seconds. This can be corrected now, in the present day, and would provide a long term solution.

29 July 2011, 11h25; Steeghs, Danny
University of Warwick, UK
Astronomy-Astrophysics
Response = 1

31 July 2011, 20h15; Ste. Marie, Paul
Amazon.com, USA
software development
Response = 1

01 August 2011, 10h05; Smith, Marlyn
British Astronomical Association, GBR
Astronomy-Astrophysics
Response = 1

01 August 2011, 10h41; Boomkamp, Henno
IAG WG 1.1.1 / ESOC, Germany
Geodesy Space-sciences
Response = 3

The essence of the current UTC definition is its sub-second offset from UT1. Take that property away, and UTC no longer exists: it becomes identical to TAI apart from the arbitrary constant offset. The discussion on stopping further leap seconds is therefore equivalent to a proposal for shifting the origin of the TAI time by the arbitrary amount of 34 seconds, and calling this shifted TAI scale "UTC", as if it is significantly different from TAI. It is not: it is exactly the same as TAI, apart from an arbitrarily different origin. The origin of TAI is already arbitrary, so what would be the point in having this UTC scale in parallel to it?

Arguments in favor of stopping further leap seconds are usually related to software issues, or to the political authority of announcing a formal leap second. We have always managed to live with these issues in the past. Furthermore, the increasingly important reprocessing activities of the scientific community imply that our software will forever have to be capable of dealing with past leap seconds (historic data often has UTC time stamps), even if no new leap seconds would occur in the future. The software argument is therefore rather weak.

Our "other preference" is therefore as follows. We introduce a new UTC definition without leap seconds, but call it "TAI2000" rather than UTC. Instead of the arbitrary shift of 34 seconds between the UTC and TAI origin, we use the offset at the J2000 epoch, which was 32 seconds. This (forever) constant offset between TAI and TAI2000 is just as arbitrary as when we would keep the current number of leap seconds frozen forever, but at least there would be some physical meaning to it. Also, it makes sense to call this new scale TAI2000 rather than UTC2000. The old UTC should then continue as it is - with leap seconds - because that is the only relevant way in which UTC is different from TAI.

01 August 2011, 13h39; Boot, Teco
Infinity Networks, Netherlands
Telecommunication
Response = 2

01 August 2011, 19h30; Gordon, David
NASA/GSFC, USA
Astronomy-Astrophysics Geodesy
Response = 1

01 August 2011, 21h33; Ruppert, Lyle
Ball Aerospace & Technologies Corp, USA
Celestial-mechanics Geodesy Space-sciences
Response = 1

UTC as presently defined is an optimal choice of timescales in many applications. Other options are already available for applications in which a leap second might be inconvenient.

---

02 August 2011, 13h26; Luzum, Brian
USNO, USA
Geodesy
Response = 2

---

02 August 2011, 15h45; Cook, Mike
n/a, France
Time-laboratory
Response = 3

The issue in hand is more than redefining UTC. There are three requirements of time transmission that are met by the current recommendation of ITU-R TF.460-6.
a) Ticks of SI seconds, used by all.
b) Current value for DUT1.
c) A civil time scale, UTC, used worldwide as a legal time scale, directly descending from and now synonymous with GMT which is still the legal definition in many countries laws. The current proposition to change ITU-R TF.460 provides for ONLY the first of the above requirements. Although there has been no consensus on change in the last 10 years, I think the whole issue should go back to ITU-R WPA7 with the remit to devise a recommendation that includes ALL of the above requirements and to postpone any change until that recommendation is finalised. The current system will be quite satisfactory out to about 2300. As there is no need for precipitation, WPA7 could start from scratch and ask what is and will be required in future for time transmission.

---

03 August 2011, 13h11; Visser, Hans
Fugro Satellite Positioning BV, Netherlands
Geodesy
Response = 2

---

04 August 2011, 01h55; Sutton, Jordan
Cascade Climatology Consulting Corp., USA
Meteorologist
Response = 1

I feel that the leap second is necessary to keep time as accurate as possible. There can be no "perfect clock" or "perfect calendar", since the earth's rotation is not constant, and therefore, the earth is not a perfect timekeeper. Due to tidal drag, the earth's rotation rate is slowing down at a very slight rate; the slowing is measured by the quantity delta-t, which is usually expressed in seconds. So, when delta-t increases by a second, a leap second becomes necessary.

Over time, leap seconds will be needed more frequently due to the fact that delta-t is proportional to the square of elapsed time. Currently, a leap second is typically added every one to two years. As time progresses, leap seconds will be needed several times a year, then every month, then every week, then every day, ..., and so forth. At this point, millennia into the future, it might be more logical to insert a leap minute, or better yet, perhaps once a century make accurate clocks that run just a bit slower, thus redefining the length of the second.

---

04 August 2011, 04h50; Channon, Tim
private, UK
various technical fields
Response = 1

The arguments in favour of a change are weak whereas longer term trust in a system which is consistent is vital. The risks of unintended consequences are considerable. (side effects) We had a good example of bad argument acting as a justification for change here in England. This was about a move of currency system to decimal. The bad argument was about newly introduced digital computers. Some years later the stupidity of the argument is forgotten, is a trivial problem for computing. Whether losing a human sized unit of measuring was good or bad is not the point here. The degree of reliance on GPS etc. is a grave concern. Any argument about the dire consequences of trouble with GPS ought to raise questions about existing safety and fixed independently. If GPS internally needs a fixed time that is a GPS problem, could for example be fixed elsewhere, time altered for display. Awkward system updates? It is their job and the job of a competent design. Design out the problem. From a design perspective, if you absolutely require local reliability you split the system, disconnect dependence. As an equipment designer I have had to do this on timing, whereas relying on an external clock as reference is cheap and asking for trouble. Plenty try to do this. Live with it.

05 August 2011, 23h15; Levandowski, Robert
MacWhiz Technologies, USA
Information technology; Finance
Response = 1
UTC reflects the reality of the universe in which we live. The proposal to redefine UTC will cause UTC to drift away from reality, making it inaccurate. In the case of the systems I maintain, it will cause immense disruption due to the need to change code and procedures that currently understand UTC to include leap seconds. Accurate, synchronized time is vital for the operation of many computer protocols. It's also vital to tracking down issues that occur between systems used by different organizations, and a sub-second accuracy can be crucial in aligning events. The possibility that some organizations will use a redefined UTC while others use DUT1 would cause massive disruption. Please stop the insanity.

---

06 August 2011, 23h14; Little, Matthew
none, USA
general public interested in metrology
Response = 3
I would like the plan to be to correct UTC to UT1 when the divergence has reached an hour (3600 seconds) as that could be implemented as no net change as countries go to Daylight or Summer Time and such correction wouldn't have to be applied for much longer periods of time than the leap second. Everyone would still subtract an hour to go to the adjusted standard time at the end of that year's Daylight or Summer time. Yes, places that don't observe Daylight/Summer Time would have to change their clocks an hour at some point, I realize. I believe it is appealing to keep the solar crossing of the zero meridian noon UTC *generally*, just that the correction can be allowed to accumulate to a quantity that can be planned for conveniently long in advance and then the next correction need not be worried about for another long period rather than worrying about "nudging" all the clocks so often. However, looking at the situation from the point of view of correcting one's clocks, I note that when corrections are applied does not make much practical difference with my computer set via Internet time server and a separate clock set automatically via WWVB as these clocks would set themselves to the updated time and involve very little effort on my part whether the leap second policy is changed or not. Those relying on UT1 would likely need to track it just as much whether UTC is being corrected by leap seconds or not and we agree UT1 is the less predictable scale that needs to be tracked and analyzed regardless of agreement with UTC.

---

06 August 2011, 23h26; Skehan, Sean
City of Los Angeles - Government Agency, USA
Transportation
Response = 1
Significant amounts of transportation infrastructure in the USA are dependent on the accuracy of UTC for communications, coordination and operation. Leap seconds play an important part in keeping this equipment in sync. Allowing UTC to diverge from UT1 will in the long term be problematic. Also, the small and predictable leap second increments are much more tolerable than larger step adjustments proposed (leap minute or leap hour) and less troubling then letting UTC drift away from UT1. Please save the Leap Second!

---

07 August 2011, 21h58; Kisselov, Ivo
n/a, Bulgaria
Telecommunication
Response = 1
As UTC is often referred to and used in numerous calculations in all sectors and activities by a really very wide, diffuse and in fact hard to reach community at present, it would be better to leave it defined as it is today for official use. After several decades when the new generation of satellite navigation and time keeping systems are fully developed and integrated, and our accuracy and modeling capabilities improve by a couple of orders, this probably would be reconsidered with a far longer perspective in mind, by a more understanding user community, and in a truly global manner. Meanwhile - not to hamper the scientific work - probably a new "UTC0" (or so) can be defined that would drift away from solar time and progress uniformly (with the inevitable frequent improvements and changes to come with our development). Competent researchers will no doubt be able to easily handle such a change and make good use of it, without throwing hundreds of thousands of non-specialists into the next half-baked "measurement system bog" just because we have noticed an imperfection and want to try to impose a fix at once. Change is inevitable and needed, but should be done intelligently, not impatiently.

07 August 2011, 12h25; King, Frank
British Sundial Society (Chairman), UK
Celestial-mechanics
Response = 1
Leap seconds are as crucial for synchronising the daily rotation of the earth to clock time as leap days are for synchronising the seasons to the calendar. Please retain the current definition of UTC and the leap second.

07 August 2011, 12h34; Mohamadi, Jahanbakhsh
Zanjan_University, Iran
Astronomy-Astrophysics Geodesy
Response = 1

08 August 2011, 14h36; Theuillon, Gwladys
SHOM, France
Geodesy Geophysics Hydrography
Response = 4

08 August 2011, 15h32; Hartmann, Wilfried
IGP, ETH Zürich, Switzerland
Geodesy Geophysics
Response = 2

09 August 2011, 14h32; Rodin, Alexander
Pushchino Radio Astronomy Observatory, Russia
Astronomy-Astrophysics
Response = 3
I prefer "leap minute" introduced every 50 or 100 years.

10 August 2011, 12h31; Sauve, Michael
Alien Technology, LLC, USA
Astronomy-Astrophysics Telecommunication Time-laboratory
Response = 1
All needs for a timescale which lacks discontinuities (leap seconds) can be fulfilled by using TAI. There is simply NO rational argument for redefining UTC, which is historically linked to earth rotation, and used for that reason, to be something it was never meant to be. There is no need for yet another time scale with a fixed offset from TAI such as GPS and SMPTE timescales.

12 August 2011, 21h30; Klepczynski, William
Global Timing Servcices, LLC, USA
Astronomy-Astrophysics
Response = 2
If UT1 is needed, it can easily be obtained from the IERS to even greater precision than that available through transmitted time signals.

In this day and age, it should be a relatively easy task to re-program computers to run at UT1 with known and predicted corrections to UTC. It is even possible to have a digital clock which keeps UT1 using the published offsets of UT1 from UTC if one is needed for guiding telescopes.
I see no technical reason for keeping the existing system other than TRADITION. Even for celestial navigation, the published corrections to UTC can easily be applied to sextant observations.

13 August 2011, 04h35; Schuh, Harald
IGG, Vienna University of Technology, Austria
Astronomy-Astrophysics Geodesy
Response = 1

13 August 2011, 04h43; Ohnuki, Tohru
Northrop Grumman Aerospace Systems, USA
Astronomy-Astrophysics Space-sciences
Response = 1

16 August 2011, 00h32; Frankston, Bob
Frankston, USA
Information Systems
Response = 2
We needn't fit all definitions of time into a single framework. Whether we call this UTC or not is secondary. We do, however, need to recognize that times in databases since 1972 have been indeterminate since we can't be sure that leap seconds were honored and, in fact, most databases can't deal with leap seconds and interval calculations can't. For this reason we need to unwind leap seconds. This would be facilitated by adopting a designation that is explicit about being uniform since 1972. We can then adopt measures appropriate to domains that need to take into account celestial objects and other considerations.

16 August 2011, 09h39; Davis, John
British Sundial Society, UK
Celestial-mechanics Sundial design and consultancy
Response = 1
Leap seconds are crucial to keeping timekeeping locked to the rotation of the earth. I (or my descendants) do not wish to have noon drift into the middle of the night.

16 August 2011, 11h27; Oja, Heikki
Helsinki University Almanac Office, Finland
Astronomy-Astrophysics
Response = 1

17 August 2011, 09h25; Gupta, Sanjeev
DCS1, Singapore
Telecommunication Network Research
Response = 1
Changing the definition of UTC will cause a discontinuity. I have no objections to a non-leap-second scale, but there is no reason to use the same name. There is no shortage of new names that can be used for such a scale, or call it TAI-34.

17 August 2011, 23h03; Cabeen, Ted
UCSB, USA
Telecommunication
Response = 1

18 August 2011, 04h52; Altman, Jeffrey
OpenAFS, USA
File system developer
Response = 2

18 August 2011, 05h32; Buhrmaster, Gary
Gary Buhrmaster, USA
Telecommunication
Response = 1

18 August 2011, 08h10; Eggert, Paul
UCLA Computer Science Department, USA
Software engineering
Response = 1
My background is software engineering. I help maintain many widely used computer programs that deal with leap seconds, including the GNU C library and the TZ (timezone) database and code. I see no real need for this change, and some reasonable arguments against it, mostly in terms of complexity of transitioning to software implementing the new system.

18 August 2011, 14h56; Zijlstra, Mark
Royal Netherlands Navy / CAMS-ForceVisio, the Netherlands
Astronomy-Astrophysics Celestial-mechanics Geodesy Defense
Response = 2

18 August 2011, 19h57; Colebourne, Stephen
OpenGamma, UK
Computing
Response = 1
I believe it is fundamentally wrong for civil timekeeping to be altered in a way that separates us from the solar day and that this has moral and ethical issues beyond science or broadcasting. I also believe that is is wrong to continue to use UTC for something different to what the UT prefix implies. TAI already provides what this change seeks.
I believe that a large part of the problem has been computer systems that are not setup to deal with leap seconds, however Java via JSR-310 is bringing full leap second support and I expect others to follow. My experience as JSR-310 spec lead indicates that developers (and humans generally) realy like the concept of 24 hours of exactly 60 minutes of exactly 60 seconds, and they would prefer to see that maintained (such as via rubber seconds) rather than having to cope with an occasional 61 second minute. I believe that the best solution to the issues here are to publish leap seconds 5 years in advance, with the understanding that DUT may exceed 0.9 seconds by a small amount if the prediction is wrong. Leap seconds should be permitted at the end of any month. I also believe that UTC-SLS (whether smoothed over 1000s, 600s or 1200s) should be more widely published as the standard mechanism for mapping TAI + leap seconds to civil time. Finally, I want to see the atomic duration of "SI seconds" renamed (to duronds?) allowing the "second" to be used for civil time. The duration of 1 second is 1 durond except near a leap, where it may be longer or shorter (see UTC-SLS).

19 August 2011, 08h27; Emanov, Alexey
Altay-Sayan branch of Geophysical Survey, RUSSIA
Geophysics
Response = 2

19 August 2011, 23h53; Storz, Mark
Air Force Space Command, USA
Astronomy-Astrophysics Celestial-mechanics Geodesy Geophysics Space-sciences Telecommunication
Response = 1
The Office of the Secretary of Defense (through a letter from ASD/NII to the State Department - June 29,2009) has already agreed to support the elimination of leap seconds, but no earlier than January 1, 2019. Although no real cost estimate for upgrading Air Force Space Command software has been performed, many subject matter experts expect costs could be in the $100s of millions. A schedule risk could also be incurred if the complexity of the software upgrades is such that they cannot be tested and implemented by 1 Jan 2018 (date suggested by ITU-R).

19 August 2011, 17h50; Vasconcelos, Manuela
Portuguese Geographic Institute, Portugal
Geodesy
Response = 1

20 August 2011, 22h29; Klein, Stanley
United States Power Squadrons, USA
Celestial-mechanics
Response = 1

22 August 2011, 15h00; Mäkinen, Jaakko
Finnish Geodetic Institute, Finland
Geodesy Geophysics Metrology
Response = 2

22 August 2011, 18h35; Fischer, Michael
Marin Amateur Radio Society, USA
Telecommunication
Response = 1
Thank you for maintaining the status quo!

22 August 2011, 20h16; Kerns, Carrol
Kerns Associates, USA
Meteorology
Response = 1

22 August 2011, 21h13; Deovlet, Benjamin
Stanford University, USA
Telecommunication
Response = 1

22 August 2011, 21h23; Abraham, James
Retired, USA
Telecommunication
Response = 1

22 August 2011, 23h30; Denny, Douglas
Ex-Institute of Measurement and Control, England
Astronomy-Astrophysics Celestial-mechanics
Response = 1
A common, globally used system, linked to the Earth's rotation/variation and adjusted to be within one second at all times, or better, adjusted to within 100milliseconds (or better) by more frequent and automatic adjustments, promulgated by the international time-standard radio systems, is the common-sense way forward for the future use by the maximum number of people with the least trouble to any of them. Scientific use is more normally restricted to a relatively miniscule number of people and systems, and can be promulgated via GPS signal embedded data and is readily obtained for specialist use by specialist receivers. Let the few have the greater trouble obtaining a continuous dynamical time system. Dynamical Time can be linked to pulsars or continue to be defined by the latest hydrogen maser, caesium fountain or other atomic systems technology.

23 August 2011, 00h26; Posick, Steven
ESPN, USA
Media and Broadcast
Response = 1
The purpose of tracking time has historically been associated to the rotation of the earth and the suns position within the sky. What's to point of a time based system that does not reflect this? Who does it benefit and why? How will this help everyday software engineers/architects like myself in making software applications that function at a global scale (remember timezones)? Currently, UTC is used; if UTC is changed then the impact will be far reaching as most developers won't even understand the divergence. If someone needs a time standard like this for scientific applications, make a new standard. Leave the one everyone has become familiar with alone.

23 August 2011, 05h09; Bliss, Gerald
Retired, USA
Telecommunication
Response = 4
Just so we all are on the same tick !

23 August 2011, 07h25; Ekne, Ignidz E
GS SB RAS, Russia
Geophysics
Response = 2
Remove leap seconds! Because it's very difficult to create programs with it support. As result today 99% of programs have no support of them. For example if we try to show graph of data in real time we should to do correction from system time (if system have support of it). And we should introduce correctly 60 second.

23 August 2011, 09h19; Stephansen, Helge
T-VIPS, Norway
Broadcasting
Response = 2
For synchronisation of Single Frequency Transmitters in DVB-T2 ETSI EN 302 755 v1.1.1 Number of UTC seconds since 1.1 2000 is used as reference for time. The handling of leap seconds adds a considerable complexity for equipment manufacturers and for operators in order to prepare and pre-program for the insertion/removal of a leap second.

23 August 2011, 05h26; Evans, Andrew
MSK, USA
Telecommunication
Response = 1

23 August 2011, 11h06; Cook, Bert
Member of American Radio Relay League, USA
Telecommunication Amateur Radio
Response = 1
Presently UTC is the world accepted time standard used in worldwide radio communication. It is also the basis for the worlds established time zones. It would be very awkward to have to refer to another time standard and further a change might make it more difficult and more costly to maintain than the current method. I realize that while the atomic time standard may be necessary for some extremely delicate scientific needs in laboratories, etc.; the present system is the easiest and less expensive system to maintain and use in my practice of Amateur Radio and Amateur Astronomy.

23 August 2011, 12h50; Reis Paulino Cascalheira, Telmo José
Geographic of Portuguese Army, Portugal
Geodesy GIS
Response = 1
Avoiding the leap seconds, will bring with time some problems that we cannot figure out at the present time the consequences that will have. I strongly recommend that a deep study into this subject is done and assess all the consequences that might have. Only then the community is ready to discuss the subject.

23 August 2011, 13h33; Schaal, Ricardo
EESC- Universitu os São Paulo, Brazil
GPS applications
Response = 1

23 August 2011, 13h43; TRABANCO, JORGE
UNICAMP, Brasil
Geodesy
Response = 1

23 August 2011, 15h17; Martins, Paulo
Portuguese Army Geographic, Portugal
Geodesy Geophysics
Response = 1
It is very well described in the reference number two table 1 of the appendix the pros and the cons of the UTC redefinition. From the perspective of our institution this will bring additional and not quantified costs with no foreseen benefits.

24 August 2011, 05h08; Calabretta, Mark
Australia Telescope National Facility, Australia
Astronomy-Astrophysics
Response = 1
I am a software engineer with over 25 years' experience in astronomical software development and maintenance. It is common for astronomical software to take advantage of the fact that UTC approximates UT1 sufficiently well to be used for purposes such as Doppler correction or low-precision calculation of apparent coordinates. If leap seconds are dropped, the approximation UTC ~ UT1 will degrade slowly over a period of years, only becoming obvious after decades. Inevitably some, possibly much, software would not be adapted to the altered definition of UTC. The error may go unnoticed at first, leading progressively to ever more erroneous results over a timescale of years to decades. The famous mis-identification of the first "pulsar planet" in the early 1990s, due to an error in calculation of the Earth's ephemeris, illustrates the potential harm that might result. It is difficult to say how much software may be affected without conducting an extensive y2k-type audit. Indeed, identifying potentially affected software, and remediating that found, would in itself would be an expensive undertaking for the astronomical community. Dropping leap seconds would fundamentally alter the meaning of UTC, effectively turning it into yet another atomic timescale offset by an integral number of seconds from TAI. It would even devalue the meaning of "UTC" for the period before leap seconds were dropped. This would affect the timestamps recorded in a vast archive of astronomical data so that a distinction would have to be made between "UTC with leap seconds" before a particular date and "UTC without leap seconds after it. Aside from astronomical considerations, there is an implicit assumption that UTC, as the basis for civil time, is tied to the motion of the Sun. The very existence of daylight savings time supports this observation. The proposal to drop leap seconds essentially ignores this. There are only vague notions of "leap hours", or resetting time zones in 600 years' time when the error has accumulated to one hour. I feel that this aspect of the proposal has not been adequately addressed. For these reasons I strongly oppose the proposal to drop leap seconds.

23 August 2011, 15h22; Goncalves Ferreira, Vagner
Hohai University, China
Geodesy
Response = 2

23 August 2011, 16h25; Rui, Dias
Instituto Geográfico do Exército, Portugal
Geodesy
Response = 1

23 August 2011, 18h14; Aranha Ribeiro, Selma Regina
University State of Ponta Grossa - PR, Brasil
Geodesy remote sensing data
Response = 2

24 August 2011, 07h58; SUAGHER, Françoise
Association Astronomique de Franche-Comt, France
Astronomy-Astrophysics Space-sciences
Response = 1

24 August 2011, 15h33; ARANA, JOSÉ MILTON
FCT/UNESP, BRAZIL
Astronomy-Astrophysics Geodesy
Response = 1

24 August 2011, 23h50; Isaacs, Michael
BSS, UK
Horology, gnomics
Response = 1
Leap seconds are as crucial for synchronising clock time to the daily rotation of the earth asleap days are for synchronising the calendar to the seasons. Please retain the current definition of UTC and the leap second.

25 August 2011, 08h20; Mele, Francesco
INGV, Italy
Geophysics
Response = 1

25 August 2011, 11h54; Maria, Escuer
Instituto de Meteorologia, PORTUGAL
Geophysics
Response = 1

25 August 2011, 14h26; Earl, Zmijewski
Renesys, USA
Telecommunication
Response = 1
A change at this point would break A LOT.

25 August 2011, 15h11; Hansen, Tony
AT&T Laboratories, USA
Telecommunication
Response = 1

25 August 2011, 15h22; Adolf, Alexander
Condition-ALPHA Digital Broadcast Techno, Germany
TV Broadcast
Response = 1
We fully understand and support the need for a non-leaping and unconstrained time source. This will be very useful in many scientific and technical use-cases.
Due to the very widespread use of UTC in commercial applications, we would however be concerned over redefining UTC. It's major role in commercial applications is providing a time-zone neutral (and hence reliable) way of specifying points in time for applications which are linked to, or depend on human activities (e.g. radio/TV broadcast or air travel) and therefore need to be aligned with day/night cycles (i.e. Earth rotation). In this role, UTC has become a brand name, and has been referred to in literally hundreds of standards, and in countless computer software interfaces.
We would hence like to kindly suggest that - instead of redefining UTC - a new time source should be defined with the described properties. This would allow commercial implementations using UTC to realize a migration path towards the new time source. Whilst existing standards and implementations could remain unchanged, new standards and application designs could make use of the new time source. This would for sure be a commercially viable solution.
Redefining UTC with a 5-year deadline for updating all implementations would imply huge investments for the commercial sector, without any perceived or visible commercial advantage though.
We are hence concerned that the de-definition approach could lead to the change being largely ignored outside the scientific community.
Hence our suggestion for defining a new time source.

25 August 2011, 15h52; Hanna, Stephen
Juniper Networks, USA
Telecommunication
Response = 1

25 August 2011, 16h28; Michael, Richardson
CREDIL, Canada
Telecommunication
Response = 1
Given that computers all over the world (except some of the toys made in Redmond), already have code to deal with leap seconds (and timezones), and we all use a standard set of TIC files maintained by NIST.gov, I see no advantage to removing leap second calculations. It isn't like we can remove that code, nor is that code particularly big.

25 August 2011, 19h01; Daniel, Christopher
British Sundial Society, UK
Astronomy-Astrophysics sundial design & delineation
Response = 1
Leap seconds are as important, indeed crucial for the synchronisation of clock time to the diurnal rotation of the Earth as leap days are for rectifying the calendar with the seasons. I see absolutely no point in abandoning the use of leap seconds, which have stood the test of time, and ask that the current definition of UTC be retained together with the leap second.

25 August 2011, 20h06; Ellermann, Frank
meta.wikimedia.org/wiki/Template:YMD2MJD, Germany
Computer Science, Mathematics
Response = 1
Whatever happens to UTC, I need POSIX timestamps based on 24*60*60 seconds per day, and Modified Julian Days counting "observed" days corresponding to various calendar dates.

25 August 2011, 20h31; McQuillan, Bill
Bill McQuillan, USA
Telecommunication
Response = 1
Making UTC exactly track TAI (with a constant offset?) is redundant. Why have two standards with the same characteristics? If a user needs an atomic time without unpredictable "leaps" use TAI. Also, for consistency TAI should NOT be expressed in terms of terrestrial motions like days and years but rather in multiples of seconds (e.g., Kiloseconds, Megaseconds,...) Computers that cannot handle leap seconds should be replaced by ones that were developed by competent engineers. Leave UTC as-is for those of us that like our sun overhead at noon!

25 August 2011, 22h17; Carpenter, Brian
Individual expert, New Zealand
Telecommunication Internet protocol design
Response = 1
UTC diverging from UT1 would become a major headache for future generations. We should continue to support the mild inconvenience of leap seconds.

26 August 2011, 04h21; Chapin, Lyman
Interisle Consulting Group, USA
Telecommunication
Response = 1

26 August 2011, 11h09; Vesely, Alessandro
TANA, IT
Telecommunication
Response = 1

26 August 2011, 11h11; Weilbier, Joerg
SIEMENS AG, Germany
Telecommunication Energy distribution
Response = 2
I think, the proposal will reduce unexpected malfunctions in computer networks, containing some components, can't handle current UTC definition of leap seconds right. It's really difficult to detect erroneous (regarding leap seconds) components of a heterogene network and to predict behavior of functions, depending from exact time. It seems to me, that almost all current implementations - erroneous _and_ well made - of UTC handling devices will work right with the new proposal.

26 August 2011, 12h47; Vicente, Raimundo
Faculty of Sciences, Lisbon, Portugal
Astronomy-Astrophysics Celestial-mechanics Geodesy Geophysics
Response = 1
In order to avoid the introduction of another discontinuity in the actual system of units, constants and parameters, employed in astronomy, geodesy and geophysics, which already presents lack of consistency. I am therefore satisfied with the current definition of UTC which includes leap second.

26 August 2011, 13h43; Pereira, Jorge
Servicio Hidrografico y Oceanografico de, Chile
Hydrography and Oceanography
Response = 1
Many users of our products, as well as surveyors support point 1 and indicate they are satisfied with current definition of UTC

28 August 2011, 16h20; Heard, Charles
Private Consultant, USA
Telecommunication
Response = 1
For applications where leap seconds are significant nuisance, direct use of TAI would be appropriate. It is not necessary to redefine UTC for that purpose. On the other hand, if UTC were redefined to have a constant offset from TAI, the lack of a time scale tied to the Earth's rotation would necessitate inventing something very similar to the current UTC to replace it. So my strong preference is to leave it as is. By way of background, I have in the past written software for test equipment to convert UTC (as reported from a commercial GPS receiver) to something with a constant offset from TAI for time-stamping purposes, so I am aware of the issues involved.

28 August 2011, 17h25; Eichenstein, Yisruel
Jewish Calendar Institute - Scientist, Israel
Astronomy-Astrophysics Celestial-mechanics Space-sciences Time-laboratory Jewish Time-Studies
Response = 1
My opinion is that the proposed change in the concept of UTC in a way it will not keep pace with the diurnal rotation of Earth is a significant deviation from the way humanity is measuring time along the history and the way we always perceived the concept of Day and Second.
I think that just like there is an accepted consensus that the Calendar is synchronized with the annual revolution of the heavenly bodies, and there is no idea to change it, the same is with the UTC concept which is not different, and for sure it's not something which could be changed in a narrow assembly, but it needs a worldwide referendum which it's currently not possible, because of this my opinion is to strongly oppose such a redefinition in the UTC concept.
About the problem of the increasing amount of machines which depends on UTC, I think that's very easy to create a universal protocol which should be programmed in such a way it should easy accept the Leap Second, also regarding the Julian Day when measured according to UTC, it's possible to program that at the day a Leap Second will be added, already in the beginning of that day the Second of that day should be measured a 86,401 part of the Day and not a 1/86,400.
Regarding the problem that the number of Leap Seconds that's will be required to inset will increase during the next tens and hundreds of years, it's possible to introduce a system that every hundred years (or any other time span) should the length of the second be determined anew according the LOD at that time, so it will be avoided the need of frequent insertion of Leap Seconds.

28 August 2011, 17h28; Genut, Mordche
Jewish Calendar Institute - President, Israel
Astronomy-Astrophysics Celestial-mechanics Space-sciences
Response = 1

28 August 2011, 22h21; Bortzmeyer, Stéphane
AFNIC, France
Telecommunication
Response = 1
After reading the two excellent papers mentioned as reference, and after discussing the matter with several persons, I tend to think that the fundamental problem is the existence of several user communities, each with different (but perfectly legitimate) requirements. Because these requirements are different, there is zero chance to find a way to satisfy them all with one time scale. The only solution is therefore to have several scales and to let each community choose the one which fits its requirements.
Of course, there are already several time scales. I feel that most of the needs of people who want the end of leap seconds would be satisfied by TAI (a very regular time scale, without "steps" and without link with the solar time). If, for one reason ot the other, TAI is not perfect for them, and there is no existing time scale suitable, it may be interesting to develop a new time scale (I'm not convinced it will be necessary: many proposals, such as the one for the "new UTC", are YATSCOT - "Yet Another Time Scale with a Constant Offset to Tai"). But using the term UTC for a new time scale seems confusing because people and software are now used to the existing definition of UTC.
The root cause of the dispute, I believe, is that too many people would like to have a "primary" time scale, one which is "more equal than others", hence the fight over UTC (actually, over the name "UTC"). The proper framework of thought would be, not only to have several time scales, but also to recognize them are "equal" and chosen at will by the different communities.

28 August 2011, 20h51; Vincent, Fiona
University of St. Andrews, Scotland, UK
Astronomy-Astrophysics
Response = 1
The abstract idea of "Time" is a concept which humans find difficult to deal with, but its embodiment in the cycle of day and night allows us to feel comfortable with it. I think it would be morally undesirable to divorce our time-keeping system from the basic rhythms of life.

28 August 2011, 21h54; Stapleton, Roger
University of St. Andrews, UK
Astronomy-Astrophysics
Response = 1
My work is on the control system for a 1m class telescope (built about 1960). Its pointing system is not super-accurate so I only need UT1+/-1sec to get the pointing accuracy needed. This is at present supplied by UTC via the internet and the NTP service. If UTC is abandoned by the removal of leap seconds I require easy access to the error between UT1 of the new timescale. I have seen no suggestion that such a service will be part of the change. There will also be a cost in programming time to incorporate this service - assuming that it is created.

28 August 2011, 23h05; Arnold, Mathieu
Absolight, France
Telecommunication
Response = 1

28 August 2011, 23h24; Rascanu, Theodor
Institut für Kernphysik Frankfurt, Germany
Astronomy-Astrophysics
Response = 1
In my opinion there is no logic behind setting UTC constant to TAI, since in that case one could directly use TAI.

29 August 2011, 11h05; Ferreira, Rui
MOG Technologies, Portugal
Broadcasting
Response = 2

29 August 2011, 12h15; Da Silva Costa, Paulo
LNEG, Portugal
Space-sciences
Response = 1

29 August 2011, 18h48; OGDEN, Andrew
Worshipful Company Of Clockmakers, Republic Of Ireland
Geophysics
Response = 1
Leap seconds are crucial for synchronising clock time with the daily rotation of the earth. Please retain the current definition of UTC and the leap second.

29 August 2011, 21h30; Auerbach, David
Technical University Eindhoven, Netherlands
Geophysics
Response = 1

29 August 2011, 22h45; Mann, Christopher
Michael Fields, USA
Agriculture
Response = 1

29 August 2011, 22h49; Leneweit, Gero
Carl Gustav Carus-Institute, Germany
Nanotechnology
Response = 1

29 August 2011, 22h52; Winiwarter, Verena
IFF Social Ecology, Austria
Environmental History
Response = 1
The symbolic significance of giving up our attachment to natural rhythms should not be underestimated.

29 August 2011, 22h58; Schaerer, Alec
Geography -- U of Basel, Switzerland
Integral Methodology
Response = 1

29 August 2011, 23h19; Weigl, Herwig
Dept. of History, Univ. Vienna, Austria
history
Response = 1

30 August 2011, 00h04; Jacobi, Johanna
University of Berne, Switzerland
Space-sciences
Response = 1

30 August 2011, 00h27; Giorgini, Jon
NASA/Jet Propulsion Laboratory, USA
Astronomy-Astrophysics Celestial-mechanics
Space-sciences solar system ephemerides
Response = 1
Redefining UTC and dropping leap-seconds must not be imposed on those already successfully using it for the mild or theoretical convenience of others. Perpetually troublesome communication problems with the ephemeris-using public would be introduced; it is not simply a numerical and software modification issue.

30 August 2011, 04h06; Woo, Wang Chun
The Hong Kong Observatory, Hong Kong, China
Astronomy-Astrophysics Geophysics Time-laboratory
Response = 2
1. Designing, operating and testing time service equipment for leap seconds require tremendous efforts, yet they are still error-prone as leap seconds are introduced only occasionally.
2. The possibility of leap seconds makes it impossible to compile calendar valid for decades/centuries.
3. It is difficult to explain to the public why leap seconds are necessary, given that the time shift of sunrise/transit/sunset occur over hundreds of years.

30 August 2011, 04h39; Koehler, Reinhard,
Carl Gustav Carus-Institut, Germany
Geophysics Hydrodynamics in Pharmazeutics
Response = 1

30 August 2011, 07h32; Heertsch, Andreas
Verein für Krebsforschung, Switzerland
Medical devices
Response = 1

30 August 2011, 08h06; FICHMEISTER, Hellmut
Graz University of Technology, AT
Astronomy-Astrophysics
Response = 1

30 August 2011, 08h30; Auerbach, Raymond
Nelson Mandela Metropolitan University, South Africa
Agriculture
Response = 1
It is important in my field to have the connection with the actual astronomical events to which time is related in terms of daylight, planetary cycles and seasons. Please keep it as it is!

30 August 2011, 08h40; RAMM, Hartmut
Hiscia, Switzerland
Cancer research
Response = 1
The Sun is the source of life - let's keep connected.

30 August 2011, 09h16; SCHWARZ, Reinhard
Ordination, Austria
Medicine
Response = 1

30 August 2011, 09h27; Sutter, Christine
Institut für Stroemungswissenschaften, France
Astronomy-Astrophysics
Response = 1

30 August 2011, 10h03; Lukas, Dostal
none, Czech Republic
Astronomy as hobby / medicine
Response = 1

30 August 2011, 10h07; Baur, Felix
private, CH
medicine
Response = 1

30 August 2011, 10h51; Moser, Max
Institute for Physiology, Med Uni Graz, Austria
Medicine, Physiology
Response = 1

30 August 2011, 11h08; Schmitt, Tobias
Gabriel-Tech, Germany
Telecommunication
Response = 1

30 August 2011, 11h10; Seifert, Georg
Charité UniversitätsMedicine Berlin, Germany
Medicine
Response = 1

30 August 2011, 11h17; Kranz, Christoph,
none, Austria
pedagogic
Response = 1

30 August 2011, 11h21; Landl, Richard
Bund der Freien Waldorfschulen, Germany
Pädagogical Science
Response = 1

30 August 2011, 12h23; Haberl, Helmut
Institute of Social Ecology, AAU, Austria
Social Ecology
Response = 1

30 August 2011, 12h49; Van der Wal, Jacob
Dynamension, Netherlands
Medicine
Response = 1

30 August 2011, 12h50; McKeeen, Claudia
Fachschule/Kindergartenseminar, Germany
Medicine und Pädagogik
Response = 1

30 August 2011, 12h50; Els, Ruitenberg
Dynamension, Netherlands
none
Response = 1

30 August 2011, 13h18; Crawford, Athalie
Quaker Peace Centre, South Africa
Diversity
Response = 1

30 August 2011, 13h36; Gelinek, Oskar
STENUM, Austria
Environmental Consulting
Response = 1

30 August 2011, 13h45; Barrett, Paul
US Naval Observatory, USA
Astronomy-Astrophysics
Response = 2

30 August 2011, 14h03; Kiene, Helmut
IFAEMM, Germany
medicine
Response = 1

30 August 2011, 14h08; Raderschatt, Bert
Praxis, Germany
General Practitioner
Response = 1

30 August 2011, 14h24; Merimaa, Mikko
Centre for Metrology and Accreditation, Finland
Time-laboratory National Metrology Institute
Response = 2
Considering seasonal deviations of the apparent solar time and quantization caused by time zones, leap seconds introduced to UTC are of minor importance to the general public. In a modern society, there are relatively few applications that require time synchronized to UT1, while leap seconds create problems in data logging, time stamping, telecommunication systems and time distribution services. Thus, a serious consideration should be given to stop corrections to UTC while the published difference between UTC and UT1 could be used in applications where UT1 is needed.

30 August 2011, 14h57; Mayer, Helmut
IPMR, Austria Vienna
Rehabilitation Medicine
Response = 1

30 August 2011, 14h57; Moravansky, Johann
Private Practice, Austria
Medicine
Response = 1

30 August 2011, 15h05; Liess, Christian
HTWG Konstanz, Germany
Fluid dynamics engineer
Response = 1

30 August 2011, 15h21; Seelbach, Dr. Volker
Waldorfschool at Wangen / Allgäu, Germany
Biology- and Chemistry-Teacher
Response = 1

30 August 2011, 15h44; Volkmann, Juern-Hinrich
DGI, Germany
Literature
Response = 1

30 August 2011, 16h33; Rabethge, Helga
NaturheilMedicine, Germany
Astronomy-Astrophysics Geisteswissenschaft
Response = 1

30 August 2011, 17h01; Albonico, Hans-Ulrich
Cosmic Intuitive System Promotion CISP, Switzerland
Biochronology
Response = 1

30 August 2011, 17h24; Bauer, Hermann
School, Germany
Astronomy-Astrophysics
Response = 1

30 August 2011, 17h27; Cimino, Giancarlo
ASL8 Cagliari, Italy
Physician
Response = 1

30 August 2011, 17h32; Malicky, Michael
Oberösterreichische Landesmuseen, Austria
Informatics
Response = 1

30 August 2011, 17h35; Allen, John
Edinburgh Univeristy, Scotland
Astronomy-Astrophysics
Response = 1
I acknowledge that UTC needs to follow the atomic time for astronomical and computing needs but it means changes to the way we read time from the sun (sundials being one example). Also it means that the meridian, now at Greenwich will effectively move slowly eastwards - which is confusing. Can astronomers and computing people (the minority of the population) use another system that keeps in step with atomic time? (I believe there are some already available.) Let us not confuse the ordinary folk with this change and keep the status quo!

30 August 2011, 18h24; Weaver, Nicholas
ICSI, USA
Telecommunication
Response = 1
The leap-second addition, when it occurs, is transparent to most computer users, programmers, etc., as systems are synced using NTP (Network Time Protocol) to UTC. But if UTC, by removing leap-second addition, is allowed to diverge from Earth rotational time, when the accumulated divergence is over >1 minute, there will be pressure to redefine local times in terms of UTC - 60s, which will significantly disrupt a large number of computers, programs, etc, which rely on twin assumptions:
a) That UTC represents human-scale time
b) That the offset between UTC and local time doesn't suffer discontinuities.
The proposed change in definition of UTC will cause significant disruptions in the future on effectively every computer on the planet, as these assumptions about UTC ~= UT1 is baked into all these devices we use today.

30 August 2011, 19h02; Schwarz, Anneliese
Physik, Austria
Astronomy-Astrophysics
Response = 1

30 August 2011, 19h11; Schröter, Astrid
GCC, China
Industry and Trade
Response = 1

30 August 2011, 20h16; Schmidt, Martina
IPSUM, Germany
physician
Response = 1

30 August 2011, 20h36; Noest, Ingrid
privat, Austria
Floristic
Response = 1

30 August 2011, 20h51; Dickson, Brian
VeriSign, USA
Telecommunication Networking protocols, equipment, service
Response = 1
Since TAI exists, any need for a fixed offset from TAI can be achieve by... a fixed offset from TAI. UTC is different from TAI because it is different than TAI. (Duh.) UTC exists for many reasons, and is an accepted standard, which is the basis for an entire category of time-related functions:
- astronomy (consistent and accurate measurements require consistent and accurate time)
- GPS
- GPS-derived super-accurate clocks for synchronized network transmission equipment
- GPS-derived super-accurate clocks for networking protocols
- GPS-derived super-accurate clocks for security logs
- GPS-derived super-accurate clocks for satellite communication buffering (Doppler effect cancellation)
- GPS-derived super-accurate clocks for keeping computers synchronized for inter-machine communication/coordination (file systems, schedulers, etc.)
All of these require that UTC be consistent, and have not much to do with TAI-UTC drift. All systems that derive nanosecond-level clocking from GPS, do so with knowledge of leap seconds, and do not experience frequency-shift off of TAI nanosecond-level clocking. All systems that derive clock-face-time do so with knowledge of leap seconds, and maintain their synchronization across leap-second events. Changing UTC to not implement leap-seconds can obviously be easily implemented, by not counting leap seconds. However, this achieves nothing of value, and does so at a significant detriment to every human activity that currently relies on UTC and GPS. Please reject this, permanently.

30 August 2011, 20h58; Jacobi, Michael
Institut für Strömungswissenschaften, Germany
Astronomy-Astrophysics
Response = 1

30 August 2011, 21h23; Kestel, Tobias
White Elephant Design Lab, Austria
Industrial Design
Response = 1

30 August 2011, 21h26; Dr. Kindt, Reinhard
Anthroposophical Society, Germany
Medical Doctor
Response = 1

30 August 2011, 21h55; Schwarz, Valentin
Weleda AG, Germany
Life Science/Microbiology
Response = 1

30 August 2011, 22h06; Kröswagn, Armin
private, Österreich
Pediatrician
Response = 1

30 August 2011, 22h14; Varga, Marta
TU Budapest, Hungary
Celestial-mechanics Geodesy Space-sciences
Response = 1

30 August 2011, 22h17; Jacobi, Martin
Sozialtherap. Gemeinschaften Ww e. V., Germany
musician (a=432 Hertz)
Response = 1
No doubt to me - by shifting time one hour ahead what happens every year end of March time is spoiled enough. "Summer Time" to me means more worrying, more distress, less enthusiasm. Redefining the second in the above way would mean wrong tone system; music would not have to do with human feelings any more. Never can I accept that.

30 August 2011, 22h22; Conradt, Oliver
Section for Mathematics and Astronomy, G, Switzerland
Astronomy-Astrophysics
Response = 1

30 August 2011, 22h27; Miller, Gary
Rellim, USA
Time-laboratory
Response = 1
Please do NOT do this! There are many GPS in the field that are over 20 years old and have no chance of a firmware update. Incompatible changes to a long established standard would lead to many problems.
Ain't broke, don't fix it.

30 August 2011, 23h07; Jacobi, Freimut
Schwarz.Jacobi Architekts BDA, Germany
Architecture
Response = 1

31 August 2011, 02h30; Wright, Frederick
Google, USA
Software Engineering
Response = 1
The effort to eliminate leap seconds seems to be the beast that won't die. At least the Julian calendar had the excuse that it was trying to do the right thing and merely wasn't accurate enough. Here the proposal is to knowingly break the correspondence between the time scale and the Earth, leaving it for future generations to clean up the mess when the error becomes sufficiently large. Note that very few modern computer systems have difficulty with the much larger one-hour step adjustments of the local time scale that occur as Daylight Time goes on and off. This is not because local time has been eliminated or redefined, but because computer systems have learned not to expect local time to be a well-behaved time scale, while continuing to use it in appropriate contexts. The only reason leap seconds pose problems is that the move away from local time didn't go quite far enough. The correct solution is to use TAI (or TAI-K) for internal timestamps, while converting to and from UTC and/or LT as needed. This is precisely what GPS does, including not only using TAI-K for internal purposes, but also tracking the UTC offset and thereby making current UTC available. The use of UTC with leap seconds for NTP synchronization does pose a couple of difficulties, but both can be dealt with:
1) The parties involved need to agree on the timing of leap seconds, in order to avoid apparent glitches in the internal time scale, which should be a "smooth" leap-free time scale.
2) The last UTC second of a day in which a leap second occurs is ambiguous.
No room for the answers. :-)

30 August 2011, 23h11; Jacobi, Georg
BSO, Switzerland
Musician
Response = 1

30 August 2011, 23h30; Pechmann, Heidi
Arztpraxis, Deutschland
Medicine
Response = 1

30 August 2011, 23h43; Pechmann, Michael
Novalisgesellschaft, 37351 Dingelstädt
Geodesy
Response = 1

31 August 2011, 01h23; Killian, Gotthard
Music-Healing-Space, Australia
Musician
Response = 1

31 August 2011, 07h27; Vogt, Jürgen
Freie Waldorfschule Kassel, Germany
Teacher
Response = 1

31 August 2011, 08h10; Taylor, David
SatSignal Software, UK
Space-sciences
Response = 1

31 August 2011, 08h32; Ziegler, Renatus
Verein für Krebsforschung, Switzerland
research scientist, mathematician
Response = 1

31 August 2011, 08h58; Seaton, Daniel
Royal Observatory of Belgium, Belgium
Astronomy-Astrophysics Space-sciences
Response = 1

31 August 2011, 09h01; Hart, Dave
ntp.org, USA
Software Developer maintaining ntpd
Response = 1
Redefining technical terms is the wrong way to tackle the perceived problem(s).

31 August 2011, 10h01; Rang, Matthias
Research institute at the Goetheanum, Switzerland
Optics
Response = 1

31 August 2011, 10h13; Nicula, Bogdan
Royal Observatory of Belgium, Belgium
Astronomy-Astrophysics Celestial-mechanics
Response = 2

31 August 2011, 10h29; Hobbensiefken, Sönke
CERES, South Africa
agriculture
Response = 1

31 August 2011, 10h37; Daniele, Antonio
Italian Institute Of Navigation, Italy
Air Navigation
Response = 3
Thank you for having sent this questionnaire to all IAIN members. It has been a very good and appreciable initiative. In Italian Institute of Navigation, after wide and long discussion, we agreed the UTC timing and TAI timing should be the same (as, it was before 1972).This sentence from the consideration that time measuring in hours, minutes, seconds and other fractions is a convention originally starting from measuring the position of the Sun in relation with the Earth surface due to its rotation. From a practical point of view, there are no reasons to maintain two different timing scales just because of the yearly revealed difference in Earth rotation by the strong precision of the atomic clocks. We see the problem by a practical approach that means: which use would be done of such a precise but different timing on respect of Earth surface positioning to the Sun every day? And, which kind of use may we do in the next hundred or thousand years of a timing progressively diverging from the upper indicated position relationship between Earth and Sun? And also to maintain the current definition of UTC including the leap second, always on a practice point of view, would be unsatisfactory, because it will become without any significance as the years will pass. The space here do not permit the complete exposure of our discussions, anyway, if you need more information, we are ready to send them to you anytime.

31 August 2011, 11h07; Dr. Rose, Ernst
Freie Waldorfschule Graz OG, Austria
biology; chemistry
Response = 1

31 August 2011, 12h27; Ferrandiz, Jose M.
University of Alicante, Spain
Celestial-mechanics Earth rotation, satellite dynamics
Response = 1
I cannot appreciate no real advantage in changing the definition of UTC but a lot of associated problems and risk of computation flows together with a large overhead to prevent them. Therefore my opinion is to keep UTC in present form. Of course new time definitions may be introduced, but not representing an alternative of UTC in the short term but with research purposes to avoid un-useful, costly changes

31 August 2011, 12h30; Schmit, Scott
N/A, USA
Telecommunication Software development
Response = 1

31 August 2011, 12h57; Steiner, Bernhard
Institut für Gegenwartsfragen, Germany
Space-sciences
Response = 1

31 August 2011, 13h39; Massey, Robert
Royal Astronomical Society, UK
Learned society (on behalf of the RAS)
Response = 3
a. If the definition is changed, then the name of UTC should also change. While the proposed re-definition will make only minor differences over the next few decades, it is a major conceptual change. It would decouple UTC from Earth rotation as represented by the UT1 timescale - and thereby break away from the original concept of universal time introduced by the IAU in 1925. We believe that it is poor practice to make changes that invalidate existing text books, especially at the conceptual level. Good practice demands a name change – to stimulate people to probe into definitions of terms.
b. Whatever is decided at the ITU-R meeting in January 2012, there needs to be an easily accessible source of information on current and historical values of dUT = UT1 – UTC (or whatever succeeds UTC). This is a fundamental requirement for anyone pointing observing systems at objects away from Earth, whether astronomers with telescopes or engineers tracking spacecraft. This information needs to be freely and easily available to all, including the amateur astronomical community. Until now, both amateur astronomers and professional scientists have relied on this and if the change is implemented, future generations should continue to have equivalent access.
c. In the UK there is also a specific political issue: the proposed re-definition would necessitate primary legislation to change the basis of UK legal time from GMT to the new system derived from UTC, something the British Government has been reluctant to do in the past.

31 August 2011, 14h45; Galvin, James
eList eXpress, USA
Internet networking
Response = 1
My network applications and services depend on the current UTC definition.

31 August 2011, 14h46; Soma, Mitsuru
Natl. Astron. Observatory of Japan, Japan
Astronomy-Astrophysics
Response = 1
We live with the Sun. If no leap seconds will be introduced, the time we use will diverge from the ideal one which is in harmony with the apparent position of the Sun, so in any case we will need to make some adjustment to the time in the future. If there is no rule for the adjustment, there will be serious confusion in the future. When one calculates the times of past and future astronomical phenomena, such as solar eclipses, sunrise, sunset etc., we need the value of TT-UT1 (for precise calculations one also needs UT1-UTC, but for most cases it is not needed). If UTC diverges from UT1, we will always need both of TT-UT1 and UT1-UTC, which complicates such calculations, and I do not like that situation.

31 August 2011, 17h52; Kirby, John
none, USA
Telecommunication
Response = 1
This change could have far reaching consequences that affect public safety and public services. Time changes will affect radio system configuration and end user service management, geographical positioning navigational aids, computer aided dispatch systems, SCADA systems, and event logging and recording systems. In the event of severe solar weather, time changes could render systems such as navigational aids inaccurate or ineffective in helping to direct the delivery of emergency services, gaining and maintaining situational awareness, or coordinating with multiple agencies.

31 August 2011, 14h58; Parker, Terry
KCOM, UK
Telecommunication
Response = 1

31 August 2011, 17h07; Schulthess-Roozen, Marjolein
Ita Wegmanklinik, Switserland
Medicine
Response = 1

31 August 2011, 20h54; Malone, David
School of Mathematics, Trinity College D, Ireland
Public NTP Server Operator
Response = 1
We have not experienced any difficulties during leap seconds. Legal time in Ireland still seems to depend on GMT and consequently I think there would be a preference here for keeping UTC close to historical definitions of GMT.

31 August 2011, 21h37; Cornec, Jean-Paul
Retired, France
Astronomy-Astrophysics Telecommunication
Response = 2

31 August 2011, 21h47; Drury, Luke
Dublin Institute for Advanced Studies, Ireland
Astronomy-Astrophysics
Response = 2

31 August 2011, 23h34; William, Thompson
NASA Goddard Space Flight Center, USA
Astronomy-Astrophysics
Response = 1

01 September 2011, 09h44; Gelinek, Christian
N/A, Australia
Electronics
Response = 1

01 September 2011, 10h45; Monstein, Christian
ETH Zurich, Switzerland
Astronomy-Astrophysics
Response = 1

01 September 2011, 12h23; Escapa, Alberto
University of Alicante, Spain
Astronomy-Astrophysics Celestial-mechanics Space-sciences
Response = 1

01 September 2011, 12h55; Gary, Dale
New Jersey Institute of Technology, USA
Astronomy-Astrophysics
Response = 1

01 September 2011, 16h10; Dominique, Marie
Royal Observatory of Belgium, Belgium
Astronomy-Astrophysics
Response = 3
I do not like the idea of having UTC increasingly diverging from UT1. Nevertheless, leap seconds considerably complicate the processing of data that must be accurately time-tagged. Confusion and mistakes are frequent. Although there is no perfect solution, I think that the situation would be simpler if time correction was applied on a deterministic date, and more rarely. I would be in favor of a correction applied, for example, on January 01 00:00 every 10 years. Or even better, apply them on each Feb 29.

01 September 2011, 16h16; Parenti, Timothy
University of Pittsburgh, USA
Student
Response = 1

01 September 2011, 16h18; Gamby, Emmanuel
Belgian Institute for Space Aeronomy, Belgium
computer science
Response = 3
I do not like the idea of having UTC increasingly diverging from UT1, but I think that the offset between both could be bigger than one sec. In addition, leap seconds considerably complicate the time-stamping of data. Although there is no perfect solution, I think that the situation would be simpler if time correction would be applied on a deterministic date: for instance, on Feb 29 every 10 years or so.

01 September 2011, 17h55; Pechmann, Johanna
Ridterapi Novalis, Sverige
Horseback Riding
Response = 1

01 September 2011, 22h19; Allen, Steve
UCO/Lick Observatory, USA
Astronomy-Astrophysics
Response = 3

I appreciate the problems faced by systems which have to handle conflicting requirements under the current implementation of leap seconds. Nevertheless, the current problems with leap seconds are largely one of representation. A better representation can preserve the existing and traditional meaning of UTC as civil time while also alleviating the problems faced by software systems. I have written a detailed description of an alternative to the draft revision of ITU-R Rec. 460. This alternative is truly a compromise. It makes use of an existing, deployed, and routinely-exercised mechanism. It also changes leap seconds into a form which is easily-testable by systems and engineers. The description can be seen here
http://www.ucolick.org/~sla/leapsecs/right+gps.html
I urge that this scheme be presented for wide consideration.

02 September 2011, 01h14; Homeyer, Gernot
Dr.med., Germany
Astronomy-Astrophysics
Response = 1

02 September 2011, 03h45; Tschannen, Ruth
Cascadia Society, Canada
Eurythmist
Response = 1

02 September 2011, 13h47; Scott-Stapleton, Graham
British Sundial Society, Britain
Astronomy-Astrophysics
Response = 1

To disconnect time keeping from the earth's rotation will be to render time an entirely theoretical entity. To discontinue leap seconds is, in the long term, as ill-advised as discontinuing leap days.

02 September 2011, 19h53; Fischer, Gwendolyn
Christian Community, Germany
Telecommunication
Response = 1

02 September 2011, 20h46; Stuart, Robin
., USA
Astronomy-Astrophysics
Response = 1

Surely the mandate of UTC is to provide a measure that is closely aligned with the principal driver of civil activity, namely time of daylight hours and the position of the Sun in the sky. While it may be argued that the existence of time zones and artificial lighting make the leap second of relatively little practical importance, at least in the near term, the drift will eventually become unacceptable on time scales that will depend on the particular application. Celestial navigation will likely be amongst the first disciplines to be affected but I suspect that even casual sky watchers will find it profoundly disturbing to know that the time of sunrise on midsummer's day will vary over time, they cannot reliably specify the date of earliest sunrise from their location and that sundials can no longer be relied upon. As scientists we seek uniform operating principals wherever possible. It may be argued that modern life and time keeping is no longer regulated by the exact position the Sun in the sky. But since few of us sow and reap the same arguments apply to the Gregorian calendar. If we do away with leap seconds then we should also revert to the Julian calendar. Of course astronomers do make use of Julian date which is fine for the conduct of science but I think that few would attempt to inflict it on the public at large just for their own convenience. This is what is effectively what is being done with regard to UTC. What a tragedy years from now when sundials no longer work and our hard won mastery over the clockwork of the universe can no longer be demonstrated and accessible to the common man.

02 September 2011, 23h01; Miguel, Martinez-Falero del Pozo
Other, Spain
Medical Doctor
Response = 1

03 September 2011, 02h04; Mischanko, Edward
None, USA
Time-laboratory
Response = 1

03 September 2011, 18h29; Saltzwedel, Gerhard
Praxis für AllgemeinMedicine, Germany
Medicine
Response = 1

04 September 2011, 04h19; Senturia, Philip
None, USA
Interested Layperson
Response = 1

04 September 2011, 19h26; Neuwirt, Rudolf
Institute for Geometry; TU Graz, Austria
Mathematics and Geometry
Response = 1

---

05 September 2011, 04h37; Excoffier, Denis
Ecole Normale Supérieure, France
computer science
Response = 3
UTC to be defined by IERS (or equivalent) instead of radio ticks. abs(UTC - UT1) to be kept small (like currently) TI-TAI is fixed. UTC-TI is an integer number of (SI) seconds, 0 at time of transition (2022, see torino/closure.pdf). Use the "right" branch in the zoneinfo database. Rename the "leap second" posterior to 2022 into another name ("intercalary second" would not be my preferred choice) Now a question: if transition is to occur on 2022-01-01, is the last possible leap second: 2021-12-31Z23:59:60 or 2021-12-30Z23:59:60 ?

---

05 September 2011, 08h04; Kozisek, Frantisek
National Institute of Public Health, Czech Republic
Environmental Health
Response = 1

---

05 September 2011, 09h54; VOLK, Gerhard
LVermGeo RP, Koblenz, Germany
Geodesy
Response = 1

---

05 September 2011, 12h46; Bos, Mara
none, Netherlands
religion
Response = 1

---

05 September 2011, 22h34; Schmidt, Thomas
Waldorf-School, Germany
Astronomy-Astrophysics Teaching
Response = 1

---

06 September 2011, 10h38; Arias, Elisa Felicitas
Bureau International des Poids et Mesure, --
Astronomy-Astrophysics Celestial-mechanics time metrology
Response = 2
UTC was defined in 1972 when UT1 which had limited ways of dissemination. Celestial navigation and astronomical observations were the most concerned applications. The methods of dissemination of time, of information, the communications in the seventies were not significantly affected by intentional unpredictable discontinuities of UTC, and this remained the case until the advent of GNSS and the development of the various communication networks. Different possible ways of accessing UT1 appeared, and scientists were able to improve the uncertainty of its prediction. Real-time predictions are calculated, and their dissemination through different networks is possible today; even we can think of dissemination via satellite navigation messages. Who is today using UTC because it represents a "unique" access to UT1? Who is using dUT1 as regularly published by the IERS? Celestial navigation is no more the case; astronomers can have rapid access to UT1 by its predictions. The leap seconds represent a nuisance for the modern applications requiring time synchronization. For avoiding the leap second, internal timescales are constructed (case of GNSS), offset of several integral seconds. Inconsistencies within a system using different references (with and without leap seconds) in different components have an impact in security. UTC without leap seconds will increase its offset with respect to UT1, not significantly affecting human activities, but it will positively impact and enhance modern applications. The IERS will increase visibility, disseminating real-time UT1.

---

07 September 2011, 04h01; Erdi, Jacobi
N/A, Canada
complementary Medicine
Response = 1

---

07 September 2011, 09h51; Coll, Guillermo
Hydrographic Office, Spain
Geodesy Nautical Chart production
Response = 1
The Official Position of the Spanish Navy Hydrographic Office is to maintain the current definition of UTC which includes leap second.

---

07 September 2011, 10h44; Husemann, Armin
Eugen - Kolisko - Akademie, Deutschland
Medicine
Response = 1
The system of Human Physiology is coordinated by several "Biorhythms" which depended all on the Earth rotation. Earth rotation, transformed in Sun - Light - intensity, is transferred via the eye and Melatonin - Response of the Epiphysis in the whole System of human biorhythms. So the connection of Biorhythms in man with Earth rotation is a result of evolution and in no concern arbitrary.

07 September 2011, 10h58; Kühl, Johannes
Science Section, Goetheanum, Switzerland
Physics
Response = 1
The system of human Physiology is coordinated by several "Biorhythms" which depended all on the Earth rotation. Earth rotation, transformed in Sun - Light - intensity, is transferred via the eye and Melatonin - Response of the Epiphysis in the whole System of human biorhythms. So the connection of Biorhythms in man with Earth rotation is a result of evolution and in no concern arbitrary.

07 September 2011, 20h36; Barlier, Francois
Observatoire de la Côte d'azur, France
Geodesy
Response = 2
Today, it is extremely easy to forecast DUT1 (Internet and space navigation and telecommunication). On the contrary, it will be extremely useful and more simple to have a uniform time for dynamical studies and ephemerides in space geodesy and space mechanics. I fully approved the position on the future status of UTC and UT1 adopted by the Bureau des longitudes in Paris in May 2007.

08 September 2011, 10h08; Bonnefond, Pascal
OCA-GéoAzur, France
Celestial-mechanics Geodesy Space-sciences
Response = 2

08 September 2011, 15h49; Achkar, Joseph
Observatoire de Paris, France
Time-laboratory
Response = 2
- As a scientist involved in the Time metrology, I prefer that UTC be redefined as a uniformly increasing atomic timescale without leap seconds and constantly offset from TAI.
- The UTC system with leap seconds was essentially introduced to give access to UT1 within the necessary approximation for astronomical navigation. This astronomical navigation has almost completely disappeared. For scientific applications, the use of an accuracy uniform timescale (atomic timescale) is required.
- It is sometimes said that the present form of UTC does not present any inconvenience and that users of continuous time are able to cope with leap seconds without encountering major problems. The low frequency of occurrence of leap seconds in the last few years might support this opinion. But the general behaviour is the increase of this frequency. Due to decadal fluctuations of the rotation of the Earth, this frequency may reach two leap seconds per year in a few years' time. This will make the probability of omitting or of making errors non negligible.

09 September 2011, 15h07; Grob, Herbert
Freie Waldorfschule, Germany
Education
Response = 1

12 September 2011, 18h13; Gambis, Daniel
Observatoire de Paris, France
Astronomy-Astrophysics Geodesy
Response = 1
The present system is a good compromise between Earth rotation and atomic time scale. Leap seconds introductions could be a nuisance for some restricted scientific communities but the system works well. Arguments to change are not sufficient compared to the advantages of a coordinated UTC time scale linked to the earth rotation. Few problems were reported after the 2009 leap second introduction. The issue is not only scientific, all scientists are able to adapt to any definition of UTC. A majority of UTC users are not aware of the difference between UT1 and UTC. If the new definition is adopted, they should. When the difference DUT1 increases, 30s, 10 min, 1 hour, a lot of problems will arise. There is too much software with the assumption of UTC being coordinated with the earth rotation. The costs of change would be important. Unforeseen problems could happen. Why having another timescale in addition to UT (GPS) parallel to TAI without leap seconds? The idea of suppressing TAI and to entrust the task of deriving a new continuous UTC by BIPM does not solve anything unless UTC be operational. The possible adoption of a continuous time UTC time scale with the introduction of leap hours putting off to future generations is much worse than the present system. The ITU does not appear to be the correct international body to change the definition of the worldwide system of civil time. There is no strong justification to adopt a time scale no longer related to the rotation of the Earth. In any case, more time should be needed to evaluate the consequences of such a change.

12 September 2011, 18h20; Moshuber, Jöran
private, Austria
Medicine
Response = 1

08 September 2011, 22h45; McBurnett, Neal
Boulder Community Network, USA
Systems software
Response = 1
The worst approach is to redefine UTC so that the basic meaning changes (i.e. no longer linked to rotation of the earth) without changing the name "UTC". This would fundamentally confuse the name, require endless clarifications for the rest of time, and be a huge waste. For people that want a timescale without leap seconds, let them simply use TAI, or if really necessary some variation on TAI like GPS time. If the goal is to redefine a legal notion of time, this should be undertaken by a different body than ITU-R, which has no remit to disassociate clock time from solar time. E.g. the United Nations, or individual countries.

15 September 2011, 11h28; Bizouard, Christian
Observatoire de Paris, FRANCE
Astronomy-Astrophysics Celestial-mechanics Geodesy Geophysics Space-sciences
Response = 1
There is no practical requirement for changing the definition of UTC. If for some practical issues, continuous time scale is required, one has already at hand UT GPS or TAI. Moreover this definition appears to be recent (the 1970's) in light of the long astronomical tradition going back to Sumerian civilisation. The current UTC concept is the fruit of a long scientific ripening, combining technological progress (atomic clock) and the natural, biological rhythm, founded on the succession of days and nights. Changing a definition too often has the same effect as to permanently produce new laws without fundamental reason: few people will note it, and this will diminish its force.

15 September 2011, 14h48; Lefebvre, Pierre,
None, France
Astronomy-Astrophysics (not professionally)
Response = 1
The proposal to remove leap seconds from UTC appears contradictory with the definition of UTC: if UTC is not synchronized with Earth's rotation (within a 1 second accuracy), why maintain it ? What we will be the meaning and interest of UTC in a few decades, when it will be 34 seconds behind TAI but, say, 10 seconds ahead of UT1 ? Should the proposal be adopted, I would recommended keeping only 2 time-scales:
- TAI (as base for civil time around the world), introducing a one-time 34 seconds shift in all clocks worldwide
- UT1 (for astronomical applications)

16 September 2011, 00h54; Glaser, Thorsten, Mir-Solutions, Germany
Telecommunication computing
Response = 3
I have a strong preference for the current system with leap seconds and keeping UTC an integral offset to TAI aligned with the real earth rotation. Computing systems have coped for decades; changing things now will introduce more new breakage than can ever be saved by changing systems. Astronomically, it's the only thing that makes any sense, too.

16 September 2011, 21h19; Tobin, William, (retired from University of Canterbury), France
Astronomy-Astrophysics
Response = 1
If I was setting up UTC again, I would decouple it from the Earth's rotation, because it has the same flaw as the French revolutionary calendar, i.e. you cannot tell how many seconds there will be from now to the end of the decade, just as the Revolutionary Calendar could not tell you how many dates until some date several millennia hence. But as presently defined, UTC is a standard, and so should not be changed lightly. The consequences on many pieces of hardware are far from clear...for example where UTC and UTC1 are hard-wired/programmed under the assumption that there can never be more than 1 second between them. Further UTC is specifically referred to in many countries' legislation. If there is to be a redefinition, it *must certainly* be given a new name and not called UTC. In fact, what I'd say is that we should just jump 34 seconds and start using TAI for civil timekeeping (but I believe there is some flaw in this because TAI is not known in real time, so something similar to TAI). Finally, of course, with the spread of computer networks we are moving to a point where it would be appropriate to abandon time zones and have everyone use a common time wherever they are on the planet. Already this is what some of the banks do when I pay on-line with my credit card. Any change to UTC should be coordinated with a change to a common time everywhere on the planet.

18 September 2011, 05h57; Gerstman, Larry, Long Beach Schools, USA
Astronomy-Astrophysics
Response = 1
Sorry for my late response, but I only just found your questionnaire. I feel that the current system of defining UTC and adding leap seconds when needed is ideal and accurate. Please do NOT abolish the current system which works well and is vital to astronomical calculations throughout the world. No other system is even adequate. A proverb we live by, "If it ain't broke, then don't fix it."

18 September 2011, 13h57; Citro, Gary, Elmont U.F.S.D., USA
Astronomy-Astrophysics
Response = 1

18 September 2011, 14h29; Kozma, Michael, CUNY, USA
Astronomy-Astrophysics Telecommunication
Response = 1
The time scale approach was tried in the past using atomic oscillation frequencies. It quickly lost favor since it was impossible to remember the constants. The current definition is more than adequate.

23 September 2011, 17h19; Coy, Robert, -, UK
Telecommunication Transportation
Response = 1

25 September 2011, 14h35; Dawson, Hylton, Britsh Sundial Society, England
Celestial-mechanics
Response = 1
Leap seconds are as crucial for synchronising clock time to the daily rotation of the earth as leap days are for synchronising the calendar to the seasons. Please retain the current definition of UTC and the leap second

29 September 2011, 20h47; Novosielski, Gary, Fort Lee (NJ) Board of Education, USA
Education
Response = 1
I believe that the current definition, which is useful to all persons as long as they do not require access to earth rotation time more precisely than the nearest second, will be useful to many more people than a definition that is permitted to drift to an undefined degree. Those who require UT1 precision closer than one second presumably already have access to such a standard, but changing the definition would require many more people to arrange access to one. UTC is currently far more widely available than UT1. It can be determined to sub-second precision over any internet and radio sources nearly anywhere. Changing its precision from sub-second to indeterminately sloppy is, in my view, unwarranted.

06 October 2011, 11h55; Maltin., Michael, Navigation., UK.
surveying.
Response = 1
The present system should remain. It is best suited for the purposes of Navigation.

11 October 2011, 12h45; Vultaggio, Mario, Italian Institute of Navigation, Italy
Astronomy-Astrophysics Celestial-mechanics
Response = 1
We focus our attention on the consequences of possible changes in the definition of UTC from the navigation point of view. Currently UTC timescale is constrained to Earth rotation, by the introduction of leap seconds such that the difference between UTC and UT1is maintained within 1 second. GPS is currently the most common navigation system and its timescale is related to UTC; GPS time and UTC differ for an integer number of seconds (the leap seconds accumulated since the GPS turn on) and the difference between GPS and UTC(USNO) (the UTC maintained by US Naval Observatory) is continuously sent to users. A change in the UTC definition, omitting the leap seconds correction, would not affect directly the navigation performance with GPS; this change would only affect the time reference of navigation, not more linked to GMT (whose UTC is an approximation).
The main problem related to the proposed change to UTC definition is that the output time form GPS is not related to legal timescale with consequences for all the application based on GPS time dissemination. Similar problems are present also in the other satellite navigation systems as GLONASS and Galileo.
For these reasons the change to UTC is not recommended.

07 October 2011, 22h26; Lefebvre, Christine, None, France
Individual
Response = 1
Personally, I am satisfied with the current situation. I would be very disappointed to see the Earth's rotation totally absent from the usual time definition, after having been used by humanity for thousands of years to measure time. If it is decided to redefine UTC as a constant offset from TAI, it should not be called UTC anymore:
- it will not be UT because not related to the Earth's rotation
- it will not be "Coordinated", because there will not be any coordination anymore between an atomic time and the Earth's rotation.

Possible Changes to Co-ordinated Universal Time (UTC)—RIN's Response, 30 September 2011

1. UTC is a man-made, atomic timescale that is constrained to approximate the Earth's rotation to within one second by the inclusion of periodic adjustments known as "leap seconds."
2. The technical reason for the proposed change, eliminating these adjustments, appear esoteric and the benefits extremely limited—bearing in mind that a main means of dissemination of time is GPS.
3. UTC (approximating to GMT) is the legal timescale in many countries and so a change of definition is likely to require
   a. a legal impact assessment to understand which laws need to be modified, and
   b. a technical impact assessment to understand how the change will affect existing systems.
4. In the UK UTC is disseminated using many different mechanisms including: satellite navigation, (*e.g.*, GPS System Time automatically corrected to UTC), low frequency radio (*e.g.*, 60 kHz MSF), and the Internet (NTP servers), hence there is a need for a technical assessment of the proposed change.
5. We note the role of timing in distributed systems (including transport, finance, communications, energy), many of which are safety or mission critical and impact on the critical national infrastructure.
6. We note the existing GPS vulnerability concerns and the almost impossible task of understanding fully the impact of a loss of GPS-based timing because many users (*e.g.*, defence, transport, finance communications, energy) are simply unaware that they are using GPS timing in their systems.
7. We note that a UTC impact assessment will be far more complex given the different dissemination techniques.
8. The technical impact is likely to require a lot of
   a. significant public relations activity (e.g., similar to that of the "Millennium Bug")
   b. time, and
   c. systems engineering activity.
9. One top of this there may be a need to upgrade or replace existing subsystems or components. The cost could be very significant for a single country, let alone globally, at a time when many national economies are still in recession or at best fragile.
10. In summary making this change to UTC has a rather esoteric rationale, limited benefits and potentially significant costs. Many governments will require a formal business case comparing this change scenario with a 'do-nothing' scenario and the change scenario is likely to fail at this point.
11. For these reasons the imperative for change is not compelling to the Royal Institute of Navigation.

DISCUSSION CONCLUDING AAS 11-668

Near the conclusion of Daniel Gambis' presentation, Dennis McCarthy interjected that this was not an IERS questionnaire. McCarthy said this was Gambis' questionnaire; the IERS "had not backed" or sanctioned the questionnaire. Gambis clarified that the survey answers were not his, but came from users.

Tyson said that McCarthy's tone "implied that they wouldn't have backed it," to which McCarthy replied "that's correct." George Kaplan asked who McCarthy meant by "they." McCarthy replied "they" meant "the President [of the Directing Board] and head of the Central Bureau of the IERS," adding that "there was no consultation with the IERS before this questionnaire was sent out." Tyson noted that lack of consultation does not mean that the IERS wouldn't "back" the questionnaire. McCarthy added that they would not have accepted it having read the questionnaire.

Kaplan asked McCarthy, "isn't this the largest response anybody has ever gotten on this question? Over 400 is not a trivial response." McCarthy was not sure. Seaman asked why "they" would not have backed the phrasing of the questionnaire. McCarthy replied, "If you looked at the questionnaire you can see what the answer is supposed to be." Seidelmann said one might make the same claim against the questionnaires sent out by the ITU-R.[1,2] Seaman said that Gambis' questionnaire language looked rather balanced, and wondered if McCarthy was saying that "they" would have backed a questionnaire worded the way "they" wanted. McCarthy replied, "This questionnaire does not supply information."

The discussion was halted so that Gambis could conclude his presentation. Gambis offered that, as the head of the Earth Orientation Center, he seemed best positioned within the IERS to conduct such a survey.[*] Additionally, the questionnaire records the direct responses of UTC users; the results therefore speak for themselves and are not supposed to represent the opinions of the IERS Directing Board. For this reason, an earlier 2002 survey was not coordinated with the IERS Directing Board either. John Seago asked Gambis if anyone from within the IERS had objected to the latest survey being conducted. Daniel Gambis said that the issue put on him was that it should be clear that the survey was conducted by Earth Orientation Center of the IERS and that the decision to conduct a survey was not made by the IERS, because the position of some members of the IERS Directing Board were completely different.

Seago said that it was unclear just how many surveys had been conducted on this topic, but a smattering of survey activity took place about a decade ago among various scientific groups. He said it was interesting that the ITU-R Special Rapporteur Group (SRG) was dismissive of that survey activity which suggested that UTC users were overwhelmingly satisfied with the *status quo*, instead reporting that these surveys "did not provide any clear resolution." Understanding

---

[*] *Editors' Note:* Daniel Gambis was the previous Director of the IERS Central Bureau and is a member of the IERS Directing Board.

that the SRG or any other ITU-R group could have conducted a more satisfactory survey at any time, Seago wondered why there had been no polling of end users by the ITU-R after so many years. Seago also wondered why simply *attempting* to gather feedback on end-user preferences now seemed to be a controversial activity. Gambis explained that a survey is not only a way to collect opinions, but it is a way to inform people and create awareness. Many people outside the IERS seemed to especially lack awareness of this issue. Gambis therefore thought that perhaps more time is needed before a decision is made. He is involved with a working group within the United Nations Educational, Scientific and Cultural Organization (UNESCO). UNESCO has many responsibilities and it would seem that they may have a stake in the issue of global time-keeping as well. Seidelmann added that the ITU-R Study Question explicitly asked for a determination of user requirements, but apparently that determination was never done.[3]

Seaman was surprised by the number of survey responses received (over 400), especially in relationship relative to the size of *Bulletin C* distribution list (about 1600). Gambis said that they were still receiving occasional responses to the survey, and mentioned that the day before he received very detailed correspondence from the English Royal Institute of Navigation (RIN). Seaman asked if there was any way to determine if a survey response came from a recipient of IERS products, to which Gambis said no. Frank Reed added that he knew ten people who responded to the survey who were not IERS product subscribers.

Steve Malys said that every survey he had seen on this issue—including an internal survey conducted within the US Department of Defense—had been characterized as being "informal"; that is, conducted by working level people and scientists who use the information. Malys was not aware of any surveys directed at decision makers who would be procuring the money necessary to affect changes to operational systems. Malys said answers will be different from working-level people versus those who control money and have to determine how much a proposed change will cost. Organizationally coordinated responses were not being requested, and different government agencies worldwide would likely to give different answers based on how these agencies use time and what the changes would cost. The informal nature of past surveys suggested that they are not being coordinated through any recognizable chain of command, and nobody seemed to be exploring questions related to cost.

McCarthy said that polling can be tricky and survey questions have to be carefully worded to elicit the desired information. He said that "we have done I don't know how many surveys, formal or informal—I would call them all informal—and it is very difficult to get responses." He said if the survey is "formal", then it has to go up various chains of command and "it ends up being more trouble than it's worth to even respond, so we don't get responses." Malys replied that if national decisions are to be made, then we want those decisions to be well coordinated. McCarthy replied that "the business of making surveys is not easy and you have to be careful when you do that. That's the problem. That's the problem with this [IERS Earth Orientation Center] survey and that's the problem with every survey." McCarthy said that perhaps the best survey he ever saw on this issue was the one commissioned through the American Astronomical Society.

Wolfgang Dick asked how the 2011 IERS Earth Orientation Center survey compared with the 2002 survey. Gambis said the number of people aware of this issue in 2002 was more limited. The number of responses in 2011 was greater and the percentage of responses from outside the IERS was greater. Seago noted that in the earlier survey 88% favored the status quo, whereas in the later survey 75% favored the status quo. McCarthy speculated that if the question were rephrased to ask "Were you satisfied if a change were made?" then such a phrasing would have also received the most favorable response. Gambis clarified that survey responders included comments, and the wording of the question does not change the detail of the comments. McCarthy responded that a survey will get comments from both sides, but he learned that within the astronomical community it makes a difference regarding how the question is asked. Seidelmann thought that the difference with the astronomical community was that only a small percentage actually understood the difference between UT1 and UTC, but IERS product users should have a more inherent understanding of what the underlying issues are—why else would they be responding to an IERS survey? McCarthy thought we might not want to make that assumption. Seidelmann thought it should safe to assume that the IERS survey was targeting people who know what UT1 is. Reed said that the people responding to the survey didn't have to be involved with the IERS, adding that he knew people who responded to the survey who didn't know what UT1 was.

REFERENCES

[1] Timofeev, V. (2010), "Questionnaire on a draft revision of Recommendation ITU-R TF.460-6, Standard-frequency and time-signal emissions." ITU-R Administrative Circular CACE/516, July 28, 2010.

[2] Racey, F. (2011), "Questionnaire on a draft revision of Recommendation ITU-R TF.460-6, Standard-frequency and time-signal emissions." ITU-R Administrative Circular CACE/539, May 27, 2011.

[3] Jones, R.W. (2001), Question ITU-R 236/7, "The Future of the UTC Time Scale." Annex I of ITU-R Administrative Circular CACE/212, March 7, 2001.

Session 4:
TIME SCALE APPLICATIONS

AAS 11-669

TRADITIONAL CELESTIAL NAVIGATION AND UTC

Frank E. Reed[*]

Traditional celestial navigation depends on accurate knowledge of some standard mean time. While transitioning to a time gradually decoupled from the Earth's rotation should present no serious problems for users who actively practice celestial navigation at sea, there are issues concerning possible confusion in the literature, textbooks, and other educational materials already published. Celestial navigation, though a backup of last resort, is still widely taught at maritime academies worldwide and in less formal classes. If DUT continues to be published and disseminated and identified as a simple "watch error," the continuity of textbooks and navigational practice will be maintained.

INTRODUCTION

Traditional celestial navigation is the science or art, or perhaps more properly the *craft*, of finding the position of a vessel at sea using handheld angle-measuring instruments to observe the positions of the Sun, Moon, and brighter planets and stars relative to the observer's horizon. Until about 1927, this subject was known as "nautical astronomy". Following the extension of the methods to air navigation, the less maritime and arguably more poetic name "celestial navigation" has become the preferred name for the subject.

This paper will address the impact of time standards and accurate time on "traditional celestial navigation" employing handheld sextants and non-electronic tools which reached a state of perfection and standardization circa 1960. It has declined drastically in importance in the past twenty-five years, and the overwhelming majority of marine navigation is now done by GPS. Note that automated electronic systems which use astronomical observations, usually in conjunction with inertial guidance systems (including classified military systems), can achieve much higher accuracies and depend more critically on exact time, but these are not the topic of this paper. Also note that astronomical surveying using theodolites and similar instruments achieved a level of accuracy an order of magnitude higher than traditional celestial navigation. While there are still a few practitioners of astronomical surveying, it is generally considered obsolete.

FUNDAMENTAL PRINCIPLE

The fundamental principle of traditional celestial navigation rests on the simplest connection between terrestrial coordinates and the celestial sphere. Any star or other celestial body at some given instant of time marks a specific location on the Earth's surface where that celestial body would be found in the observer's zenith. If we are fortunate enough to observe a bright star or planet, *e.g.* Jupiter, exactly at the zenith, and if we have ephemeris data giving us access to the

---

[*] ReedNavigation.com, Navigation Instructor, Treworgy Planetarium, Mystic Seaport Museum, Mystic, Connecticut 06355, U.S.A.

object's Declination and GHA (Greenwich Hour Angle = the longitude, always measured west, where the celestial object is in the zenith at some instant of time) then we have determined our position without further calculation. The topocentric Declination is equal to the observer's latitude. The topocentric GHA is equal to the observer's longitude (Figure 1).

118			1994 JUNE 12, 13, 14 (SUN., MON., T				
UT (GMT)	ARIES	VENUS −4.0	MARS +1.2		JUPITER −2.3		SATURN
d h	G.H.A. ° ′	G.H.A. Dec. ° ′ ° ′	G.H.A. ° ′	Dec. ° ′	G.H.A. ° ′	Dec. ° ′	G.H.A. D ° ′ °
12 00	260 05.6	141 35.3 N22 51.0	218 12.1	N15 28.5	46 34.8	S12 06.4	275 44.8 S 8
01	275 08.0	156 34.6 50.4	233 12.7	29.0	61 37.4	06.3	290 47.2
02	290 10.5	171 33.9 49.9	248 13.4	29.6	76 40.0	06.3	305 49.7
03	305 13.0	186 33.2 ·· 49.3	263 14.0	·· 30.2	91 42.6 ··	06.3	320 52.1 ··
04	320 15.4	201 32.6 48.8	278 14.7	30.7	106 45.2	06.2	335 54.5
05	335 17.9	216 31.9 48.2	293 15.4	31.3	121 47.8	06.2	350 56.9

Figure 1. GHA and Declination. Note also that UT and GMT were formerly treated as synonyms.

In practice, for the visual observer using a hand-held instrument, like the traditional marine sextant, stars are never observed exactly in the zenith, and instead zenith distances are acquired by measuring altitudes from the sea horizon. Observed altitudes are corrected for dip and refraction (and a few other corrections specific to the object in question) and then these corrected altitudes may be subtracted from 90 degrees to yield zenith distances. The zenith distance is the distance of the observer away from the tabulated location where the object is in the zenith. So an altitude measurement places an observer on a "circle of position". At all points along the circle, the observed star would have the same zenith distance. A single altitude observation places the observer at some location along that circle of position. If we observe a pair of stars, then the locations where the circles cross yield the observer's position fix. There is a minor ambiguity since two circles cross in two points, but these are usually separated by thousands of miles so this is not a practical problem; the navigator is never that lost (Figure 2). Note that the radius of the circle expressed in nautical miles is exactly equal to the corrected zenith distance expressed in minutes of arc. Thus, an error of one minute of arc in a measured altitude yields an error in position of the edge of the corresponding circle of position equal to one nautical mile.

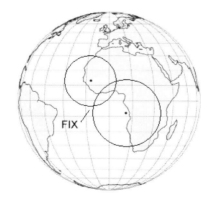

Figure 2. Two circles of position yield a fix.

The idea of the circle of position took hold relatively slowly in the history of celestial navigation. Although we can see that it is fundamental today and though mathematicians understood it from the mid-nineteenth century, practical navigators continued to picture latitude and longitude sights as distinct, separate operations. By contrast, the concept of the circle of position applies to all altitude measurements (Figure 3). Practical navigators began to think in terms of circles of position beginning at the end of the nineteenth century thanks in large part to one of the most influential books in navigation education, S. T. S. Lecky's "Wrinkles in Practical Navigation".[1] According to a reviewer of the third edition, "What Shakespeare is to the player, so is Lecky to the navigator."[2] The emphasis on circles of position was not entirely due to Lecky himself. There was a genius behind the curtain. Lecky was in close correspondence with an accomplished mariner and yachtsman who emphatically believed in the modernization of British navigation methods: the great physicist William Thomson, Lord Kelvin. Kelvin encouraged Lecky to use the language of circles of position, already standard among mathematically-inclined navigators, and to abandon the older explicit separation between latitude and longitude sights. Lecky, thanks to his light-hearted prose, made it palatable to working navigators.

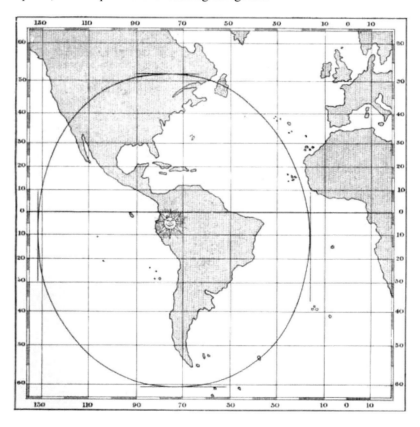

Figure 3. A circle of position from Lecky's "Wrinkles" (1885).

Clearly traditional celestial navigation explicitly depends for its operation on the connection between terrestrial coordinates and therefore Earth orientation and the coordinates of the celestial sphere. As the Earth rotates, the locations where the stars are directly overhead—the centers of those circles of position—travel west at a rate of one minute of longitude in four seconds of time. At the equator, this is equivalent to one nautical mile in four seconds:

$$4 \text{ seconds of time} \rightarrow 1 \text{ nautical mile} \tag{1}$$

ACCURACY AND ERRORS IN TRADITIONAL CELESTIAL NAVIGATION

Traditional celestial navigation never evolved into a high accuracy method. One nautical mile of accuracy has been considered a reasonable expectation of accuracy, and two or three nautical miles of uncertainty are not unusual or problematic. The accuracy of the method is limited by several factors, primarily terrestrial refraction and instrument errors.

The great majority of celestial navigation observations require the sea horizon as a reference. The sea horizon is typically some minutes of arc below the true horizon. The true horizon is the great circle 90 degrees distant from the observer's zenith while the sea horizon, where the sky meets the sea visually, is depressed below the true horizon by an amount which, in minutes of arc, is very nearly equal to the square root of the observer's height of eye in feet (a case where English units yield a simpler rule than metric units). While introductory textbooks often introduce this as a problem in pure geometry involving a simple plane triangle, this is really a physics problem. Rays of light travelling for some miles near the Earth's surface are deflected downward by refraction. Known as "terrestrial refraction", this has the effect of making the Earth seem larger in radius for all optical experiments performed near the Earth's surface. The terrestrial refraction is variable. It depends on the rate of change of the density of the atmosphere with height in the lowest level of the atmosphere and this in turn depends almost entirely on the temperature *lapse rate* --the rate at which the temperature in the lower atmosphere falls with altitude. The lapse rate varies with the weather. Therefore the depression of the visible horizon, known to navigators as the "dip" of the horizon, cannot be known in advance. This is the single greatest limiting factor in the methodology of traditional celestial navigation. Altitudes can be measured to an apparent precision of 0.1' of arc using a sextant, but the accuracy is limited to about +/-0.5' depending on the weather. This variability is itself variable. Close to shore, where the navigator is most concerned about an accurate position, the variability of dip can be even greater. Dip anomalies of several minutes of arc are not uncommon close to land. These uncertainties in dip translate directly to uncertainties in position. They are unavoidable.

While the modern metal sextant can be read to a precision of 0.1' of arc, the practical accuracy is frequently lower. The instrument has to be properly adjusted. Mirrors should be set perpendicular to the frame, telescopes parallel to the frame. There are simple, well-known methods to make these adjustments. Many sextants, especially those dusted off in an emergency, will not be properly adjusted, but we will assume for now that they are properly adjusted.

One adjustment always required of any sextant is the "zeroing error" known in celestial navigation as "index error". The sextant combines light from a direct and a reflected pathway allowing the observer to look in two directions at once. This reflecting principle makes the instrument immune to pitching motions, and is responsible for its remarkably high accuracy for a handheld instrument, despite the small errors under discussion here. When the two pathways, direct and reflected are pointed at the same object (distant enough to avoid parallax effects from the distance between the two mirrors) and the direct and reflected views coincide, the instrument should read zero. If it does not, the difference is applied as an "index correction" to all sights. Typical methods for estimating this index correction are accurate to approximately +/-0.25' based on group experiments that I have conducted with experienced students in navigation classes. There are methods for reducing the error in the index correction almost to zero, but they are not widely used.

The modern sextant also suffers from arc error. After the zeroing error or index error has been accounted for, observations may still show small variable errors along the arc. For example, all observations at 30 degrees may show a bias of +0.7' while all observations at 60 degrees may show a bias of -0.4 '. Formerly, these arc errors were tested on optical benches and recorded on certificates kept with each sextant affixed in its case, but today they are usually considered insignificant for practical use. Sextant certificates for some forty years have been ritually marked with strings of zeros and labeled "free of error for practical navigation." There is no meaningful statistical data on the extent of arc error. Unquestionably, it depends on the manufacturer. Anecdotal evidence and tests I have performed suggest that most modern metal sextants have arc errors less than 1.5' for most points along the arc. If measured and tabulated, arc error is a correctable error, not unlike the standard index correction.

Errors in dip and the instrument errors described above together guarantee that the standard deviation error in celestial navigation sights is around 1' leading to errors in the plotting of circles of position of 1 nautical mile. Most practitioners of traditional celestial navigation are happy with errors two or three times larger than this. In the middle of the ocean, such errors are inconsequential. Close to land, the navigator can see that far (when the weather is good, which is, in any case, a pre-requisite for traditional celestial navigation). Practical safety and economy are not compromised. In addition, the "system error" of celestial navigation might be considered much larger since the astronomical methods are only available when the sky is relatively clear and the horizon is distinctly visible. At all other times, the vessel's position is traditionally determined by dead reckoning. Dead reckoning or *DR* is simple, manual, two-dimensional integration of course and speed between fixes (*N.B.*: while the etymology is uncertain, this is not derived from "deduced reckoning" which has become a popular "folk etymology"). Given that skies may be cloudy for days, the combined method of celestial navigation + dead reckoning has total average "system error" at least several times larger than the purely celestial portion of traditional navigation.

HISTORY OF TIME STANDARDS IN CELESTIAL NAVIGATION

Given the typical errors expected from traditional celestial navigation, errors in UT of one second or even somewhat more do not significantly affect the position fix. A one second error adds an error of 0.25' to the longitude or 0.25 nautical miles for points near the equator. Keeping or carrying time to this level of accuracy once it has been determined by an Internet time check or GPS of shortwave radio time source is a trivial matter today. Common watches can maintain accuracy for months without special care. In this sense, there is no such thing as a "marine chronometer" in the modern world. The care and feeding of the timepiece is a matter of purely historical interest. Cumulative error in a modern watch, once selected, can be ignored. The only error of concern to the modern navigator is drastic watch failure: a dead battery, saltwater in the case, etc. And here the modern navigator is advised to follow the traditional practice dating back to the early nineteenth century: bring two. With two good watches, the navigator can at least detect if one watch is in trouble. And of course, with three or more watches, a bad watch can be ignored or multiple watches can be averaged.

If leap seconds are dropped from UT, that time standard or its descendant under a different name, call it UT*X*, will gradually drift from the alignment required by celestial navigation. For a navigator, this could be counted as a simple "watch error". Before addressing this 21st century issue further, we can consider some of the changes in timekeeping standards that have impacted navigators in the past two hundred years.

Traditional celestial navigators once mastered a daily computation known as "calculating the true time." Today we recognize this as the calculation of Local Apparent Time or LAT. That it was known even in the middle of the nineteenth century as the *true time* reflects the long-standing importance of apparent solar time. This calculation, by the way, determined LAT from the altitude of the Sun measured with a sextant. In effect, it turned a sextant into a highly accurate sundial.

The ancestor of both the modern *Nautical Almanac* (NA) and the *Astronomical Almanac* (AA) was the *Nautical Almanac and Astronomical Ephemeris* (NA&AE) which was rushed into publication starting in 1767 to introduce and facilitate the new *lunar distance* method of determining longitude which had recently become practical thanks to the lunar ephemeris tables of Tobias Mayer and the indefatigable enthusiasm of Nevil Maskelyne. The NA&AE, remarkably, listed all ephemeris data, and especially the lunar distance tables, in terms of Greenwich Apparent Time, GAT (Figure 4). A navigator could directly compare the Greenwich Time derived from lunar observations with the LAT or "true time" observed in his longitude. The difference between the two was the longitude in units of time, and no correction for the Equation of Time was necessary. The tables did not switch to Greenwich Mean Time, the time naturally suited to marine time-keepers or chronometers, until 1834, though this was considered embarrassingly late in the era. The dominance of time by machine gradually led to the Sun being counted as *early* or *late* (as tabulated in the Equation of Time) rather than clocks being considered *fast* or *slow*.

[24] FEBRUARY 1767.

Diſtances of ☽'s Center from ☉, and from Stars weſt of her.

Days.	Stars Names.	12 Hours. ° ′ ″	15 Hours. ° ′ ″	18 Hours. ° ′ ″	21 Hours. ° ′ ″
1	The Sun.	40. 10. 52	41. 54. 11	43. 37. 9	45. 19. 46
2		53. 47. 13	55. 27. 32	57. 7. 27	58. 46. 58
3		66. 58. 20	68. 35. 24	70. 12. 3	71. 48. 18
4		79. 43. 17	81. 17. 4	82. 50 27	84. 23. 28
5		92. 2. 59	93. 33. 49	95. 4 18	96. 34. 28
6		104. 0. 15	105. 28. 28	106. 56. 24	108. 24. 4
7		115. 38. 27	117. 4. 35	118. 30. 29	119. 56. 7
6	α Pegaſi.	71. 18. 35	72. 50. 18	74. 21. 50	75. 53. 9

Figure 4. All ephemeris data in the *Nautical Almanac and Astronomical Ephemeris* were tabulated by Greenwich Apparent Time from 1767-1833.

Navigators through the middle of the nineteenth century frequently dealt with another peculiarity of time-keeping known as the "nautical day" or the "sea day". At sea, the work day and the logbook day began at noon, and that was when officers and crew adjusted the day and date. Monday, June 7 turned to Tuesday, June 8 at noon as observed from the deck of the vessel. Upon arriving in any port for any length of time, the vessel's time-keeping would be switched to the civil standard leading to the curiosity of days of "twelve hours" and days of "thirty-six hours" duration, and these were sometimes recorded in logbooks.[3] Note that this day beginning at noon is very similar to the old concept of the "astronomical day" but one calendar day offset. The nautical day overlapped the corresponding civil day in the morning hours while the astronomical day overlapped in the afternoon and evening hours. The "nautical day" was eliminated by decree in the British Royal Navy in 1805, but it continued in common use on commercial vessels until it faded out after 1850.

The concept of the nautical day created some curious issues in daily life at sea. Respecting the Sabbath among crews which, in American and British shipping, were almost universally Christian until the modern era, was a source of enduring confusion. In general, the captain or master of the vessel had the last word on declaring the date. Similarly in the Pacific, when the date was changed from western hemisphere reckoning to eastern hemisphere reckoning (known in modern euphemism as "crossing the date line," though this is a recent notion), ships' crews had to decide when to observe the Sabbath if Sunday was doubled in the account of days or, worse yet, dropped from the calendar. Here, too, the captain had the last word. Nathaniel Bowditch of Salem, Massachusetts, famous as the author/editor of the *New American Practical Navigator* was one of the first Americans to face this issue when his ship sailed into Manila in 1796. At that time, the Philippines and other Spanish outposts in the western Pacific maintained the date consistent with the western hemisphere in order to remain synchronized with Mexico. Travelling via the Atlantic and Indian Oceans, Bowditch arrived with the western hemisphere accounting of days. As good Yankee Protestants, Bowditch and the other New England traders made do. They kept the date on-board unchanged (since they would be sailing back to Massachusetts the way they came rather than circumnavigating). And so they worked on "our Sunday", as he put it, in order to make the most of their short time in Manila.[4] I would add with respect to the idea of dropping leap seconds from civil time-keeping that if word gets out that civil time will be allowed to drift away from Earth rotation and might, in a few thousand years, effectively turn Saturday into Sunday, there will be confusion at best, and potentially outrage, among the faithful around the world.

The introduction of standard time zones in the late 19th century (in 1883 in the US) had little significance for celestial navigation since navigators already dealt with these issues (time zones do, however, cause lingering confusion for modern navigation students). Following the introduction of time zones, jurisdictions have generally tended to shift their time-keeping east. In the US, Ohio, Michigan, most of Indiana, most of Georgia, were not originally on Eastern Time but have legislated themselves east for economic and cultural reasons. With Daylight Saving Time now in effect for 238 days out of the year in the US (since 2007) equal to 65% of each year, the Eastern US and much of the Midwest now maintain a civil time which would correspond to actual solar Mean Time far out in the Atlantic, east of Bermuda, at 60 degrees West longitude. We are already culturally disconnected from the time implied by the Earth's orientation to an extraordinary extent.

In the early twentieth century, the "big bang" in almanac time-keeping occurred on January 1, 1925 when the astronomical day was abandoned and the almanacs adopted the civil standard of initiating each calendar day at midnight instead of noon. Anecdotal evidence suggests that this may have caused some brief confusion for navigators at sea, uncertain whether to take, e.g., the Declination of the Moon, from the column for 0 hours or 12 hours. I have found no direct evidence of such confusion in actual primary-source materials such as logbooks.

The official almanacs used by most celestial navigators in the first half of the twentieth century were the British *Abridged Nautical Almanac* (AbNA) and the *American Nautical Almanac* (AmNA). Both of these had been spun off, originally as extracts, from the more complete and progressively more astronomically-oriented *Nautical Almanac & Astronomical Ephemeris* and the *American Ephemeris and Nautical Almanac*. By mid-century, the publication known widely as the *Nautical Almanac* (NA&AE) was not nautical except in name. Economy and international good-will suggested the unification of the AbNA and the AmNA and eventually a renaming of these as the international *Nautical Almanac*. This unification of content occurred in 1958 close to the plateau of the standardization of modern celestial navigation. This publication has been remarkably stable in its run. In terms of time-keeping the time standard for the Nautical Almanac changed from GMT to UT *c*.1980, but navigators take little notice of this. Indeed, it is normal among celestial navigators to use GMT and UT interchangeably. The difference has no practical significance, and the world's best navigation schools continue to use the terms interchangeably.

Will the Nautical Almanac exist in another fifty years? As I noted in my presentation on the history of the Nautical Almanac at Mystic Seaport in 2008, I would bet against it. But we can, with some confidence, assert that "nautical almanacs" published by various sources will still exist, even if only to serve a dwindling community of enthusiasts. There are a number of online sources for nautical almanac equivalent data, including my own. Geoffrey Kolbe, of Scotland, publishes a fifty-year *Long-Term Almanac*.[5] As he noted recently on the "NavList" message boards, "In the second edition of the LTA, it is specifically noted that the time system to which the ephemerides are referenced is UT1, and that a decision may be taken sometime during the period of validity of the tables to cease inserting leap seconds into UTC to keep it aligned with UT1. So, if DUT continues to be published as it currently is on the Internet, there should be no problem."[*] Unlike the official *Nautical Almanac*, his almanac already addresses this issue.

NAVIGATION EDUCATION AND SOLUTIONS

As noted, traditional celestial navigation reached a state of near perfection and high standardization some fifty years ago. It is stable, almost ritualistic. Small, even insignificant changes in authoritative data can cause confusion and lead to cynicism if not handled properly. In 2004, the refraction tables in the *Nautical Almanac* were recomputed. The changes are minor and unnecessary (Figure 5). For practical computation, they present no problem, but for pedagogic purposes, they created the impression that some published examples and sample problems were slightly incorrect.

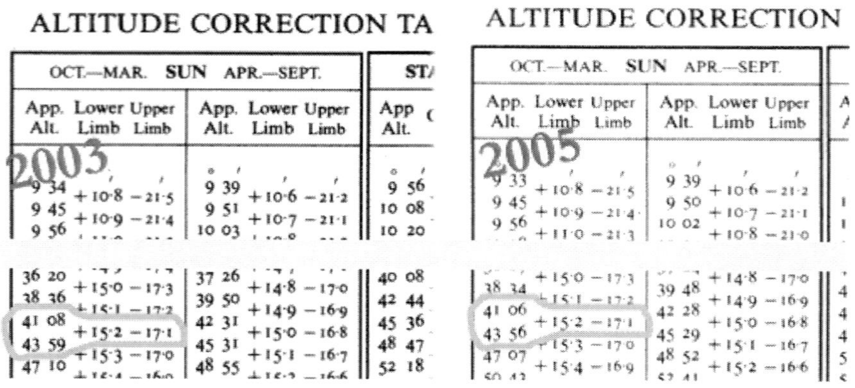

Figure 5. Small changes can lead to confusion: a minor change in the refraction tables in 2005.

---

[*] http://fer3.com/arc/m2.aspx?y=201108&i=116958

Navigators who practice celestial navigation tend to be highly independent to the point of distrusting authority, despite their dependence on the published tables. Small changes cause consternation. Celestial navigators tend to have strong aesthetic attitudes with respect to maintaining the link between mean solar time and civil time. More importantly, very few navigators remain in practice. This is the conundrum of the ultimate backup. GPS and other electronic systems work with great reliability, but eventually a navigator may need to fall back on traditional tools. If the navigator only vaguely remembers the methodology, any changes in the rules, such as a change in the meaning of the tabulated time, could lead to crippling confusion in a crisis situation. In the event that a time standard replacing UTC gradually drifts from Earth orientation, after fifty years the accumulated drift might amount to some forty seconds, equivalent to a ten nautical mile error in a celestial navigation fix near the equator. This problem can be avoided by the formulation and publication of extremely concise and clear rules carefully cleared of pedantic terminology irrelevant to the *craft* of celestial navigation. Celestial navigators are smart; they will figure it out. But they are not scientists, by and large, and any solution must respect the nature of rapid, non-technical education.

I have recently surveyed some of the most prominent textbooks, almanacs, and other resources used in traditional celestial navigation including recent editions of the *American Practical Navigator* (1962, 1995, US National Geospatial-Intelligence Agency, formerly authored by US Navy Hydrographic Office),[6] *Celestial Navigation in the GPS Age* by John Karl (2006),[7] the *Long-Term Almanac* by Geoffrey Kolbe (2000)[5], *Practical Celestial Navigation* by Susan P. Howell (1981),[8] and the *Primer of Navigation* by Mixter (1943)[9] seeking out exceptional cases or other problems that might arise if "leap seconds" are dropped. I am convinced that the most economical solution for traditional celestial navigation is the continued publication of DUT, identified as a cumulative "watch error", and inserted on an annual basis as a prominent separate page with brief instructions in the official *Nautical Almanac* and similarly in any un-official nautical almanacs. Such instructions might read, for example, "For the year 2025, subtract 8.0 seconds from broadcast time before entering these tables". Ephemerides for navigation would continue to be calculated on the basis of UT (UT1). Navigators would acquire the leap-second-free civil time, which I am calling for the purpose of this paper UTX, from the Internet, GPS receivers, shortwave time signals, etc. and then apply this simple correction, equivalent to a traditional "watch error", before entering the tables. All textbook problems and examples would remain valid even in publications that are decades old.

Navigators and navigation enthusiasts that I have emailed and interviewed (all informally) appear convinced that there should always exist a non-Internet, non-digital, plain language, in-the-clear announcement of the time offset from the established international civil time standard if leap seconds are dropped. The system of doubled ticks presently used by WWV is neither obvious nor necessary, and few celestial navigators are even aware of the significance of the doubled ticks. Announce it in plain English. For example, WWV in 2025 might announce, "UT1 for navigation today differs from UTX by 8.3 seconds." An announcement once an hour, similar to the tropical weather updates, should be more than sufficient for the needs of celestial navigators. The watch error or DUT need not be more accurate than the nearest second though there could be some small benefit in terms of confidence-building by providing this offset to the nearest tenth of a second.

CONCLUSION

The impact of dropping leap seconds on traditional celestial navigation would be small and manageable. The expected accuracy of celestial navigation, about one nautical mile under good conditions, is low enough that an error of a second or two would not be counted as significant.

The greater risk arises from the potential for confusion in a subject that is, at best, a rare backup. A single, concise, clearly-labeled page published in the annual Nautical Almanac and a simple, plain-language, hourly shortwave radio announcement will satisfy the needs of celestial navigation. The tools and the textbooks will continue to work. Beyond that, it is merely a matter of education.

ACKNOWLEDGMENTS

Many thanks to the members of the *NavList* navigation community who offered their opinions and advice. *NavList* discussions related to all aspects of traditional celestial navigation including issues of time standards, UTC, and leap seconds, spanning over fifteen years, are archived online at *http://www.fer3.com/arc*.

REFERENCES

[1] S.T.S. Lecky, *Wrinkles in Practical Navigation*. New York: John Wiley & Sons, 1885.

[2] *The Marine Review*, Vol. 37, No. 9. New York: Feb. 1908.

[3] Logbook, Bark Mary & Louisa, 1858-59 (Aug 25, 1859), Mystic Seaport Collections Research Center, Mystic, CT.

[4] Mary C. McHale, ed., *Early American-Philippine trade: the journal of Nathaniel Bowditch in Manila, 1796*. New Haven, Yale Univeristy, 1962.

[5] Geoffrey Kolbe, *Long-Term Almanac*. Pisces Press, 2000.

[6] US Navy Hydrographic Office, *American Practical Navigator*. 1962.

[7] John Karl, *Celestial Navigation in the GPS Age*. Paradise Cay Publications, 2007.

[8] Susan P. Howell, "Practical Celestial Navigation." Mystic Seaport Museum, 1970.

[9] George W. Mixter, *Prime of Navigation*. D. Van Nostrand, 1943.

DISCUSSION CONCLUDING AAS 11-669

Steve Allen asked how historically significant changes to almanacs were perceived by practicing navigators, such as the changeover from apparent solar time to mean solar time in 1833, and the changeover from the astronomical day (beginning at noon) to the civil day (beginning at midnight) in 1925. Frank Reed responded that he had found no evidence in the primary-source logbooks available to him that these changes caused serious problems, but secondary source information suggested that there may have been issues.[*] Reed thought that the personal memoirs of D.H. Sadler (Superintendent of HMNAO from 1936 to 1971) expressed more anger with the Admiralty than was deserved, but admitted that is the nature of a personal memoir as the recollections of someone at the end of their career. But there was no known primary source documentation of anything bad happening with ships at sea on January 1, 1925, for example.

David Terrett asked if WWV still included "double-ticks" (the encoding of DUT1 in the audio time signal). Reed and others replied affirmatively. Rob Seaman noted that part of the proposed Recommendation TF.460-7 was to eliminate the requirement for DUT1. Reed said he did not think anyone used those anymore, and was not sure he knew anyone who was aware of them. Reed had suggested in his talk that a verbal announcement of DUT1 could instead be declared at some point within the hourly broadcast, but Allen wondered who might have the time or patience to listen to WWV for an hour. Reed clarified that he meant that a verbal announcement could be made at regularly scheduled minutes so that people would know when to tune in to get the information. Reed said that a page within future editions of the nautical almanacs could include a note on how to get UT1-UTC information. Seaman noted that a single page could get ripped out; Reed replied that much of the information within the almanacs is duplicated to a degree, so a page in front and a page in the back should provide desired redundancy.

Wolfgang Dick commented that sailors familiar to him all use GNSS, so it was unclear to him if anyone still used sextants. Reed clarified that he is discussing celestial navigation in the context of providing the "ultimate back-up". Reed said that if one somehow lost their GPS receiver overboard, one could make it back to port with a sextant in hand without the embarrassment and complications of having to be rescued by the Coast Guard. Although celestial navigation is not actively used by most navies today, all of the leading maritime academies still teach it extensively.

Reed suggested that the amount of navigation taught within the US Navy today is remarkably trivial because surface vessels tend to have both GPS navigation in combination with inertial navigation systems. Thus, the crews of large surface ships can no longer estimate their location by

---

[*] *Editors' Note*: Mr. Reed provided the following addendum to his remarks following the colloquium. He spent several hours examining two logbooks in the library at Mystic Seaport having hundreds of extensively worked lunars. Both logbooks had worked examples from January and February of 1834 where the navigators were initially confused because the lunars were providing longitudes that did not match the chronometer or the dead reckoning by a wide margin. Logbook notes added later explained that the discrepancy was due to the change from Greenwich apparent time to Greenwich mean time in the Nautical Almanac tables.

observing buoys and lighthouses, and they can literally drive a vessel upon a reef if both systems happen go out at the same time. Modern navigation systems are instead computer driven and the computers combine information from both GPS and inertial navigation. These systems are very advanced and precisely tell a helmsman how to drive a boat, but unfortunately operators know how to turn off the software that is protecting them if it seems to complain too much. Reed said this actually happened a few years ago when a US Navy cruiser was driven on to a reef off of Honolulu International Airport, with the city lights and the airport plainly visible. The navigation system was a mile off course because the crew had turned off the system that could have alerted them to the fact that there was no GPS fix and the inertial guidance system had not been re-calibrated for three days.

Reed said this is just an example of just how little celestial navigation is practiced now; today's navies don't really use it much at all anymore and in some respects the practice of celestial navigation is regarded as not much more than a hobby to some. But it still functions as the "ultimate back-up"—useful when all else fails. Tyson offered the caveat that celestial navigation only works as a back-up in fair weather.

Reed continued that celestial navigation still has its place, but mostly in automated systems. For example, Reed said that if GPS were somehow disabled by war, celestially navigated drones could operate at a very high altitude within a theater of operations, providing all-weather navigation signals to specialized receivers on the ground. Rob Seaman asked if this was an operational capability or a concept; Reed replied that to the extent of his knowledge it was still a concept. Seaman said that such capability, if operational, would seem to involve "a heck of a lot of software" that required precise knowledge of UT1. Reed agreed.

Kaplan clarified that the practice of celestial navigation within the US Navy is actually highly variable. While the practice has gone down drastically over time, Kaplan said that some captains still demand that their crews know how to do it, and conduct exercises by establishing a fix in the middle of the ocean with GPS receivers turned off. Yet, because of the high amount of automation within modern navies, there is much gadgetry that demands training. So there is a question as to how much time should be devoted to crew training for celestial navigation versus "the latest gizmo on the bridge." Kaplan therefore reiterated that celestial navigation has not disappeared completely from the Navy; a lot of quartermasters still practice it.

Kaplan continued by saying that if celestial navigation is not part of one's daily assignment, then one gets rusty, because it requires a certain level of practiced skill to navigate accurately. Reed said that issue leads to the other problem with changing the time standard; namely, if a navigator hasn't practiced celestial navigation in a long time and the time standard has been changed out from under him, it could be very confusing in a crisis. To avoid risks, word would need to get out about any such change. This is why Reed suggests that a bold notice should be put be a permanent part of future almanacs if a change occurs.

AAS 11-670

THE CONSEQUENCES OF DECOUPLING UTC ON SUNDIALS

Denis Savoie[*] and Daniel Gambis[†]

Severing the link between the rotation of the Earth and time-signal broadcasts will require a fourth correction to convert between solar time and standard legal time in the long term. For sundials, this represents an additional complexity, both for people using these instruments as educational tools, and for those who design them. Knowing the precision that can be expected of sundials, how long will it take for the decoupling of UTC to become noticeable?

INTRODUCTION

Though sundials may seem old-fashioned today in the time of the GPS, they are nevertheless part of our scientific and cultural inheritance. The first astronomical constants like the latitude, the ecliptic obliqueness or the inequality of the seasons were determined during Antiquity with a simple gnomon. The study of the gnomon's shadow trajectory represented a powerful tool in mathematical research on conics.[1] Considered as objects of prestige during the Greco-Roman period, their accuracy improved. The works of Vitruve (*De Architectura*, book IX) and especially of Ptolemy (Treaty of *De l'analemme*) greatly improved the mathematics of sundial layout. Under the influence of Arabo-Persian astronomers, sundials began to indicate the hours of prayers, while retaining their aesthetic qualities.

Let us highlight two important points: firstly, from antiquity to the end of the Middle Ages, the hour is the twelfth part of the interval of time between the rising of the Sun and its sunset.[2] The duration of this hour, called "temporary", varies according to the date and place; this hour is equal to 60 uniform minutes only at the equinoxes. The second point is related to the equation of time. Thanks to Ptolemy, one knows at least since the IInd century of our era that solar time doesn't flow uniformly because of the obliqueness of the ecliptic and the non-uniform speed of the Sun along the ecliptic. The equation of time was thus known to Greek astronomers and their astronomical observations (made in apparent solar time) were corrected to mean solar time.[3]

The arrival of the clock industry during the XIIIth century partly stimulated the development of sundials in the West. These sundials became references for the adjustment of clocks and pendulums until the XIXth century.

FORESEEABLE IMPACT ON ACCURACY

In Europe during the XVIth and XVIIth centuries, the first indoor meridians followed the reform of the Gregorian calendar for the determination of the tropical year length or the decrease of

---

[*] SYRTE, Observatoire de Paris, CNRS, UPMC, 61, avenue de l'Observatoire, 75014 Paris and Palais de la découverte, France.

[†] Earth Orientation Center of the IERS, Observatoire de Paris, 61 av. de l'observatoire, 75014 Paris, France.

the obliqueness of the ecliptic. It is estimated that the accuracy of a meridian such as the one Cassini built in 1680 at the Paris Observatory, is about a few seconds of time.

Two crucial factors diminish the accuracy of a sundial: firstly and most importantly there is the effect of the gnomon penumbra (or spot of light), and secondly there is refraction.[4] Let us discard the decrease of the ecliptic obliquity, which affects only the reading of the diurnal arcs indicating the date. In general, a meridian is much more accurate than a simple ornamental sundial for which an accuracy of about one minute of time is acceptable, even if many sundials made during the last decades are accurate to 10 or 20 seconds of time and even better.

The idea of incorporating the equation of time into the calculation of a meridian, which allows a direct reading of the local mean solar time, is due to the astronomer Grandjean de Fouchy in 1730.[5] Sundials with a figure of eight (analemma) became increasingly numerous during the XVIIIth and XIXth centuries. The acceptance of the international Greenwich meridian in 1884, the unification of the hour in each country at the end of the XIXth century, not to mention the advance of Universal Time (UT) for economic reasons, lead the meridian and sundial users to introduce two additional corrections: one in longitude and one due to advance of the legal time with respect to UT.

It is obvious that much of the time, the public is not aware of all these relations that link these different times scales, and that sundials are generally considered as old-fashioned and inaccurate; it even happens that standard sundials' gnomon are manually twisted by their owner in order to indicate the hour given by the clock.

Hence, since almost a century, to convert the solar time read on a sundial to legal time, one has been taking three corrections into account: the equation of time, the longitude of the place and the advance with respect to Universal Time. Of course, these corrections can be partly or entirely integrated on a sundial, in order to make a direct reading. What does the decoupling of UTC entail for a sundial? If one calls ΔE the difference between UTC-without the leap seconds and UT1, then in the long term this fourth correction that is ΔE is going to be necessary in order to always be able to compare the apparent solar time with the Sun:

$$\text{Legal time at the clock} = \text{apparent solar time} + \text{equation of time} + \text{longitude} + (1h \text{ or } 2h) + \Delta E \qquad (1)$$

Let us be realistic: except for specific sundials (either very accurate meridians or exceptional sundials), only an accuracy of about one minute of time is expected from these instruments. Some time will elapse before ΔE reaches this order of magnitude (approximately 140 years if only the secular deceleration of Earth rotation is taken into account).

However, let us consider some particular cases. The drawback of sundials indicating the mean solar time is that the equation of time slowly varies, causing an error which increases over centuries. Over a century, the maximum variation of the equation of time is a little more than 20 seconds, which is rather low (the main factor of variation of the equation of time is the displacement of the perigee).[6] Assuming that the only source of error is the variation of the equation of time, the error after a century on a sundial indicating the mean solar time will thus amount to 20 seconds in absolute value. This is perfectly acceptable for most sundials featuring an analemma. But obviously, the difference between UTC-without-leap-seconds and UT1 has to be added to this variation of the equation of time. Let us note that, for the past 40 years, 23 leap seconds were added for the difference between UT1 and UTC to remain less than 0.9 second of time. Thus, the drift due to the Earth's rotation is more important than the variation of the equation of time: the sum of the two can predictably reach 1 minute of time within much less than a century!

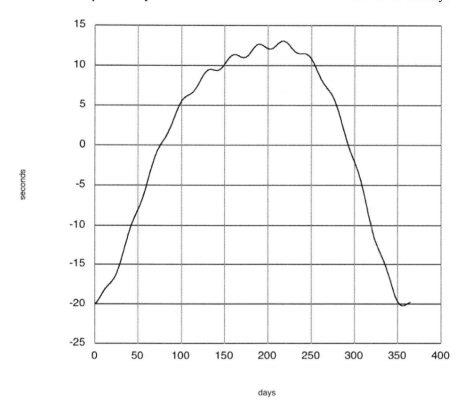

Figure 1. Variation of the Equation of Time between 2001 and 2101.

With regard to the accurate analemma of midday Universal Time, one will no longer be able to observe the perfect coincidence between the shadowing of this figure and the "beep" of the speaking clock: there will be an increasing offset over time. Let us underline that the ΔE correction is not foreseeable a long time in advance, contrary to the equation of time. Thus, it is going to be impossible for the sundials' calculation to take it into account: does it mean that sundials indicating the mean solar time with an analemma are to disappear? It depends on the desired accuracy.

Let us note that sundial manufacturers use the Sun to determine the astronomical orientation of walls and to trace meridians on the ground. It will be necessary to take the ΔE term into account since all these operations are based on Universal Time. Neglecting these few seconds would lead to annoying consequences: for instance, considering that the azimuth of the Sun varies most quickly close to the summer solstice, a 1-second error in the determination of the time of passage on the meridian line at 30° of latitude would lead to an error in azimuth greater than 0° 02 '.

CONCLUSION

From a pedagogical point of view, the decoupling from UTC can be seen as an additional complexity that is not easy to explain. Until now, sundials have been wonderful teaching objects, usable as early as elementary school to explain the seasons or the inequality of the days; and then later at college, Kepler's laws; without forgetting that they are also a useful practical application of trigonometry.[7] But experience shows that the most difficult point to understand, for the pupils or the general public, is the conversion from solar time to legal time, in particular the equation of time. So, if it becomes necessary to explain an additional correction that corresponds to the difference between a uniform scale of time and the solar time, sundials could be considered as complex objects indeed!

However, it seems unlikely that the decoupling of UTC will prevent the general public or astronomers from enjoying sundials, because their interest goes well beyond the problem of the conversion of time, which is known only by a very few. But sundials are almost the last link between the measure of time and the Sun, and it would be somewhat afflicting to note the ending of a link going back to Antiquity.

REFERENCES

[1] J. Evans, « Gnomonike Techne : The Dialer's Art and its Meanings for the Ancient World », *The New Astronomy : Opening the Electromagnetic Window and Expanding our View of Planet Earth,* ed. Springer, Printed in the Netherlands, 2005, p. 273-292.

[2] S. L. Gibbs, *Greek and Roman Sundials*, New Haven and London, 1976.

[3] O. Neugebauer, *A History of Ancient Mathematical Astronomy*, Springer-Verlag, Berlin-Heidelberg-New York, 1975, p. 61-68.

[4] D. Savoie, *La Gnomonique*, Les Belles Lettres, Paris, 2007, p. 481-490.

[5] D. Savoie, « L'aspect gnomonique de l'œuvre de Fouchy: la méridienne de temps moyen », *Revue d'Histoire des Sciences*, t. 61-1, janvier-juin 2008, p. 41-61.

[6] J. Meeus, D. Savoie, « L'équation du temps », *L'Astronomie*, février 2008, p. 16-22.

[7] D. Savoie, *Sundials, Design, Construction, and Use*, Springer, Berlin-Heidelberg-New York, 2009.

DISCUSSION CONCLUDING AAS 11-670

The presentation was made by Daniel Gambis on behalf of Denis Savoie, who could not attend the meeting in person.

David Simpson asked if anyone was familiar with a so-called *digital sundial*. This dial is comprised of two opaque sheets separated by a transparent spacer. A set of slits are cut into each one of the sheets, and are arranged such that if the Sun is at a certain angle, light will go through the correct slits to provide a digital readout of time. Simpson wondered what the consequences of a redefinition of UTC might be on such a dial: could one simply reorient the dial, or would the slits need to be remanufactured? John Seago thought the precision of this type of dial was perhaps 15 minutes. Simpson said that he thought the precision might be much finer, perhaps one to five minutes.[*]

Frank Reed said that sundials are really good indicators of *local apparent* time, noting that every useful public sundial will have a table next to it to account for the equation of time plus an additional correction for the static longitude offset. If the placard containing the value of longitude correction is replaced every few years then standard time can still be recovered from even the oldest sundial. Seago commented that the group would be visiting a public sundial on Friday designed to keep standard time *directly*, without any added tabular corrections for the equation of time or longitude.[1]

REFERENCES

[1] Seidelmann, P.K. (2011), "The Longwood Gardens Analemmatic Sundial." Paper AAS 11-682, this volume.

---

[*] *Editors' Note*: The actual precision of a digital sundial is ten minutes. URL http://www.digitalsundial.com

AAS 11-671

TIME SCALES IN ASTRONOMICAL AND NAVIGATIONAL ALMANACS

George H. Kaplan[*]

Commission 4 (Ephemerides) of the International Astronomical Union (IAU) includes astronomers from many countries responsible for the production of printed almanacs, software, and web services that provide basic data on the positions and motions of celestial objects, and the times of phenomena such as rise and set, eclipses, phases of the Moon, etc. This information is important for pointing telescopes, determining optimal times for observations, conducting night operations, and also for celestial navigation. Commission 4 also includes researchers involved in the more fundamental tasks of determining the orbits of solar system bodies based on a variety of observations taken from the ground and spacecraft. We assume that the data we produce are used by a variety of people that have a broad range of scientific sophistication.

In the almanacs, software, and web services that Commission 4 members produce, data that are independent of the rotation of the Earth, such as the geocentric celestial coordinates (right ascension and declination) of the Sun, Moon, planets, and stars, are generally provided as a function of Terrestrial Time (TT). In practice, TT is based on atomic time (TT=TAI+32.184s) and as such, it can be extended indefinitely into the future without ambiguity or error.

On the other hand, data that depend on the rotation of the Earth, such as Greenwich hour angles or the topocentric coordinates (zenith distance and azimuth) of celestial objects, have traditionally been provided as a function of Universal Time, specifically UT1. UT1 is inherently unpredictable because of natural irregularities in the length of day, but the current international time protocol guarantees that UTC, the basis for civil time worldwide, is never more than 0.9 seconds from UT1. For many users and software applications, the approximation UT1=UTC is adequate and is assumed. Many users, particularly navigators, are probably not even aware of the distinction between UTC and UT1.

A change in the definition of UTC that allows it to diverge from UT1 without bound therefore creates a challenge as to how to provide future data that are a function of the rotational angle of the Earth, and how to educate users on the change. Several ideas for how to proceed that have been circulating among Commission 4 members will be explored.

WHAT IS IAU COMMISSION 4?

The International Astronomical Union (IAU) was founded in 1919 and Commission 4, Ephemerides, was among the first group of commissions formed within the new organization. Its purpose was to facilitate international cooperation in the computation and distribution of information on the coordinates of celestial bodies, and related information such as rise and set times, predictions of eclipses, moon phases, etc., to facilitate astronomical observations, timekeeping, surveying, the comparison of dynamical theory with observations, and celestial navigation.

---

[*] President, IAU Commission 4 (Ephemerides), 35 Oak Street, Colora, Maryland 21917, U.S.A.

It's worth remembering that at the time, computers were people doing arithmetic, the most accurate clock was the rotating Earth, the only distribution mechanism was print, and celestial navigation was the only means of determining position at sea. None of that is true today, of course, but the mission of Commission 4 is remarkably unchanged. Two years ago the organizing committee of the commission re-wrote our mission statement (or, in IAU terminology, our "terms of reference"), as the IAU periodically requires. Here is the latest version:

1. Maintain cooperation and collaboration between the national offices providing ephemerides, prediction of phenomena, astronomical reference data, and navigational almanacs.

2. Encourage agreement on the bases (reference systems, time scales, models, and constants) of astronomical ephemerides and reference data in the various countries. Promote improvements to the usability and accuracy of astronomical ephemerides, and provide information comparing computational methods, models, and results to ensure the accuracy of data provided.

3. Maintain databases, available on the Internet to the national ephemeris offices and qualified researchers, containing observations of all types on which the ephemerides are based. Promote the continued importance of observations needed to improve the ephemerides, and encourage prompt availability of these observations, especially those from space missions, to the science community.

4. Encourage the development of software and web sites that provide astronomical ephemerides, prediction of phenomena, and astronomical reference data to the scientific community and public.

5. Promote the development of explanatory material that fosters better understanding of the use and bases of ephemerides and related data.

There are two broad categories of work that the commission supports. The first type of work is the computation of fundamental solar system ephemerides, that is, using gravitational theory along with observations of many types to determine the orbits of bodies in the solar system. The second kind of work uses these fundamental ephemerides to compute practical astronomical data, such as the geocentric or topocentric coordinates of the Sun, Moon, planets and stars for any given time; the prediction of times of astronomical phenomena, such as the times of rise, set, and transit, and eclipse phenomena; the parameters that describe the apparent orientation and illumination of solar system objects at specific times; and various quantities that allow knowledgeable users to transform coordinates or vectors between standard reference systems.

In the U.S., the fundamental solar system ephemerides are computed by the Jet Propulsion Laboratory (JPL); their Development Ephemeris (DE) series, which has been the de facto international standard since 1984, dates back to the 1960s. Before that, the U.S. Naval Observatory (USNO) was the primary U.S. source for such information. High quality solar system ephemerides comparable to those produced by JPL are also now available from the Institut de Mécanique Céleste et de Calcul des Éphémérides (IMCCE) in Paris and the Institute for Applied Astronomy (IAA) in St. Petersburg. There has been a transition in the last half century from ephemerides based on analytical theories to those based on N-body numerical integrations, and that is an interesting story in itself, which parallels that of the development of electronic computers.

The organizations providing authoritative almanac data are more numerous, and these usually are national institutions of various countries. These include USNO and JPL in the U.S.; Her Majesty's Nautical Almanac Office (HMNAO) in the U.K. (one of the remnants of the Royal Greenwich Observatory); the aforementioned IMCCE in France and IAA in Russia; the National Astronomical Observatory of Japan (NAOJ); and the Spanish Naval Observatory. Obviously the distribution of almanac data has also undergone tremendous change in recent decades, first with the proliferation of personal computers, and later the Internet. Although printed publications are still produced, software and web services are now in the mix.

True to the purposes of Commission 4, there is quite a bit of cooperation and data exchange among the institutions involved. Certainly the solar system ephemeris work at IMCCE and IAA would not be of the high quality that it is without the active assistance of E. Myles Standish (now retired) and Bill Folkner of JPL's solar system dynamics group. There is now, within Commission 4, a number of people involved in analyzing the differences in the ephemerides produced by these three groups, in order to improve all of them. A formally established Commission 4 working group is currently investigating a common data format for these ephemerides, so that users can more easily switch among them. Among the almanac producers, USNO and HMNAO will, in two more weeks, celebrate 100 years of cooperation and joint publication arrangements, and information sharing among all the organizations active in Commission 4 is common.

THE TIME SCALES WE USE

I do not presume to be able to provide a definitive history of astronomical time scales; there are people at this symposium much more qualified than I am for that, and I refer interested readers to their papers. A little background is useful, however. From ancient times through the middle of the 20th century, time was defined by the rotation of the Earth; there was no better clock. Significant irregularities in the rotation of the Earth were first definitively established in the 1930s.[1] But it wasn't until after World War II that the astronomical community officially recognized the need for two time scales for ephemerides: one that is a function of the rotation of the Earth and one that is independent of it. The latter time scale, called Ephemeris Time, was introduced by the IAU in 1950s. At that point, astronomical information could be expressed in either Universal Time (UT) or Ephemeris Time (ET), as appropriate. These time scales were based, respectively, on the rotation of the Earth on its axis (the mean solar day) and the revolution of the Earth around the Sun (the tropical year).

At the same time, atomic clocks were being developed, leading to a revolution in practical time-keeping and eventually to the establishment of International Atomic Time (TAI). The length of the Système International (SI) second, the unit of atomic time, was set equal to that of the ephemeris second, the unit of Ephemeris Time. Formally, the SI second is 9,192,631,770 cycles of the radiation corresponding to the ground state transition of Cesium 133, which Markowitz and collaborators established in 1958 as the best estimate for the length of the ephemeris second, which is 1/31,556,925.9747 of the length of the tropical year at 1900.[2]

Things got more complicated in the 1960s, when a number of new high-precision measurement techniques were introduced into astronomy and geodesy, including radar ranging to the nearby planets, very long baseline radio interferometry (VLBI), lunar laser ranging (LLR), and various kinds of observations of and from spacecraft. These new types of data required that relativity be integrated into our dynamical theories and our data analysis algorithms. This has led to a proliferation of specialized time scales, based on general relativity, some of which are entirely theoretical — that

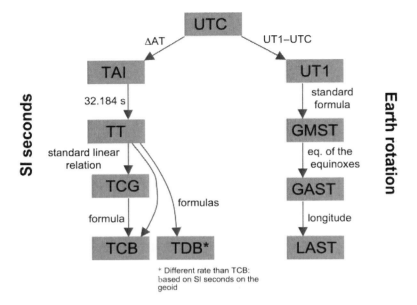

Figure 1. Relationships between time scales used in astronomy, from a user's perspective, starting from Coordinated Universal Time (UTC). The SI-based times scales (left side) are: International Atomic Time, Terrestrial Time, Geocentric Coordinate Time, Barycentric Coordinate Time, and Barycentric Dynamical Time. The time scales based on the rotation of the Earth (right side) are: Universal Time 1, Greenwich Mean Sidereal Time, Greenwich Apparent Sidereal Time, and Local Apparent Sidereal Time.

is, by their definition, they cannot be realized by an actual clock anywhere. (In the terminology of general relativity, they represent *coordinate time* rather than *proper time*.) This has been a rather messy business, with a confusing evolution of ideas about time and timekeeping over the last few decades, with time scales being defined then dropped altogether or redefined.

The current definition of Universal Time, specifically the relationship between the time scales UTC and UT1, was introduced at the beginning of 1972. That is when leap seconds as we currently know them began. UT1 is tied directly to observations of the rotation of the Earth, so it is continuous but irregular in rate; for example, the ≈25 ms annual term in UT1 corresponds to a fractional variation in rate of order 10^{-9}. Some would argue that UT1 is not a time scale at all, but a measurement of the Earth's rotational angle. UTC, the basis for the worldwide system of civil time, is a hybrid time scale, with the SI second as its unit but with occasional leap-second adjustments to keep it within 0.9 s of UT1. Some would argue that UTC is also not a real time scale because, although its rate is fixed and well defined, a day of UTC can comprise either 86,400 or 86,401 SI seconds (or, in theory, 86,399 SI seconds, although this has never happened) and there is no deterministic algorithm that provides the number of seconds between two UTC epochs. The difference between UT1 and

UTC is measured and published daily by the International Earth Rotation and Reference Systems Service (IERS Bulletin A*). The measurements are now based primarily on VLBI observations.

Figure 1 shows a schematic of the relationships between the currently used astronomical time scales, from a user's perspective, starting with UTC. Figure 2, borrowed from a paper by Seidelmann and Fukushima,[3] shows the differences between these time scales, in seconds, as a function of year. Readers interested in a more thorough explanation of modern astronomical time scales, along with the formulas for relating one to another, may refer to Chapter 2 of USNO Circular 179.[4]

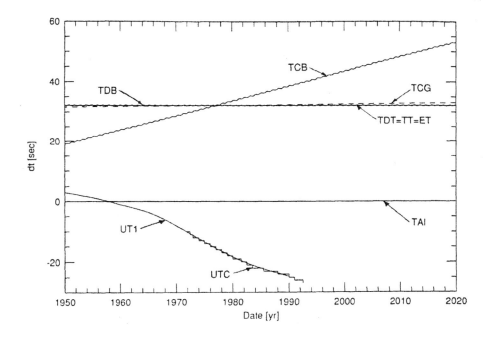

Figure 2. Difference, in seconds, between time scales used in astronomy as a function of year. Note especially the "stair-step" appearance of UTC, due to the leap seconds inserted in that time scale so that it approximates UT1. TT, TAI, and the "steps" of UTC are parallel because they are all based on the SI second on the geoid. The figure is taken from Reference 3.

In modern astronomical software, it is not unusual to have to keep track of three, four, or even five time scales. However, the almanac offices strive to keep this rather untidy situation hidden from our users. In modern almanacs, only two time scales commonly appear. For tabulations of astronomical coordinates and phenomena that do not depend on the rotation of the Earth, such as the celestial coordinates of the planets, Terrestrial Time (TT) is used. TT is the successor to ET and is considered continuous with it. For coordinates and phenomena that are a function of the rotation of the Earth, including the data for celestial navigation, UT1 is used, sometimes labeled simply UT.[†] For practical purposes, TT is equal to TAI + 32.184 s, and TAI can be obtained from UTC if one

---

*http://www.iers.org/nn_10968/IERS/EN/DataProducts/EarthOrientationData/eop.html?__nnn=true

[†]UT is also sometimes used for data that do not depend on the rotation of the Earth but are of public interest and are not given to high precision, such as the times of the phases of the Moon.

knows the leap-second count. Similarly, UT1 can be obtained from UTC if one knows the current value of UT1–UTC published by the IERS. So, both almanac time scales are related to UTC in a straightforward manner, but it is up to the user to do that transformation. The difference TT–UT1 (formerly ET–UT) is called ΔT and is tabulated separately; it is a smooth and (so far) monotonically increasing function that is free from leap-second discontinuities.

A third time scale, TDB, is the basis of heliocentric or barycentric ephemerides of solar system bodies but it differs from TT by less than 0.002 s. Most users can assume TDB=TT with negligible error for their applications.

Figures 3 and 4 show examples of modern almanac data.[5,*]

34 2010 FEBRUARY 6, 7, 8 (SAT., SUN., MON.)

UT	ARIES	VENUS −3.9		MARS −1.2		JUPITER −2.0		SATURN +0.7		STARS		
	GHA	GHA	Dec	GHA	Dec	GHA	Dec	GHA	Dec	Name	SHA	Dec
d h	° ′	° ′	° ′	° ′	° ′	° ′	° ′	° ′	° ′		° ′	° ′
6 00	136 01.5	170 06.9	S15 08.5	5 24.6	N22 53.2	159 27.6	S10 46.4	311 13.5	N 0 35.7	Acamar	315 20.1	S40 16.0
01	151 04.0	185 06.3	07.4	20 28.1	53.4	174 29.5	46.2	326 16.1	35.8	Achernar	335 28.6	S57 11.3
02	166 06.4	200 05.7	06.4	35 31.5	53.6	189 31.4	46.0	341 18.6	35.8	Acrux	173 11.8	S63 09.2
03	181 08.9	215 05.1	. . 05.4	50 35.0	. . 53.9	204 33.3	. . 45.8	356 21.2	. . 35.9	Adhara	255 14.1	S28 59.3
04	196 11.4	230 04.5	04.4	65 38.4	54.1	219 35.2	45.6	11 23.8	35.9	Aldebaran	290 52.0	N16 31.8
05	211 13.8	245 03.9	03.4	80 41.9	54.3	234 37.1	45.3	26 26.3	36.0			
06	226 16.3	260 03.3	S15 02.4	95 45.4	N22 54.5	249 39.0	S10 45.1	41 28.9	N 0 36.0	Alioth	166 22.3	N55 53.9
07	241 18.8	275 02.7	01.4	110 48.8	54.8	264 40.9	44.9	56 31.4	36.1	Alkaid	153 00.5	N49 15.4
S 08	256 21.2	290 02.0	15 00.4	125 52.3	55.0	279 42.9	44.7	71 34.0	36.1	Al Na'ir	27 47.1	S46 54.8
A 09	271 23.7	305 01.4	14 59.4	140 55.7	. . 55.2	294 44.8	. . 44.5	86 36.5	. . 36.2	Alnilam	275 48.6	S 1 11.8
T 10	286 26.2	320 00.8	58.4	155 59.2	55.4	309 46.7	44.3	101 39.1	36.2	Alphard	217 58.2	S 8 42.3
U 11	301 28.6	335 00.2	57.3	171 02.6	55.6	324 48.6	44.1	116 41.6	36.3			
R 12	316 31.1	349 59.6	S14 56.3	186 06.1	N22 55.9	339 50.5	S10 43.8	131 44.2	N 0 36.3	Alphecca	126 13.1	N26 40.5
D 13	331 33.6	4 59.0	55.3	201 09.5	56.1	354 52.4	43.6	146 46.7	36.4	Alpheratz	357 46.3	N29 08.9
A 14	346 36.0	19 58.4	54.3	216 13.0	56.3	9 54.3	43.4	161 49.3	36.4	Altair	62 10.9	N 8 53.6
Y 15	1 38.5	34 57.8	. . 53.3	231 16.4	. . 56.5	24 56.2	. . 43.2	176 51.9	. . 36.5	Ankaa	353 18.3	S42 15.2
16	16 40.9	49 57.2	52.3	246 19.9	56.7	39 58.1	43.0	191 54.4	36.5	Antares	112 29.4	S26 27.3
17	31 43.4	64 56.6	51.2	261 23.3	56.9	55 00.0	42.8	206 57.0	36.6			

Figure 3. Data from the Nautical Almanac. The time argument is UT1.

```
##########################################################################################
                          EPHEMERIS OF SOLAR SYSTEM BODIES
##########################################################################################

 Planet  2 Venus
 Planetary theory INPOP10
 Apparent coordinates (true equator; equinoxe of the date)
 Frame center: geocenter
 Relativistic perturbations, coordinate system 0
 Equatorial coordinates (R.A, Dec.)

##########################################################################################

         Date TT                 R.A             Dec.            Distance     V.Mag   Phase   Elong.  muRAcosDE      muDE      Dist_dot
              h  m  s        h  m  s           o  '  "             au.                  o       o        "/s          "/s        km/s

10  5  2011  0  0  0.00    13 32 49.11386 -08 48 53.7535      1.658916553   -3.90   18.61    13.50  0.477E-01 -0.202E-01   -4.5267
10  5  2011  6  0  0.00    13 33 58.67518 -08 56 10.2049      1.658261626   -3.90   18.70    13.56  0.477E-01 -0.202E-01   -4.5448
10  5  2011 12  0  0.00    13 35  8.28068 -09  3 25.9160      1.657604085   -3.90   18.79    13.63  0.477E-01 -0.202E-01   -4.5629
10  5  2011 18  0  0.00    13 36 17.93094 -09 10 40.8746      1.656943939   -3.90   18.88    13.69  0.478E-01 -0.201E-01   -4.5809
10  6  2011  0  0  0.00    13 37 27.62655 -09 17 55.0687      1.656281196   -3.90   18.97    13.75  0.478E-01 -0.201E-01   -4.5988
10  6  2011  6  0  0.00    13 38 37.36809 -09 25  8.4865      1.655615865   -3.90   19.05    13.82  0.478E-01 -0.200E-01   -4.6167
10  6  2011 12  0  0.00    13 39 47.15617 -09 32 21.1158      1.654947955   -3.90   19.14    13.88  0.478E-01 -0.200E-01   -4.6346
10  6  2011 18  0  0.00    13 40 56.99134 -09 39 32.9447      1.654277472   -3.90   19.23    13.95  0.478E-01 -0.200E-01   -4.6524
10  7  2011  0  0  0.00    13 42  6.87422 -09 46 43.9612      1.653604426   -3.90   19.32    14.01  0.478E-01 -0.199E-01   -4.6701
10  7  2011  6  0  0.00    13 43 16.80537 -09 53 54.1534      1.652928825   -3.90   19.41    14.07  0.479E-01 -0.199E-01   -4.6878
10  7  2011 12  0  0.00    13 44 26.78539 -10  1  3.5091      1.652250675   -3.90   19.50    14.14  0.479E-01 -0.199E-01   -4.7054
10  7  2011 18  0  0.00    13 45 36.81485 -10  8 12.0165      1.651569986   -3.90   19.58    14.20  0.479E-01 -0.198E-01   -4.7229
```

Figure 4. Data from a web service provided by IMCCE. The time argument is TT.

*http://www.imcce.fr/en/ephemerides/formulaire/form_ephepos.php

THE CHOICES WE FACE IF UTC IS REDEFINED

As we all know by now, the proposal before the International Telecommunication Union (ITU) for decision early next year is to cease inserting leap seconds into UTC by the end of the decade. If that happens, then the difference UT1–UTC will no longer be restricted to the range ± 0.9 s; UT1–UTC will grow (negatively), presumably without bound. On the other hand, the difference between UTC and TAI will become fixed at some integral number of seconds; therefore the difference between UTC and TT will also be frozen.

How will this affect almanac producers and users? Since almanac data are generally computed and displayed as a function of either UT1 or TT, the initial, naive, answer is that nothing changes; UTC is not used in the computation of the data and UTC is not the independent argument of any of the tabulations. It has been the user's responsibility to convert between UTC (or whatever his external time scale is) and the time scale(s) used for the almanac data, and we can simply assert that this responsibility remains unchanged. In fact, the basic conversion formulas from UTC to UT1 or TT won't change if leap seconds no longer occur.

On further reflection, though, we encounter both challenges and opportunities. I base the ideas in this section on a meeting of the almanac office personnel of USNO and HMNAO that took place earlier this year in which the consequences of a change in UTC were discussed. A few months ago, I inquired, as president of IAU Commission 4, whether the other almanac offices have had similar conversations, but with little response. One office replied that it had not done any specific thinking about the issue so far, and the others never responded at all.

In the previous section, I mentioned that in the almanacs, UT1 is sometimes labeled "UT" or "Universal Time", although the explanatory material always states that this means UT1. It is a safe bet that fewer than one astronomer in ten, and a similarly small fraction of navigators, know the difference between UTC and UT1, or even that there are two kinds of Universal Time. That's because, up to this point, the difference has been bounded at 0.9 s, and for the vast majority of purposes, we can set UT1=UTC to sufficient accuracy. The error in celestial navigation resulting from such an approximation is never greater than 0.225 nautical mile (0.4 km) and is more typically about half of that (the error is always only in longitude). These errors are less than the uncertainty of a typical position fix from hand-held sextant sights. Similarly, assuming UT1=UTC when pointing a telescope currently results in a pointing error of not more than 13.5 arcseconds and more typically less than 10 arcseconds (always in right ascension). That is less than the size of Saturn's disk; and on average, only stars fainter than about V magnitude 20 are separated by arcs this small, so confusion of targets is unlikely. A radio or radar dish operating at X band would have to be about 500 m in diameter to have resolution this good. So we can now get by with the UT1=UTC approximation very well for many ordinary astronomical applications.

Obviously, this will not be the case if leap seconds end and UT1–UTC becomes unbounded. At a minimum, we will have to re-label our time argument to be explicitly "UT1" and make sure that the explanatory text is very clear about the conversion from UTC, something that most users have not had to worry about. That is, people who have never heard of the IERS or the difference between UT1 and UTC will have to come up to speed on these time conversions. In the U.S. Navy, celestial navigation is generally performed by enlisted quartermasters, and we expect that some new training will be required. (It may come as a surprise that celestial navigation is still practiced by the Navy, but it is regarded as a basic seamanship skill, like piloting by visual navigation aids, that must be available in case of emergency.)

Aside: I suspect that there is a great deal of software out there in the broader astronomical community — beyond the influence of Commission 4 — that assumes that UT1=UTC, whether explicitly or not. If the change to UTC is made, these applications will simply slowly degrade in accuracy, unnoticed at first. This will be worse than the Y2K problem, because it is much more subtle; in many cases, the original coder may not even have been aware of the distinction between UT1 and UTC. This problem is outside the scope of this paper.

John Bangert of USNO has suggested that if leap seconds are dropped from UTC, we could switch our Earth-rotation-dependent tabulations from UT1 to UTC. This would require, in preparing the printed publications, that we predict UT1–UTC to sufficient accuracy a little more than two years into the future. The IERS publishes UT1–UTC predictions for a year in the future, and their estimated accuracy at the end of that time is currently 0.02 s. (This must be an underestimate; a recent authoritative paper[6] gives the prediction accuracy after one year as 0.04–0.15 s.) The IERS also provides a linear extrapolation formula for dates beyond those in the published table. Regardless of the expected accuracy of such predictions, the printed almanacs would probably have to provide correction tables so that users could, if necessary, adjust the data to be consistent with the better values of UT1–UTC that would be available closer to the time that the data are actually needed.

Online services would fare better with UTC as the time argument, because the value of UT1–UTC is known for past and current dates; but if users request data for dates more than a few months in the future, the same considerations apply as for the printed almanacs. In such cases, the web pages could, perhaps, supply a projected value of UT1–UTC that the user could change if she wants.

If we change to UTC as the time argument for Earth-rotation-dependent data, we might also do it for the other data. As I mentioned, TT would be simply a constant offset from UTC, so, aside from historical continuity arguments, there would be little point in *not* switching from TT to UTC. The heliocentric or barycentric tabulations now given as a function of TDB should probably remain that way, because these data are most appropriately expressed in the same time scale in which the fundamental ephemerides are computed. (Another time scale, TCB, shown on Figures 1 and 2, is likely to become more common for computing these data in the future.)

CONCLUSION

If leap seconds are removed from UTC, there is at least the possibility that most almanac data could in the future be tabulated as a function of UTC. That would undoubtedly be a great convenience for users. In the case of Earth-rotation-dependent data, however, the convenience would come at the price of some degradation in accuracy. It seems likely that the increased error (for tabulations computed about two years in advance) would not be worse than that from assuming UT1=UTC now, which is probably a common assumption. Furthermore, if necessary, the error could be removed by application of corrections based on the measured value of UT1–UTC for the date on which the data are needed.

On the other hand, if these tabulations continue to be provided as a function of UT1, a rather intensive user-education strategy will be required, because the approximation UT1=UTC will no longer apply. Users will have to understand the difference between UTC and UT1 and how to obtain and apply the UT1–UTC value from the IERS or other sources. This will be new to many current users of these data.

My personal view on the proposed change to UTC, which may not be shared by other members of Commission 4, is that it has not been widely enough discussed outside of the specialized communities that are most directly involved. I find it astonishing that an obscure technical organization like the ITU has the responsibility for defining the worldwide system of civil time — essentially, people who normally decide on radio spectrum allocation, satellite orbits, and communication protocols are being asked to decide the meaning of what our clocks say. I believe that the issue has so far been couched in terms (e.g., UTC, UT1, leap seconds) that are too technical even for many astronomers, and the low response to surveys and other inquiries reflects that. I think the title of this meeting has it about right: Should we decouple civil time from the Earth's rotation? I think a question like that needs more discussion, among a much wider audience.

Our concept of public time has evolved considerably over the centuries, and the U.S. experience has been well documented in the very readable book, *Keeping Watch: A History of American Time*.[7] Every change to our system of timekeeping has been a contentious issue for the general public, and I expect that if the proposed change to UTC is made, it will be equally contentious once the popular media really get hold of it. The recent article on UTC in the *American Scientist*[8] is one of several good contributions to a more public discussion, but I believe that much more education and debate needs to take place before a decision is made on changing something so fundamental as the system of civil time.

REFERENCES

[1] Spencer Jones, H., "The Rotation of the Earth, and the Secular Accelerations of the Sun, Moon, and Planets," *Monthly Notices Roy. Astr. Soc.*, Vol. 99, No. 7, 1939, pp. 541–558.

[2] Markowitz, W., Hall, R. G., Essen, L., & Parry, J. V. L., "Frequency of Cesium in Terms of Ephemeris Time," *Phys. Rev. Letters*, Vol. 1, No. 3, 1958, pp. 105–107.

[3] Seidelmann, P. K., & Fukushima, T., "Why New Times Scales?" *Astron. Astrophys.*, Vol. 265, 1992, pp. 833–838.

[4] Kaplan, G. H., *The IAU Resolutions on Astronomical Reference Systems, Time Scales, and Earth Rotation Models*, USNO Circular 179 (Washington: USNO), 2005.

[5] Her Majesty's Nautical Almanac Office & U.S. Naval Observatory Nautical Almanac Office, *The Nautical Almanac for the Year 2010* (Washington: GPO), 2009, p. 34.

[6] Gambis, D., & Luzum, B., "Earth Rotation Monitoring, UT1 Determination and Prediction," *Metrologia*, Vol. 48, 2011, pp. S165–S170.

[7] O'Malley, M., *Keeping Watch: A History of American Time* (New York: Viking Penguin), 1990.

[8] Finkleman, D., Allen, S., Seago, J., Seaman, R., & Seidelmann, P. K., "The Future of Time: UTC and the Leap Second," *American Scientist*, Vol. 99, No. 4, 2011, pp. 312–319.

DISCUSSION CONCLUDING AAS 11-671

George Kaplan ended his presentation by noting that if leap seconds are removed from UTC, then the UT1 ≈ UTC assumption is no longer valid. This requires either re-labeling "UT" as "UT1" and educating users about the conversion from UTC to UT1, or changing "UT" tabulations to be explicitly based on UTC and using predicted values of UT1–UTC to compute the data. Steve Allen asked whether that choice would be made by evaluating user preferences or by deciding on behalf of users. Kaplan said that thinking had not progressed that far, but major almanac changes tended to elicit user input. Unfortunately, the users do not always provide sufficient feedback.

McCarthy speculated that by the year 2020 (after the possible retirement of the leap second), electronic versions of the almanac may largely replace printed almanacs. Kaplan said that situation exists already, noting that the *Nautical Almanac* is available in electronic form now, and internet web services get an enormous number of users. McCarthy said that the value of UT1-UTC could be updated very frequently in that situation; Kaplan agreed that it could be updated daily if the user preferred.

Neil deGrasse Tyson asked about the degree of differences in various national almanacs. Kaplan said that the data in various national almanacs are consistent to a very high degree (on the order of milliarcseconds, because they all originate from the same astronomical ephemerides), but that different national almanacs may choose to display different content depending on the expectations of their customer base. Steve Malys added that there are obvious language differences.

With regard to educating users about the differences between "UT" and "UTC", David Simpson shared an anecdote about an acquaintance who had heard about the proposal to redefine UTC, and believed that in 2018 we would not only stop adding future leap seconds from UTC but we would retroactively eliminate all past leap seconds from the definition of UTC as well. Kaplan said that it is worth pointing out that if future leap seconds were eliminated from UTC then we would have three (3) time scales approximately offset from each other by a constant amount: UTC, TAI, and TT. All are basically realized from the *SI* second on the geoid. Rots commented that one could add GNSS time scales to this list, such as GPS time.

McCarthy noted that there is a technical difference between TT and TAI that is more than a constant offset of 32.184 seconds. Kaplan concurred that the definition is different. McCarthy noted that plots of TAI-TT differences exist.[1] Seidelmann clarified that TT is a more ideal version of TAI, where TAI reflects how well the TAI second was realized.[2] McCarthy added that TAI had historically varied due to slight changes in its realization, such as blackbody corrections, *etc*. Kaplan noted that TT is now defined relative to Geocentric Coordinate Time (TCG), which almost seems backwards in terms of how it is practically realized today by most people. McCarthy conceded that most people find TCG from TT = TAI + 32.184 although these are defined in the opposite way.

John Seago noted that the options to change "UT" in almanacs to either "UTC" or "UT1" seemed like a rather subtle matter for a user base that may not have ever understood that there was a distinction. Seago asked whether there might be some benefit to retiring the name "Coordinated Universal Time" and the acronym "UTC" and replacing it with something that would not be as easily confused with Universal Time. Kaplan replied that if UTC didn't exist, it might raise legal issues wherever national timescales refer to something called "UTC". McCarthy offered that a British colleague had proposed the name "Global Mean Time" to preserve the acronym "GMT".

Seaman pointed out that the more immediate discussion seemed "focused on the solution." In comparison, the discussion of the subtle differences between TT and TAI, was more of an "exploration of the problem space." Seaman's preference was that the discussions should sufficiently explore the problem space instead of tweaking solutions. The colloquium's title referred to a more fundamental question—*civil* timekeeping—although many presentation topics were somewhat esoteric and had little to do with *civil* timekeeping. Seaman suggested that we should first explore the design of the civil timekeeping system before addressing the implications that design might have on leap seconds.

Storz commented that he was surprised by the magnitude of the relativistic rate difference between Barycentric Coordinate Time (TCB) and the terrestrial scales. Allen offered that TCB still keeps "*SI* seconds". Kaplan said that the figure he used for his presentation originated from Seidelmann and Fukishima.[3] He also clarified that the chart did not reflect what would happen if leap seconds were terminated; UTC would just continue parallel to all other terrestrial atomic scales, diverging from UT.

Malys noted that with regard to the issue of the definition of "civil timekeeping" and nomenclature, one misconception he had faced within the US DoD is that a redefinition of "civil timekeeping" would not affect the Defense Department, which does not see itself as "civilian". He therefore noted that terminology is important, and that terms can mean different things to different people.

If leap seconds are completely ignored, Tyson asked if humanity was supposed to wait until |UT1-UTC| is approximately 86400 seconds and then add an unscheduled leap day into the calendar? Seidelmann responded that there was a suggestion to wait until |UT1-UTC| was approximately an hour, but that idea had since been discarded.[4] He added that other suggestions had been offered, such as making adjustments at the end of each century, or making an adjustment whenever the difference reaches some other tolerance (such as a minute), but none of these schemes are contained within the proposal coming before ITU-R Radiocommunications Assembly in January 2012. Seaman quipped that the proposal to make adjustments at the end of each century goes well with Paul Gabor's discussion of celebratory days, where we invent renewed chaos at a predetermined epoch and have a big party.

Tyson was also willing to gamble that no software would be in use one-hundred years later, although some in the room were not convinced this gamble was a sure thing. Several people commented that they have familiarity with systems that are still operating software from the 1960's, which is approaching a half-century already. As a point of reference, Seaman noted that one of the Kitt Peak telescopes tracked backwards during their Y2K remediation tests because it was using code written in Forth from the late 1960's. Tyson noted that bugs are fixed all the time in software, so fixing bugs should not be a concern. Kaplan clarified that the effect of UTC redefinition on software is not a bug in the usual sense of someone having coded an isolated mistake. It is also unlike Y2K where mistakes and fixes can be rather obvious; rather, UT1 ≈ UTC could be an undiscoverable working assumption. Variables may be generically called "UT" in software and clear distinctions are often not made. Kaplan also raised a point about NORAD two-line orbital elements having an ambiguous time scale. Mark Storz replied that for the SGP-4 theory commonly used today, there is an assumption that UTC is approximately UT1.[*] Kaplan continued that the issue is therefore hidden and there might not be a comment anywhere that discloses the assumption that UTC ≈ UT1 in code; the programmer may not have even appreciated that there was a difference.

Because the issues seemed to be limited to mostly astronomers, Tyson asked what the worst scenario could be. Kaplan said that, for problematic astronomical applications, things would slowly go awry. Frank Reed said that this should only result in some additional programming work that might take a little while to discover, but all problems would be eventually discovered and fixed. Kaplan agreed that the issue creates job security for programmers.

Tyson said he liked the idea of making adjustments to the end of each century, because most people don't live that long and therefore people would have to deal with the problem either once or never. However, Seidelmann added that software is being written all the time so it is not simply a once-per-century matter. Tyson said that the foresight offered from an agreeable adjustment plan allows software designers to take anticipate centurial adjustments, and the knowledge that something will happen at the end of the century puts everyone halfway there. Seago responded that the leap second was introduced 40 years ago, and now we are talking about removing them because they were not universally adopted. He said there is a real risk that very long-term adjustments may not be reliably implemented regardless of planning. The issue with any long-term adjustment scheme is that our posterity may not agree to it or may realize it improperly. Tyson countered that the Gregorian calendar has worked well since 1582, but Seidelmann offered that a major software company incorrectly programmed the year 1900 as a leap year and still maintains the error for compatibility.[†]

Seidelmann added that the discussion point could be turned the other way: because we've had leap seconds for 40 years, everybody should already know about them and therefore they shouldn't be a problem now. Terrett said that they are a problem nonetheless in practice, as all sorts of software, including computer operating systems, do not deal with leap seconds properly. Tyson therefore asked what has happened with the two dozen leap seconds so far—do operators accept failure or do they retrofit the programming once failures are noticed? McCarthy said that some systems simply shut down before a leap second and restart afterward, rather than risk any failure, which is the easiest approach. Seaman likened this to a finicky pressure valve: an occasional restart seems to release a little bit of pressure and thereby introduces only a small amount of chaos every few years. However, alternative schemes allow increasing pressure to build up.

In response to Tyson's question about the worst possible scenario, Malys clarified that this issue not just limited to astronomers, but also affects the defense communities. If the issue of UTC redefinition is not properly understood and dealt with, he said the worst possible scenario is that critical systems could go off-line, such as missile warning systems. Tyson said that astronomers shouldn't focus on their own consequences if the issue also affects missile warning or other defense capabilities, because safety of life is a much bigger concern. Kaplan said that we still should not assume that most astronomers are aware of the issue and will know how to deal with it.

---

[*] *Editors' Note*: General-perturbation theories for artificial satellites predate the creation of UTC with leap seconds.
[†] *Editors' Note*: Quoting Microsoft Support Article ID: 214326 (URL http://support.microsoft.com/kb/214326), "Although it is technically possible to correct this behavior so that current versions of Microsoft Excel do not assume that 1900 is a leap year, the disadvantages of doing so outweigh the advantages."

David Simpson expected software developers in the year 2075 or 2080 to not think about what will happen in the year 2100; just like today, their focus will be to address more immediate problems. Also, they may not expect their software will still be in use 20 years into the future. Tyson commented that such software might very well become entrenched as other code becomes built around it.

Returning to Tyson's queries about leap days replacing leap seconds, Reed foresaw a potential civil issue if a situation existed where, say, a Saturday could turn into a Sunday. There were only about 440 people who replied to the IERS Earth Orientation Center survey, but outrage would result if an internet survey asked "Is it okay for science to turn Saturday into Sunday?" Tyson noted that the power to change everyone's perception of the name of a day of the week would be extraordinary, as that convention follows no other measure except mutually agreed tradition. Reed agreed that the seven-day week is a really big deal, not just for the general populace but for millions of faithful across the globe. Reed said that matter is really a fundamental issue; if an adjustment is not applied then the reckoning by the calendar and the days of week will become jeopardized. Allen suggested that particular issue seemed to be beyond the ability of this meeting to address. Reed clarified that it is simply a fundamental point of debate that will get the attention of people that might not otherwise pay attention. Allen quipped that riots would still not erupt.

Tyson enjoyed the earlier discussions regarding cultural significance for the reckoning of time, but noted that today's culture is not strictly agrarian or overly concerned about whether the celestial bodies will suddenly depart their courses. If time is to be considered within the service of today's culture, then the most susceptible activity seems to be the programming of computing devices that run modern society. Like it or not, that activity has decades of legacy code embedded deep within it, some of which is inaccessible or compiled so long ago that it could not be recompiled even if the original high-level code were still available. These operate as black boxes. At some point there needs to be recognition that this is the way we run our world today. Allen said that his presentations would also address some of these issues.

Rots noted that the application of daylight saving time is often deliberated, but its existence is evidence that elements of society prefer to correlate their clocks with the time of sunrise. McCarthy said that China was a notable exception to this hypothesis. Tyson asked for clarification. McCarthy said that "China has six time zones compressed into one." McCarthy asked what percentage of the world's population "lives in a single time zone which essentially takes the space of six," and yet "they somehow have learned to live with that" situation. Tyson noted that his wife was from Alaska, where the variation in daylight hours over the course of the year far exceeds one hour. He was surprised that anyone there would even care about daylight-saving-time adjustments at that latitude; its practice there is hardly noticed. Tyson asked McCarthy if the example of China was also an argument to put the entire United States onto a single time zone. McCarthy said that he was not aware of any such arguments within the USA. Kaplan said that the Chinese might complain about their situation "if they could", with others agreeing.[*]

Reed noted that Percival Lowell proposed that the world should adopt Greenwich Mean Time globally as far back as 1915, so the idea of maintaining the same time over a large area is not new. Terrett said that there have been proposals to consolidate all of Western Europe into a single time zone; presently Portugal is based on UTC / GMT, while the rest of Western Europe (excluding the UK) is one hour later.

---

[*] *Editors' Note*: The borders of mainland China span from about 73.5° E to 134.8° E, or about four hours. Before 1949, China employed 5 time zones.

REFERENCES

[1] McCarthy, D.D. (2011), "Evolution of timescales from astronomy to physical metrology." *Metrologia*, Vol 48, p. S138.

[2] Petit, G. (2004), "A New Realization Of Terrestrial Time." Proceedings of the 35th Annual Precise Time and Time Interval (PTTI) Meeting.

[3] Seidelmann, P.K., T. Fukushima (1992), "Why new time scales?" *Astronomy and Astrophysics*, Vol. 265, No. 2, pp. 833-38.

[4] Beard, R. (2011) "Role of the ITU-R in time scale definition and dissemination." *Metrologia*, Vol. 48, p. S130.

AAS 11-672

ISSUES CONCERNING THE FUTURE OF UTC

P. Kenneth Seidelmann[*] and John H. Seago[†]

Historically, civil timekeeping has been based on mean solar time. With the discovery that the rotation of the Earth was not perfectly uniform, time scales based on the rotation of the Earth were differentiated from more uniform scales, with astronomical time still serving as the basis of calendars and time of day. UT1 is now the observationally determined time based on the rotation of the Earth, whereas International Atomic Time (TAI) is a precise uniform time scale determined from atomic clocks. Coordinated Universal Time (UTC) was introduced in 1972 as an atomic time scale referenced to TAI, but with epoch adjustments via leap seconds to remain within one second of UT1 for the purposes of civil timekeeping. A family of dynamical times was further established to satisfy the theory of relativity and the requirements of solar system ephemerides. A proposal to redefine UTC without leap seconds has been forwarded for final consideration by the Radiocommunications Assembly of the International Telecommunications Union (ITU) without having reached a consensus within the study group commissioned to resolve the study question. The question of whether to redefine UTC has been discussed, surveyed, and studied for over a decade, yet there is no public record of an analysis of requirements and no cost estimates of the various alternative options. The status of the leap second issue, user considerations and perspectives, and the unresolved issues concerning the proposed change to UTC will be overviewed in this paper. Due to the pervasiveness of the UTC time scale, concern is expressed that a fundamental change to UTC will require much technical activity, review, testing, and documentation changes. This will occur regardless of whether or not certain systems or applications functionally benefit from the change in definition, and may create additional work for applications which may not ordinarily deal with these technical details, or which are already satisfied and compliant with the status quo.

INTRODUCTION

Mean solar time was used for uniform-time measurement for millennia and has been used as the basis for civil time for centuries. Once the variability of the rotation of the Earth became detectable in the 20th century, Ephemeris Time was established as a theoretically uniform scale defined by the independent variable of solar-system ephemerides, which became the basis of the *Système International d'Unités* (*SI*) second in 1960. Universal Time stayed a measure of Earth rotation, serving as the global basis of civil timekeeping as it kept pace with the synodic day.

---

[*] Research Professor, Astronomy Department, University of Virginia, P.O. Box 400325, Charlottesville, Virginia 22904, U.S.A.

[†] Astrodynamics Engineer, Analytical Graphics, Inc., 220 Valley Creek Blvd., Exton, Pennsylvania 19341-2380, U.S.A.

At about the same time, precise timekeeping was developed based on atomic frequency standards. In the 1950s and 1960s, different timing centers used various technologies as frequency sources, and sought to coordinate their broadcast time scales with the rotation of the Earth by introducing small steps or changes in the length of the second.[1] The advent of spaceflight initiated more careful coordination of global timing centers, and as precision and techniques improved it became apparent that slight variations in the length of the broadcast second (or equivalently, in broadcast frequency), became increasingly inconvenient and potentially troublesome.[2]

When the *SI* second was redefined in 1967 in terms of the radiation from the hyperfine transition of the cesium-133 isotope, a background time scale called *International Atomic Time* (TAI) was maintained based on the accumulation of *SI* seconds from ensembles of atomic frequency standards.[3] However, pure atomic frequency was unsuitable for civil timekeeping because atomic frequency standards maintained a different rate than Universal Time of day. To avoid problems caused by varying the broadcast second, the epochs of the atomic scale were infrequently adjusted relative to TAI by inserting (positive) or neglecting (negative) leap seconds to remain within one second of UT1 for civil timekeeping. This system went into effect in 1972 and is called Coordinated Universal Time (UTC).[4]

By the 1970's, the operational difficulties concerning the definition and determination of Ephemeris Time were apparent, and a group of dynamical time scales based on the theory of relativity were developed. These evolved to be Terrestrial Time (TT), Barycentric Coordinate Time (TCB), Geocentric Coordinate Time (TCG), and Barycentric Dynamical Time (TDB).[5] More recently, highly specialized background time scales have been developed, each having some relationship to TAI or UTC. The reasons for their existence vary, including system security or avoiding leap seconds. Examples are GPS time, which approximates TAI plus 19 seconds, and communications systems that are intentionally offset from other scales for security purposes.[6]

Coordinated Universal Time (UTC)

Coordinated Universal Time (UTC) was adopted as a recommended means of broadcasting time signals by the International Radio Consultative Committee (CCIR)[*] after official consultation with affected scientific organizations, such as the International Astronomical Union (IAU). UTC provides the TAI frequency and time scale as an atomic realization of UT1 within ±0.9 seconds. UTC is the basis for time broadcasts by national time services and is the time distributed by other services. The predicted difference between UTC and UT1, known as DUT1, is made available to a precision of 0.1 seconds. Originally the Bureau of Longitude (BIH) was responsible for the international standardization of UTC. The International Bureau of Weights and Measures (BIPM) eventually took over responsibility for TAI, and the International Earth Rotation and Reference Systems Service (IERS) became responsible for UT1, DUT1, and leap second announcements.[7,8,9]

THE PROPOSAL TO REDEFINE UTC: 2000-2012

A proposal to redefine UTC by halting leap seconds after 2017 has been advanced from the Radiocommunications Sector of the International Telecommunications Union (ITU-R) Study Groups for the consideration by the ITU-R Radiocommunications Assembly in January, 2012. The proposal originates within ITU-R Working Party 7A, which appointed a Special Rapporteur

---

[*] predecessor of the ITU-R

Group (SRG) on the future of UTC in October 2000 to address the following ITU-R Study Question:[10, 11]

1. What are the requirements for globally-accepted time scales for use both in navigation and telecommunications systems, and for civil timekeeping?

2. What are the present and future requirements for the tolerance limit between UTC and UT1?

3. Does the current leap-second procedure satisfy user needs, or should an alternative procedure be developed?

The Study Question further decided that the "studies should be completed by 2002 at the latest" but this completion date has been continually extended up to the present (2011). User surveys and discussions have evidently taken place, but there have been no published studies that definitively answer the Study Question since its establishment.[12, 13, 14, 15] This is noteworthy because the Study Question decided that "the results of the above studies should be included in (a) Recommendation(s)." Study results would help resolve uncertainty about user requirements and provide insights regarding an optimum means for satisfying the users' needs.

The proposal to cease leap seconds is now quite long in the tooth.[16] By 2002 the SRG had already reportedly "converged to the opinion" to halt leap seconds, and called the ITU-R *Special Colloquium on the Future of UTC* in 2003 to present and discuss its judgment with interested and representative parties.[17] At the colloquium, which was advertised as "concluding" and for "drafting a recommendation to the ITU-R,"[18] the rapporteur group proffered the substitution of leap seconds with less-frequent *leap hours* to "satisfy all civil requirements and concerns" regarding potential problems with the definition of national time scales tied to Universal Time.[19, 20] However, continuing use of the titles "Coordinated Universal Time" and "UTC" was agreed to be potentially harmful and technically confusing because the label "Universal Time" is a technical term reserved for time of day based on Earth rotation.[21] The colloquium's consensus recommendation for a change of name was discounted by the SRG over concerns that it might cause "great confusion and complications in the ITU-R process." Afterward, the purpose of the colloquium was re-characterized to suggest that the SRG never had a conclusive proposal under consideration.[22, 23] ITU-R delegates from the USA and SRG later proposed a revised Recommendation within Working Party 7A calling for the replacement of leap seconds with leap hours, but the suggestion of leap hours was eventually deprecated. This resulted in the current proposal to simply discontinue leap seconds without an alternative adjustment mechanism.[24]

There have been limited studies of user requirements for time scales and accurate cost estimates are lacking. Surveys have favored the *status quo*, in most cases overwhelmingly. There has been no consensus on the subject of UTC redefinition in either Working Party 7A or its parent, Study Group 7. There have consistently been negative votes concerning the proposal. On a determination by the ITU-R that there was no technical basis for any objection, the proposed new definition was forwarded from Study Group 7 to the Radiocommunications Assembly of 2012.

ISSUES CONCERNING THE RECOMMENDATION TO REDEFINE UTC

There are a number of issues concerning the proposed redefinition of UTC that do not appear to have yet been satisfactorily addressed. There are differences of qualified opinions on the technical issues and in most cases supporting documentation is limited. There have been a number of papers written, including a special issue of *Metrologia* dedicated to the subject, which simply affirms that there are differences of opinion. The discovery of potential issues has not been exhausted due to a lack of methodology to address all relevant engineering requirements.[25] At best, here we can only attempt to increase awareness of the different issues and provide commentary with the hope that these issues could be largely resolved before any final action is taken concerning the proposed redefinition of UTC.

Significance with respect to Radiocommunications

The current recommendation to redefine UTC was forwarded for consideration by the Radiocommunication Assembly on the grounds that opposition within the study groups could cite no technical issue related to radiocommunication. However, any radiocommunication issue at this time seems unremarkable compared to other technical, legal, and public issues. The major concern supposedly relevant to radiocommunication is the claim that "advances in telecommunications, navigation and related fields are moving towards the need for a single internationally recognized time scale to regulate and provide uniformity to these systems." Thus, navigation and communication systems need a continuous time scale.[20] The implication is that the presence of leap seconds makes UTC discontinuous, or worse, that "UTC is not a time scale on account of its discontinuities."[26] However, our calendar has leap days (February 29th) yet that does not make the calendar or its definition discontinuous. The existing UTC standard with leap seconds remains capable of time-tagging events unambiguously and with full atomic accuracy for centuries to come, while also satisfying long-standing requirements for civil clocks maintaining mean solar time.

Involvement of International Standards and Scientific Organizations

Currently there is a shortage of unified responses by many major stakeholder organizations. According to a summary by the former chairman of the SRG and Chairman of WP 7A...

> Studies and information gathering on the potential future of the UTC time scale have been conducted over the past ten years by special groups from the ITU-R, the IAU, the IERS, URSI, the American Astronomical Society, and others. The issue of a continuous time scale for general usage has been pushed aside or generally ignored by the scientific societies at large. Consequently, special study groups have been faced with little interest from the parent bodies, which has resulted in an inability for some to make informed decisions.[20]

Ordinarily, organizational abstentions would be regarded as contentment with the *status quo*, or at least evidence of ambivalence or a lack of consensus amongst professional memberships; however, ITU-R study groups have interpreted this situation as one of organizational neutrality or as "having no concern" about the subject of UTC redefinition.[27] Still, the five named organizations only represent a small fraction of the immense UTC user base, and ironically, the official positions of these named organizations are still largely indeterminate after a decade of consideration.

ITU-R. Direct discussions within ITU-R Study Group 7 and Working Party 7A have not led to a recognizable consensus after more than a decade; therefore, a questionnaire was circulated among the almost 200 member-state administrations of the ITU-R in 2010.[28] Approximately 5% of administrations responded to the questionnaire; but of those, most were already represented within Study Group 7 and Working Party 7A. This situation triggered the issuance of yet another questionnaire.[29] However, significant abstentions seeming reflect ignorance of the technical issues and their impact across the majority of ITU-R administrations.

International Astronomical Union (IAU). A special *IAU Working Group on the Definition of Coordinated Universal Time* was established in 2000, and after extensive consultation it concluded that there was "no strong consensus within the IAU either for or against a proposed change in the definition of UTC."[30] The group dissolved in 2006 understanding that no imminent action by the ITU-R was taking place.[31] However, IAU Commission 31 (Time) announced (via its website) that the IAU General Secretary responded to the 2010 ITU-R questionnaire, suggesting that the IAU's "opinion has shifted toward eliminating leap seconds from UTC" since 2006.* The response was based on a letter from the Chair of Commission 31 who polled its nearly 100 members and received about six responses supporting the proposal and three opposing it.[32] However, after feedback from other IAU members of different Commissions, the IAU General Secretary directed the IAU's representative to the ITU-R to clarify that the IAU's imputed sector-member response to the ITU-R questionnaire did not represent the consensual opinion of its 10,000-member organization.

IERS. A 2003 survey by the IERS Earth Orientation Center suggested that the large majority of its users (88%) were satisfied by the current UTC determination method including leap second adjustments, and only 26% thought that changing the determination method would provide an improvement.[33] A 2011 survey in found that less than 20% of respondents favored the current proposal to cease leap seconds.[34] As a service of the IAU, the IERS does not have an organizational opinion on the matter of UTC redefinition; however, the IERS and the broader geosciences communities could be potentially deprived of public awareness if civil timekeeping is no longer based on Earth rotation.[35]

The International Union of Radio Science (URSI). URSI Commission J conducted its own survey in 1999-2000, with "about half the responses that were received were opposed to any change, while one-fourth were in favor of a change."[36] This survey might seem especially significant to the ITU-R having been commissioned by radio scientists; however, the ITU-R's SRG concluded that such fact-finding "did not provide any clear resolution."[37]

American Astronomical Society. The American Astronomical Society (AAS) Division of Dynamical Astronomy (DDA) produced a report to the AAS Council on the topic of UTC Redefinition in November 2005.[38] This report outlines arguments for and against change, suggested possible impact, but reached no conclusion or recommendation regarding UTC redefinition other than urging that the ITU-R take no action to allow all affected parties time to evaluate the technical merit of the recommendation.

Other scientific and international standards organizations should be officially consulted and involved with the proposal to redefine UTC. Considering the wide impact of UTC redefinition and its technical and non-technical ramifications, the ITU-R may no longer be sufficiently well positioned to broadly consider this issue. It has been recently proposed that responsibility for the definition of UTC should now be considered under the Meter Convention.[39]

Time Scale Nomenclature

When the conceptual definitions of time scales have changed, the names have also been changed to avoid confusion. In fact, the term *Universal Time* was encouraged to overcome a twelve-hour ambiguity with the previous term for mean solar time at Greenwich, *Greenwich Mean Time* (GMT).[40] In 1925 astronomical and navigational almanacs in the USA and Great Britain switched from the "astronomical day" which began and ended at noon to adopt the civil day beginning and ending at midnight; however, the British *Nautical Almanac* continued to label this new convention as GMT. Although the IAU has recommended since 1928 that "astronomers are advised not to use the letters GMT in any sense for the present," the acronym still survives as a common navigational synonym for UT1, and in non-astronomical usage as a synonym for Universal Time or UTC.[41,42] Multiple issues are therefore perceived with regard to the need timescale nomenclature.

---

* http://www.atnf.csiro.au/iau-comm31/activities.php

A civil standard decoupled from Earth rotation would be fundamentally different from existing and historical practice and the name UTC has been statutorily adopted in many countries. The lack of name change appears to alter the basis for civil timekeeping without the usual publicity required for such a conceptual and technical change. Universal Time remains a technical term reserved for Earth rotation. As was experienced with GMT, such terminology does not fall out of use easily. It would be legally, technically, and historically confusing to have a version of UTC with leap seconds and a version without leap seconds. It was concluded by the attendees of the ITU-R *Special Colloquium on the Future of UTC* in 2003 that "UTC without leap seconds" should omit any reference to "Universal Time" and that "International Time" (TI) might instead serve as a replacement label for "UTC without leap seconds."[43]

Applications dealing with historically UTC-tagged data cannot be spared from responsibly accounting for leap seconds, even if future leap seconds are abolished. It would be technically burdensome to have a historic version of UTC with leap seconds and a newer version of UTC without leap seconds—both called UTC. Many systems use an internal or background time scale (in the sense of a system of labeling epochs) such as TAI or GPS time to avoid leap seconds internally. These systems would not likely benefit from redefinition of UTC. Future systems would almost certainly be designed to use UTC as a uniform scale; these future systems would be seriously disadvantaged should they discover that UTC is not historically uniform when processing historic data. Also, scientists and analysts, both now and in the future, would be inhibited from converting archives of historical UTC data onto a new uniform civil time scale once and for all, if both the past and future scales were identically called UTC.[44] Ironically then, the lack of a change of name could encourage the retention and proliferation of so-called internal "pseudo time scales" in future systems that must process historical data.[20]

Alternative Time Scales

In the past, a new time scale was introduced whenever technical requirements dictated such. Today, there are a staggering number of technically precise time scales available, most of which were invented in the last half-century.[45] With the exception of UT1, all modern scales are functionally related to, or approximate the rate of, TAI. Therefore, it seems reasonable to suggest that TAI be made available and used as a time scale without leap seconds, and as a source of determining precision time interval whenever necessary, rather than create yet another atomic time standard parallel to TAI called UTC. The suggested use of TAI[*] as an internal reference scale for operational systems has been explicitly recommended in the past by the Director of the BIPM, the Consultative Committee on Time and Frequency (CCTF), and the ITU-R via Recommendations TF.485-2, TF.536-2, and TF.1552.[46, 47, 48] The recommendation to broadcast DTAI = TAI − UTC for this purpose is still prescribed by Recommendation TF.460-6.

The ITU-R study group responsible for these Recommendations have noted that "TAI is not an option for applications needing a continuous reference" as it has no means of dissemination and is not physically represented.[49] It has also been noted that "GPS time is not a reference time scale but is instead an internal time for GPS system synchronization."[50] Nevertheless, many operational systems rely on high-precision GPS signals to establish internal reference time scales, such as CDMA cellular telephone networks.[51] Moreover, the DTAI is easily deduced from external data and added to UTC to recover TAI, as UTC is basically TAI with leap second adjustments. Therefore, if a time scale without leap seconds is required, a time scale comparable to

---

[*] "TAI" actually refers to TAI(k) = UTC(k) + DTAI, with "UTC(k)" being a realization from a contributing timing center, k symbolizing an identifying acronym of a particular time service, and DTAI equaling TAI − UTC.

GPS time or TAI could be introduced, not as a replacement but as an addition. For applications that only need precise time interval, differences between TAI or GPS time epochs suffice. For those who need only precise frequency without regard to epoch, *status-quo* UTC provides this already.

User Preferences

The opinions of the wide range of users who will be affected by the proposed change should be sought by official means. There have not been broad studies of who is using UTC or assessments of impact of the change on these different types of users, including costs. This is not surprising, however, as there is significant expense associated with accurate cost analyses, and organizations are likely unwilling to make such investments until they are necessary. Nevertheless, user surveys thus far have indicated that the majority of the respondents prefer retaining the *status quo*. The former Chairman of the SRG and Chairman of Working Party 7A acknowledges that recently reported issues involving leap seconds are small in number and seemingly result in "only minor anomalies," that users continue to express satisfaction with the status quo, and that contingency procedures already exist in situations where leap seconds might or might not be an issue:

> The 2005 [leap second] event allowed the ITU-R to collect further documentation on leap second problems experienced in the areas of communication, navigation and other electronic systems. [...] From the small number of responses collected from international bodies, timing laboratories, satellite agencies and network engineers, it appeared that only minor anomalies occurred, mostly on GPS driven equipment and on NTP time servers. At the same time, a few of the responses indicated their satisfaction with the present UTC system. It was noted by some that the early announcement of the leap second application by the IERS allowed them to avoid or fix any potential anomaly. In one case a computer network was shut down about an hour before the leap second occurred and brought back into operation an hour afterwards. The indications were that system operators using time information have learned to cope with the irregularities by one means or another, service disruption being one method.[20]

Thus, a major concern is that there is no publicly available documentation that adequately or consistently justifies a proposed redefinition of UTC or expresses overwhelming user dissatisfaction with the *status quo*.

Software and Hardware Modifications

Global navigation satellite systems (GNSS) are often cited as applications benefiting from the elimination of leap seconds.[52] However, GPS as a system is not particularly affected by leap seconds, and UTC redefinition may require changes to end-user software, where the difference between UTC and UT1 are expected to be less than 1 second, such as spacecraft and ground observing systems that equate UTC and UT.[53,54,55] Even systems requiring no changes from UTC redefinition will still need to be thoroughly investigated and tested to determine this for a fact. This would be an unnecessary cost incurred by systems already compliant with the existing standard.

Alternate implementations of timekeeping systems in software systems which preserve Universal Time may provide compromises that could simplify the solution of the problem.[56,57] There is limited evidence that organizations and professionals with expertise on many types of software have been consulted, and the status of any formal communications with computer-science and software development organizations regarding the proposal is unclear. Computer scientists and software developers would be a useful source of information about methods to handle the current UTC and what would be involved in any possible change to UTC.[58]

Timing signals are now widely distributed by telecommunications networks with varying accuracies.[59] Because computers are not very good time pieces, many systems frequently and automatically check for time updates. The costs of changing software due to a change in the definition of UTC are not well established, whereas the distribution of UTC with leap seconds on computer networks is already being facilitated and would be expected to be facilitated, if no change were made.[60, 61] There are many software development firms that would be affected by this proposed change (if nothing else, documentation would need to be revised).

Distribution of UT1

The IERS will continue to estimate UT1. The need will remain to provide UT1 and DUT1 in an easily accessible manner to the many users.[62, 63] Also, as DUT1 becomes non-negligible, there will be increasing numbers of users of UT1 data. Throughout the long discussion concerning the proposed redefinition of UTC, but there has not been specific information as to what data would be made available and how it would be distributed after the change, even though distribution of UTC, UT1, and DUT1 is fundamentally a telecommunications matter. It would be naïve to presume that every user of UT has a computer network by which information access is unlimited, and it is unknown how robust UT1 / DUT1 servers are to network denial-of-service attacks or other service outages.

Legal Considerations

In almost all countries the official (regulatory) time is realized as a fixed offset from some national frequency standard synchronized to UTC. In some nations statutory basis for official time is specified in relation to mean solar time at Greenwich or Universal Time; in other countries it is explicitly designated as Coordinated Universal Time.[64] For nations where statutory basis is Universal Time; the proposed redefinition of UTC defines a scale that increasingly deviates from the legal prescription without bound, resulting in a *de facto* change in the legal time. There would at least be a need for statutory and regulatory changes to national legal systems for which astronomical time is the explicit standard. For countries where statute explicitly designates Coordinated Universal Time, there may be a question as to whether there is clear understanding of the consequences of decoupling civil time from Earth rotation by legislators, as representatives of the general public. They could be confusion as to whether UTC still represents Earth rotation and astronomical time as the implicit standard.

Non-Technical and Non-Scientific Applications

There may be a variety of societal practices that are linked to Universal Time, the impact of which is presently unclear. One particular issue, which has been raised but not pursued within the precision time and time-interval community, is religious activities or religious preferences.[65] Sacred holidays which are astronomically determined, and calendars which have been refined through the ages to maintain concordance with the heavens in the long term, exemplify a philosophy supported by religious texts that time reckoning by astronomical means is divinely established.[66, 67]. Local clocks and almanacs (or equivalent software) serve as intermediates for certain ritual customs that depend on actual sightings of the Sun or the Moon, whenever it is impractical for individuals to accomplish accurate astronomical sightings. For example, daily prayer times may be regulated by astronomical time of day and the apparent position of the Sun; such times are functions of Universal Time, luni-solar ephemerides, and the worshipper's location on Earth. If the definition of UTC is a consideration in the scheduling of worship activities, the degree to which religions might endorse the decoupling of clock time from Earth rotation is not well documented, and the consequences do not seem to have yet been thoroughly explored or dismissed by religious authorities, who may or may not have vested reliance or strong philosophical preferences regarding the representation and distribution of astronomical time.

Re-education

There is presumably a large amount of technical and educational literature reliant upon or citing the current definition of UTC, which would need to be revised if UTC is redefined.[68] Much literature and textbooks are dedicated to explaining the definition of UTC and its relationship with other time scales. To fundamentally different definitions of UTC and their dates of implementation will need to be clearly documented, perhaps taking many decades for understanding to propagate through user communities.

Celestial Navigation and Almanacs

Celestial navigation is no longer routinely used, but it is still widely taught and critically relied upon as a backup to electronic navigation aids.[69] There may be questions or confusion concerning the necessary corrections to a time scale not tied to the rotation of the Earth during a nautical emergency. Similarly Universal Time is used in national almanacs. A change in the definition of UTC might necessitate changes to almanacs and might present challenges on how to conveniently provide data and to educate almanac users of the change.[70, 71]

Rate of Rotation of the Earth

There is a long term slowing in the rate of rotation of the Earth that would indicate that sometime in the future the rate of leap seconds should increase. Currently the rotation rate of Earth is not closely following the long term trend, so predicting the short-term rate of leap second insertion remains inaccurate. Also, there seems to be conflicting opinions as to the consequences of increasingly frequent leap seconds. Some experts speculate that problems "will become worse when multiple leap seconds per year will be required."[72] Others suggest that a primary problem with leap seconds is their unusual rarity; if so, their increased frequency should lead to more awareness, better support, and improved infrastructure.[73] Regardless, the system of leap seconds was introduced at a time when the rate of insertion was already anticipated to be twice per year (with guidelines suggesting that these insertions take place primarily at the end of June and December), with two adjustments introduced during calendar year 1972.

Long-Term Societal Effects

Because the issue of decoupling civil time and Earth rotation has not been seriously contemplated before now, the long-term philosophical and sociological concerns do not appear to have been carefully assayed.[74] If mankind formally severs its timekeeping from the motion of the sky, it remains unclear how the two might ever be returned again. The cessation of leap seconds now would remove future expectations that timekeeping and telecommunications equipment have built-in capability to maintain intercalary adjustments, creating technological barriers for realigning global timekeeping practices back to the heavens.[75] Long-term adjustment scenarios have been contemplated, such as so-called leap hours or adding a number of seconds or minutes to the end of each century, but these alternatives have all of the drawbacks of the current definition and without the benefits.

A PROPOSED APPROACH

Based on the lack of responses from ITU-R member administrations regarding recent ITU-R Questionnaires on the issue of the UTC redefinition, it seems that most administrations are not sufficiently informed to make decisions concerning the issue involved. Hence, the 2012 Radiocommunications Assembly may not be the most appropriate time or place for a conclusive decision concerning the definition of UTC. Before any final action is taken on a proposed redefinition of UTC, the following activity would seem prudent.

1. Study Question ITU-R 236/7 explicitly decided that the study results should be included in whatever Recommendation was brought before the ITU-R. Because there is no publicly available study outcome that adequately or consistently justifies a proposed redefinition of UTC, the current Recommendation should be withdrawn and Study Question should be reconsidered.

2. Because the ITU-R Working Party 7A has not been able to establish a conclusive and consensual study outcome after more than decade of consideration, international standards and scientific organizations should become more involved to help determine what stakeholder organizations and user groups outside the ITU-R should become involved in the decision.

3. User requirements for time scales like UTC, UT1, and TAI should be well established from the study efforts. The means to satisfy established user requirements should be documented and pursued.

4. A consensus (unanimity) should formally be sought between international standards organizations, scientific organizations, and national governments before changes to existing conventions are exercised.

The 1960's paradigm for having a singly transmitted time scale has already been disrupted by worldwide exposure and easy access to high accuracy GNSS signals, so consideration should be given to making existing, more-uniform alternatives to UTC more visible. One possible solution to all concerns would be the distribution of TAI through the broadcast of DTAI. This compromise would require *no change* to Recommendation 460-6 because the transmission of DTAI is already recommended by Recommendation 460-6. Unfortunately its operational implementation has been seemingly delayed owing to the inconclusiveness of Study Question ITU-R 236/7 and ceaseless uncertainty regarding the future status of Recommendation 460. Therefore, simple defeat of the proposed revision to Recommendation 460-6 by the ITU-R Radiocommunication Assembly could retire the debate long enough for timing centers and hardware manufacturers to begin the broadcast distribution of DTAI such that users can realize $TAI(k) = UTC(k) + DTAI$ alongside $UTC(k)$ wherever warranted.

REFERENCES

[1] Seidelmann, P.K. (ed, 1992), *Explanatory Supplement to the Astronomical Almanac*, University Science Books, Mill Valley, CA.

[2] Duncombe, R., P.K. Seidelmann (1977), "The New UTC Time Signals." *Navigation: Journal of the Institute of Navigation*, Vol. 24, No. 2, Summer 1977, p. 162.

[3] 1969 *Comptes Rendus de la 13e CGPM (1967-9)* 103. Also 1968 *Metrologia*, Vol. 4, p. 43.

[4] McCarthy, D.D. (2011), "Evolution of timescales from astronomy to physical metrology," *Metrologia*, Vol. 48, p. S132-144.

[5] McCarthy D.D., P.K. Seidelmann (2009), *Time—from Earth Rotation to Atomic Physics*. Wiley –VCH.

[6] Chadsey, H., D. McCarthy (2000), "Relating Time to the Earth's Variable Rotation." *Proceedings of the 32nd Annual Precise Time and Time Interval (PTTI) Systems and Applications Meeting*, Reston, Virginia, November 28-30, 2000. p. 241

[7] Quinn, T.J. (1991), "The BIPM and the Accurate Measurement of Time." *Proceedings of the IEEE*, Vol. 79, No. 7, pp. 894-905.

[8] Arias, E.F., Panfilo, G. and Petit, G. (2011) "Timescales at the BIPM." *Metrologia*, Vol. 48, p. S145-S153

[9] Gambis, D. and Luzum, B. (2011) "Earth rotation monitoring, UT1 determination and prediction." *Metrologia* Vol. 48, S165-S170

[10] Document CCTF/01-33, "Report of ITU-R Working Party 7A in the Period 1999 to 2001 to the 15th Meeting of the CCTF (Sevres, 20 – 21 June 2001)."

[11] Jones, R.W. (2001), Question ITU-R 236/7, "The Future of the UTC Time Scale." Annex I of ITU-R Administrative Circular CACE/212, March 7, 2001.

[12] Engvold,O. (ed) 2006 *Reports on Astronomy 2002-2005* IAU Transactions XXVIA(Cambridge: Cambridge University Press) p 51

[13] Beard, R. (2009), "Report on Possible Revision of the UTC Time Scale." 49th Meeting of the CGSIC Timing Subcommittee, 22 September 2009.

[14] Bartholomew, T.R., "The Future of the UTC Timescale (and the possible demise of the Leap Second)–A Brief Progress Report," *Proceedings of the 48th CGSIC Meeting*, Savannah GA, 2008.

[15] Finkleman, D., S. Allen, J.H. Seago, R. Seaman and P.K. Seidelmann (2011), "The Future of Time: UTC and the Leap Second." *American Scientist*, Vol. 99, No. 4, p. 316

[16] Capitaine, N., Chapront, J., Hadjidemetriou, J. D., Jin, W., Petit, G., and Seidelmann, K. (2003). "Division I: Fundamental Astronomy (Astronomie Fondamentale)," in Reports on Astronomy 1999-2002, Transactions of the International Astronomical Union Vol. 25A, edited by H. Rickman (San Francisco: Astronomical Society of the Pacific), p. 8.

[17] Press Release "UTC Timescale Conference." *The Institute of Navigation (ION) Newsletter*, Vol. 12, No. 4 (Winter 2002-2003)

[18] http://www.ien.it/events/docs/web_titoli_utc.pdf

[19] Arias, E.F., B Guinot, and T.J. Quinn (2003), "Proposal for a new dissemination of time scales," in: *Proceedings of the ITU-R SRG Colloquium on the UTC Timescale*, IEN Galileo Ferraris, Torino, Italy, 28-29 May 2003.

[20] Beard, R. (2011) "Role of the ITU-R in time scale definition and dissemination." *Metrologia*, Vol. 48, p. S130.

[21] CCTF (2004), Consultative Committee for Time and Frequency (CCTF) Report of the 16th meeting to the International Committee for Weights and Measures (April 1–2, 2004), Bureau International des Poids et Mesures. p. 17.

[22] Document CCTF/04-27, "UTC Transition Plan." WP-7A Special Rapporteur Group, March 1, 2004.

[23] Beard, R., (2004), "ITU-R Special Rapporteur Group on the Future of the UTC Time Scale." *Proceedings of the 35th Annual Precise Time and Time Interval (PTTI) Meeting*, p. 327.

[24] Winstein, K.J., "Why the U.S. Wants To End the Link Between Time and Sun." *Wall Street Journal*, 29 July, 2005, p. 1 (URL http://www.post-gazette.com/pg/05210/545823.stm)

[25] Seaman, R. (2011) "System Engineering for Civil Timekeeping" Paper AAS 11-661, from *Decoupling Civil Timekeeping from Earth Rotation—A Colloquium Exploring Implications of Redefining UTC*. American Astronautical Society Science and Technology Series, Vol. 113, Univelt, Inc., San Diego, 2012.

[26] Guinot, B. (2001) "Solar time, legal time, time in use." *Metrologia*, Vol. 48, S184.

[27] Seidelmann, P.K., J.H. Seago (2011), "Time Scales, Their Users, and Leap Seconds." *Metrologia*, Vol. 48, pp. S186–S194.

[28] Timofeev, V. (2010), "Questionnaire on a draft revision of Recommendation ITU-R TF.460-6, Standard-frequency and time-signal emissions." ITU-R Administrative Circular CACE/516, July 28, 2010.

[29] Timofeev, V. (2011)), "Questionnaire on a draft revision of Recommendation ITU-R TF.460-6, Standard-frequency and time-signal emissions." ITU-R Administrative Circular CACE/539, May 27, 2011.

[30] McCarthy, D.D. *et al.* (2006), "Division I Working Group on 'Definition of Coordinated Universal Time'," from Engvold, O. (ed.) *Reports on Astronomy 2002-2005*, Proceedings IAU Symposium No. XXVIA, 2006, pp 63-66.

[31] McCarthy, D.D. *et al.* (2006), "Developments in the Possible Redefinition of UTC—Report of the IAU Working Group on the Definition of Coordinated Universal Time." (URL http://www.atnf.csiro.au/iau-comm31/pdf/ContDoc-010_IAU_060710.pdf).

[32] Commission 31 Activities (URL http://www.atnf.csiro.au/iau-comm31/activities.php), Jan 28, 2011.

[33] Gambis, D., C. Bizouard, G. Francou, T. Carlucci (2003)," Leap Second Results of the Survey made in Spring 2002 by the IERS." in *Proceedings of the Colloquium on the UTC Timescale*, 28-29 May 2003, IEN Galileo Ferraris, Torino, Italy.

[34] Gambis, D., G. Francou, T. Carlucci (2011), "Results of the 2011 Survey by the IERS Earth Orientation Center about a Possible UTC Redefinition." Paper AAS, 11-668, from *Decoupling Civil Timekeeping from Earth Rotation—A Colloquium Exploring Implications of Redefining UTC*. American Astronautical Society Science and Technology Series, Vol. 113, Univelt, Inc., San Diego, 2012.

[35] Dick, W.R. (2011), "The IERS, the Leap Second, and the Public" Paper AAS, 11-667, from *Decoupling Civil Timekeeping from Earth Rotation—A Colloquium Exploring Implications of Redefining UTC*. American Astronautical Society Science and Technology Series, Vol. 113, Univelt, Inc., San Diego, 2012.

[36] Matsakis, D., *et al.* (2000), Report of the URSI Commission J Working Group on the Leap Second, July 2, 2000. (URL http://www.ietf.org/mail-archive/web/ietf/current/msg13828.html)

[37] Beard, R.L., "The Future of the UTC Time Scale," *Navigation, Journal of the Institute for Navigation*, Vol. 56, No. 1 (Spring), 2009, pp. 1-8.

[38] AAS Division on Dynamical Astronomy Working Group on Time and Coordinate System Standards, "Coordinated Universal Time (UTC) and the Status of the Leap Second: Report to the AAS Council," November 2005 (URL http://aas.org/files/DDA-UTCreport.pdf)

[39] Quinn, T. (2011) "Time, the SI and the Metre Convention." *Metrologia*, Vol. 48, pp. S121-S124.

[40] Sadler, D.H. (1978), "Mean Solar Time on the Meridian of Greenwich." *Quarterly Journal of the Royal Astronomical Society*, Vol. 19, p. 300.

[41] IIIrd General Assembly - Transactions of the IAU Vol. III B Proceedings of the 3rd General Assembly Leiden, The Netherlands, July 5- 13, 1928 Ed. F.J.M. Stratton Cambridge University Press, p. 224.

[42] McCarthy, D.D. (2011), "Evolution of timescales from astronomy to physical metrology." *Metrologia*, Vol 48, (2011) S134.

[43] Beard, R.L., *et. al.*, "Annex A to the Colloquium Report Information Paper: Special Rapporteur Group 7A (SRG 7A) Report of the UTC Timescale Colloquium 28-29 May 2003," in: *Proceedings of the ITU-R SRG Colloquium on the UTC Timescale*, IEN Galileo Ferraris, Torino, Italy, 28-29 May 2003. (URL http://www.ucolick.org/~sla/leapsecs/torino/annex_a.pdf)

[44] Seago, J.H., M.F. Storz (2003), "UTC Redefinition and Space and Satellite-Tracking Systems." in: *Proceedings of the ITU-R SRG Colloquium on the UTC Timescale*, IEN Galileo Ferraris, Torino, Italy, 28-29 May 2003.

[45] McCarthy, D.D., P.K. Seidelmann (2009), *Time-From Earth Rotation to Atomic Physics*. Wiley-VCH.

[46] Steele, J. (1999), "Report of the 14th Meeting of the Consultative Committee on Time and Frequency (CCTF), BIPM, Sevres, 20-22 April." Report on the URSI Commission A: Scientific Activities for the Period 1997- 1999.

[47] Recommendation ITU-R TF.485 (1990), "Use of time scales in the field of standard-frequency and time services." (Suppressed October 24, 1997.)

[48] Recommendation ITU-R TF.1552 (2002), "Time scales for use by standard-frequency and time-signal services." (Suppressed February 18, 2011.)

[49] Beard, R. (2009), "Report on Possible Revision of the UTC Time Scale." 49th Meeting of the CGSIC Timing Subcommittee, 22 September 2009.

[50] Beard, R. (2010), "Report on Possible Revision of the UTC Time Scale." 50th Meeting of the CGSIC Timing Subcommittee, 20 September 2010.

[51] Schneuwly, D. (2001), "Robust GPS-Based Synchronization of CDMA Mobile Networks." Proceedings of the 33rd Annual Precise Time and Time Interval (PTTI) Systems and Applications Meeting, 27-29 Nov 2001, Long Beach, CA. pp. 191-98.

[52] Lewandowski, W. and Arias,E.F.(2011) "GNSS times and UTC." *Metrologia*, Vol. 48, pp. S219-S224

[53] Engvold,O. (ed) 2006 *Reports on Astronomy 2002-2005* IAU Transactions XXVIA(Cambridge: Cambridge University Press) p 51

[54] Malys, S. (2011) "Proposal for the Redefinition of UTC: Influence on NGA Earth Orientation Predictions and GPS Operations" Paper AAS 11-675, from *Decoupling Civil Timekeeping from Earth Rotation—A Colloquium Exploring Implications of Redefining UTC*. American Astronautical Society Science and Technology Series, Vol. 113, Univelt, Inc., San Diego, 2012.

[55] Simpson, D. (2011) "UTC and the Hubble Space Telescope" Paper AAS 11-673, from *Decoupling Civil Timekeeping from Earth Rotation—A Colloquium Exploring Implications of Redefining UTC*. American Astronautical Society Science and Technology Series, Vol. 113, Univelt, Inc., San Diego, 2012.

[56] Wallace, P.T. (2011) "Software for timescale applications". *Metrologia*, Vol. 48, pp. S200-S202

[57] Allen, S. (2011) "Timekeeping System Implementations: Options for the *Pontifex Maximus*" Paper AAS 11-681, from *Decoupling Civil Timekeeping from Earth Rotation—A Colloquium Exploring Implications of Redefining UTC*. American Astronautical Society Science and Technology Series, Vol. 113, Univelt, Inc., San Diego, 2012.

[58] Kuhn, M 2003 Leap-second considerations in distributed computer systems Proc. ITU-R SRG Colloquium on the UTC Timescale (Torino, Italy 28-29 May 2003) IEN Galileo Ferraris.

[59] Levine, J. (2011) "Timing in telecommunications networks" *Metrologia*, Vol. 48, pp. S203-S212

[60] Mills, D. 2006 Computer Network Time Synchronization: The Network Protocol (Boca Raton, FL/London CRC Press/Taylor and Francis) pp 209-11.

[61] ftp://time-b.nist.gov/pub/leap-seconds.list

[62] Terrett, D. (2011) "Automating Retrieval of Earth Orientation Predictions." Paper AAS 11-679, from *Decoupling Civil Timekeeping from Earth Rotation—A Colloquium Exploring Implications of Redefining UTC*. American Astronautical Society Science and Technology Series, Vol. 113, Univelt, Inc., San Diego, 2012.

[63] Deleflie, F. et al. (2011) "Dissemination of DUT1 Through the Use of Virtual Observatory" Paper AAS 11-680, from *Decoupling Civil Timekeeping from Earth Rotation—A Colloquium Exploring Implications of Redefining UTC*. American Astronautical Society Science and Technology Series, Vol. 113, Univelt, Inc., San Diego, 2012.

[64] Seago, J.H., P.K. Seidelmann, S.L. Allen (2011), "Legislative Specifications for Coordinating with Universal Time" Paper AAS 11-662, from *Decoupling Civil Timekeeping from Earth Rotation—A Colloquium Exploring Implications of Redefining UTC*. American Astronautical Society Science and Technology Series, Vol. 113, Univelt, Inc., San Diego, 2012.

[65] Chadsey,H., D. McCarthy (2000), "Relating Time to the Earth's Variable Rotation." Proceedings of the 32nd Annual Precise Time and Time Interval (PTTI) Systems and Applications Meeting, Reston, Virginia, November 28-30, 2000. p. 250.

[66] Genesis 1:14

[67] *Qur'an* 6:96

[68] Seago, J.H. (2011), "Leap Seconds in Literature." Paper AAS 11-664, from *Decoupling Civil Timekeeping from Earth Rotation—A Colloquium Exploring Implications of Redefining UTC*. American Astronautical Society Science and Technology Series, Vol. 113, Univelt, Inc., San Diego, 2012

[69] Reed, F. (2011) "Traditional Celestial Navigation and UTC" Paper AAS 11-669, from *Decoupling Civil Timekeeping from Earth Rotation—A Colloquium Exploring Implications of Redefining UTC*. American Astronautical Society Science and Technology Series, Vol. 113, Univelt, Inc., San Diego, 2012

[70] Kaplan, G. (2011) "Time Scales in Astronomical and Navigational Almanacs" Paper AAS 11-671, from *Decoupling Civil Timekeeping from Earth Rotation—A Colloquium Exploring Implications of Redefining UTC*. American Astronautical Society Science and Technology Series, Vol. 113, Univelt, Inc., San Diego, 2012

[71] Hohenkerk, C.Y. and Hilton,J.L. (2011) "Time references in US and UK astronomical and navigational almanacs," *Metrologia*, Vol. 48, p. S195-199.

[72] McCarthy, D.D., Fliegel, H.F., and Nelson, R.A., "Redefinition of Coordinated Universal Time." Letters to the Editor, *AAS Newsletter*, Issue 124, March 2005, p. 3.
(URL http://aas.org/archives/Newsletter/Newsletter_124_2005_03_March.pdf)

[73] Finkleman, D., J.H. Seago, P.K. Seidelmann (2010), "The Debate over UTC and Leap Seconds". Paper AIAA 2010-8391, from the Proceedings of the AIAA/AAS Astrodynamics Specialist Conference, Toronto, Canada, August 2-5, 2010.

[74] Gabor, P. (2011) "The Heavens and Timekeeping, Symbolism And Expediency" Paper AAS 11-679, from Decoupling Civil Timekeeping from Earth Rotation—A Colloquium Exploring Implications of Redefining UTC. American Astronautical Society Science and Technology Series, Vol. 113, Univelt, Inc., San Diego, 2012.

[75] Finkleman, D., S. Allen, J.H. Seago, R. Seaman and P.K. Seidelmann (2011), "The Future of Time: UTC and Leap Seconds." *American Scientist*, Vol. 99, No. 4, p. 316

DISCUSSION CONCLUDING AAS 11-672

Ken Seidelmann's ended his presentation with a list of proposed actions, suggesting these as an alternative approach to addressing the issue of UTC redefinition. Steve Allen thought that Seidelmann's proposal to first seek consensus among standards and scientific organizations and national governments was similarly proposed back in 1969 before leap seconds were introduced, but the International Radio Consultative Committee (CCIR)[*] ignored that proposal back then too. Seidelmann disagreed, replying that the CCIR officially consulted various organizations and modified their proposals based on the technical feedback received, the result being driven by different scientific requirements.

David Simpson observed that the *status quo* didn't seem to present any imminent problem. Perceived concern over two leap seconds per year is a situation which is not likely to occur for many decades. Seidelmann agreed.

Rob Seaman said that he has not seen much discussion about Seidelmann's point that DTAI, UT1, and UT1-UTC should be made more readily available. Seaman remarked that if we stop the current convention where UTC ≈ UT1, there is no obvious infrastructure currently in place to make up for that lost functionality. If we start focusing on the future infrastructure needed to deploy these now, it should become obvious that applications can switch to the scale most appropriate for their application in the future. Seidelmann recalled a paper whereby the author was interested in a uniform time scale and the availability of DTAI would have been useful for his application. Seaman asked Seidelmann to explain what he meant by "DTAI"; Seidelmann responded that DTAI meant TAI-UTC, an integral value [currently 34 seconds].

George Kaplan noted that DTAI stays constant for long periods, but Seidelmann replied that software developers could benefit from monitoring a broadcast value that could be used to convert the broadcast scale UTC back to TAI whenever necessary. John Seago suggested that the broadcast availability of DTAI and DUT1 seemed to be a telecommunications issue, yet the proposed ITU-R Recommendation 460-7 no longer supported the broadcast availability of either DTAI or DUT1. He mentioned that the broadcast of DTAI had been explicitly recommended for the past ten years, already being written into Recommendation TF.460-6. David Terrett suggested that the reason why DTAI is not broadcast is because of its peculiar behavior; because it changes instantaneously that instant must needs be known in advance. Seago replied that characteristic seems not much different than broadcast UTC with leap seconds.

Dennis McCarthy said that he would really take issue with any claim that DTAI, UT1, and UT1-UTC are not readily available. He admitted that UT1-UTC could be made *more* available, but it wouldn't become more available until people *wanted* it to become more readily available. McCarthy said that his colleagues from NIST have found no users of broadcast DUT1, and they are seriously considering stopping its broadcast. Seago wondered if there might be more demand

---

[*] *Editors' Note*: The CCIR is the predecessor of the ITU-R.

for transmitted UT1-UTC if its value were allowed to increase to the point of being non-negligible. McCarthy replied that we cannot think about now; we must think about ten years in advance when people will be presumably much more automated and electronically capable, and "this stuff will just be out there." McCarthy said that DTAI is already in bulletins right now and there is no need to broadcast it because it doesn't change every day.* Allen suggested that ΔLS in GPS signals already provides a means of acquiring DTAI and knowing when it will change.

Daniel Gambis said that if UTC is redefined, then TAI may be officially suppressed at some point in the future and the BIPM would be in full charge of UTC. Allen clarified that the suggestion to suppress TAI originated with the Consultative Committee for Time and Frequency (CCTF). McCarthy responded that the CCTF "has not been totally in favor of that" suggestion. Allen clarified that written language exists which admits that the CCTF "would consider" suppressing TAI, adding if that possibility looms, then there is no motivation to invest in an infrastructure to broadcast DTAI.[1]

McCarthy said that the suggestion of using GPS time is very unacceptable to the precision timing community because GPS time does not meet national standards for precise frequency because the frequency of GPS clocks is changed, or, steered. National frequency standards with too much frequency variation are not allowed to contribute to the formation of UTC. Seago offered that Allen was simply noting that DTAI could be backed out of a GPS signal. McCarthy clarified that he was referring to one of Seidelmann's points that GPS time could be used potentially for timekeeping purposes. He then added that when Coordinated Universal Time was redefined in 1970 there was no name change.

Seaman redirected the discussion back to civil timekeeping by noting that "99.9999% of the technologically mature clocks on the planet are layered on GPS" because GPS is the foundation of the network time protocol (NTP). Seaman argued that all of the discussion about alternative time scales doesn't mean anything unless such times can be delivered to the devices that need it, and "that genie is now out of the bottle." Allen added that engineers are very often satisfied with GPS time and do not care about turf wars between the IERS, BIPM, CGPM, ITU-R, *etc.* Steve Malys said that UTC is a convention that has certain characteristics, and it may be clearer to phrase the issue in terms of whether we are discussing a change to the fundamental character of civil timekeeping versus changing some standard for distributing time. GPS is readily available and there is every reason to believe that it will continue to be available as a time-distribution mechanism.

McCarthy added that UTC gotten out of GPS is not the same as GPS time. Seaman asked for clarification, to which McCarthy responded that GPS time is comprised from the clocks in the spacecraft and the GPS monitoring sites to create a time scale internal to that closed system. Malys added that the system is not quite closed, as GPS clocks are steered to track the rate of UTC. Seaman did not understand why UTC from GPS is acceptable, but GPS time is not acceptable. McCarthy added that there are corrections broadcast within the GPS signal that allows one to get UTC from GPS time; receivers can apply this correction in order to recover accurate UTC. Allen questioned whether the most common devices that get time from GPS signals actually implement these corrections fully. McCarthy suggested that these corrections are applied automatically, but then said that it does happen from time to time that people confuse GPS time and UTC—even within the US DoD. He had heard "anecdotal stories of planes on the deck of an aircraft carrier being 34 seconds apart in time." David Simpson wondered why there should be a difference of 34

---

* *Editors' Note*: The discussion concluding AAS 11-676 adds to this point.

seconds.[*] Seago noted that the cessation of leap seconds would not remedy confusion between UTC and GPS time. McCarthy clarified that his example was offering evidence that GPS time is sometimes confused with UTC via a GPS receiver. Seago replied that the addition of leap seconds combats that confusion by making the differences noticeably obvious. McCarthy rebutted that a 34-second difference is "good enough".[*]

To Simpson, it seemed that GPS time is less rigorously defined. Neil deGrasse Tyson agreed, who remarked that if one must index UTC to GPS time, then that step justifies a certain lack of confidence in GPS time. McCarthy replied that GPS time is monitored continuously, and corrections are provided to the GPS master control station so they can be uploaded daily. Seago asked if McCarthy could comment on the size of the corrections being discussed. McCarthy said that he thought the specification was 1 μs, but in practice the corrections were much smaller, probably on the order of tens of nanoseconds. Malys agreed that the corrections were much smaller than 1 μs. Seago therefore wondered how many people outside of metrology worry about that level of distinction. Seidelmann added that he thought some laboratories gently steer their own local realization of UTC(k) to minimize their difference relative to what they expect TAI to be. McCarthy said that GPS time provides a means of time transfer by being a common source for comparison of frequency standards.

Storz said that if DTAI becomes static and if leap seconds go away, then that would "be a big problem" for his organization (Air Force Space Command). His organization had been entertaining an idea of cloistering their systems together and running an atomic time scale that continues to track Universal Time to within one second, if UTC is redefined. They would therefore need to provide those systems with a "classic DTAI" that changes. Seidelmann said that if leap seconds cease, DTAI will become constant. Storz reaffirmed that his systems would still need a DTAI that represents the offset between TAI and "classic UTC"—a time scale that stayed within ±0.9 seconds of UT1. Allen said that if the current ITU-R proposal is adopted, no agency will be responsible for announcing when leap seconds ought to occur.

Storz said that he had been under the mistaken impression that some agency would still announce when a leap second opportunity *should* occur, even if those leap seconds were not going to be utilized by UTC. Rots said that the same information would be reflected in the growing value of UT1-UTC. Storz replied that his systems cannot handle a value whereby |UT1-UTC| is greater than one second. Allen said that no one will be providing that and Storz agreed that they would have to do it themselves. Rots suggested that the integer part of |UT1-UTC| could be applied to their "classic UTC". Storz acknowledged that they could do this but it would create additional software changes for their system regardless. Seaman envisioned a situation whereby different agencies would be performing these *ad-hoc* modifications differently: some might drop the integer part, other may round off. Storz realized that there was also a risk that they might provide data to external agencies relative to a time scale that no one else is using. Seaman said that, whatever the future holds, it "will be entertaining." Storz said it would be "a mess" and Seidelmann remarked that the situation offered "job security" for somebody.

Malys said that most people at the colloquium seemed able to present arguments that could favor the *status quo*. He asked if people "from the other side of the argument" were invited, as he was curious who those people were and what their arguments were. Seago said that the announced scope of the meeting was expressly focused to attract papers that might address adverse

---

[*] *Editors' Note*: The difference between GPS time and UTC in 2011 was only 15 seconds. The difference between UTC and TAI in 2011 was 34 seconds.

consequences of decoupling civil timekeeping and Earth rotation, because that area had not been well studied. Seago felt that most of the people in the room probably have some general ideas of the arguments favoring the cessation of leap seconds already, such as computer problems and telecommunications concerns, and the purpose of the colloquium was to present and discuss ramifications should UTC be redefined.

Allen asked McCarthy if there were descriptions of systems that have been shut down because of leap seconds, and the concerns about this, that could be distributed to help explain and answer Malys' concern about specific arguments favoring UTC redefinition. McCarthy said that a very interesting presentation was made at the US Naval Observatory by Poul-Henning Kamp, in which Kamp outlined some ramifications of continuing leap seconds.[*] McCarthy had asked Kamp how he had come into the information he was sharing, because it was McCarthy's experience that customers almost always say they have no problems with the insertion of a leap second. McCarthy hypothesized that there may be few reported issues because no one wants to admit that their system failed. McCarthy said that Kamp "gathered some of this information" and, while he may not remember the precise numbers correctly, he recalled that about one-third of the NTP systems that Kamp monitored "got it wrong" and that none of the time systems he monitored "inserted the leap second properly." McCarthy reported that Kamp presented a slide at the USNO "showing the times that when people were broadcasting during a leap second event" that the most common thing that appeared on his bank of receivers was *hang*, meaning they "failed to operate."

McCarthy said "one of the really scary things" that Kamp related was "a well-known system of air control within Europe in an unnamed country" monitors planes with a radar system which supplies information to a database. Another system reads that database and displays the information for air-traffic controllers. During a leap-second the two systems became unsynchronized such that "the air-traffic controllers were presented with the scenario at that airport one year previously. Although it was midnight, all the planes jumped and caused a great deal of consternation." McCarthy said that it was Kamp's contention that no air-traffic controller ever wants to say "I had a problem during the insertion of a leap second." We wouldn't want to say that all of the planes coming into LAX on December 31st were at risk—"that won't happen."

According to McCarthy, another problem that Kamp reportedly encountered was the complete shutdown of a production line at a drug company during a leap second because this particular company "requires time to advance" during its manufacturing. "So if time doesn't advance, something is wrong, and if something is wrong, the option is to shut down, so as to prevent an incorrect batch of drugs from being produced." Simpson said that the problem was probably due to the fact that they implemented an *ad-hoc* solution to the leap second, perhaps repeating the same second. McCarthy said possibly a time tag was repeated, or something hung, or "any one of these things." McCarthy had asked Kamp if any of the systems he had monitored broadcasted "23:59:60" and the answer was "one group did"—a Danish oil company.

McCarthy said that these stories are out there but they are almost impossible to document because they admit a failure that no one wants to confess; it would be too scary to let everyone know that they didn't do it right. David Terrett said he believed that it was common knowledge that most computers systems "don't handle leap seconds properly" and that "NTP as implemented gets it wrong" because people aren't managing the servers, *etc*. The question then becomes: "To whom does that matter?" To the vast majority of people now it does not matter. Terrett offered

---

[*] http://phk.freebsd.dk/pubs/usno_slides.pdf (date 09-Oct. 2011)
http://www.usno.navy.mil/USNO/tours-events/usno-scientific-colloquia/time-from-microseconds-to-leap-seconds

that leap seconds also happen in the middle of the night for many people but McCarthy replied that leap seconds happen during the middle of the day in Tokyo. The argument that it also happens during a holiday also falls short whenever a leap second happens in June. Seaman said he would refrain from explicitly commenting on Kamp's stories, but noted that there is a phenomenon known as *confirmation bias* whereby we hear the evidence that supports our preconceived position more favorably than other evidence. The notion that "*absence of evidence* is not *evidence of absence*" cuts both ways.

To address Malys' original question about the announced scope of this meeting, Seaman added that there has not yet been significant exploration of the issues and risks from defining UTC. To Seaman, this matter seemed like a much more major change than simply dealing with technical issues surrounding a leap second every year or two. Seaman said it would be nice to have Kamp's air-traffic-control story in a place where it could be wrestled with. Yet, as scary as the story sounded, if a display jumps as described, then a mistake would be perceived. However, if things drift slowly off into areas where they are no longer correct, then that is potentially scarier to programmers because there is a risk of acting on incorrect data not knowing that it is wrong. Seago interjected that today's system of leap seconds was actually introduced to overcome concerns about air safety; specifically, air collision-avoidance systems proposed by the early 1970's were intolerant of frequency variations caused by changes in the length of the broadcast second.[2]

Seaman commented that, as a programmer, he is "spectacularly unconvinced" by arguments that, because some software was badly written, conventions must change so that programmers can continue to write software badly. Although the air-traffic-control issues seemed more fundamental, the consequences of the drug-company story were simply evidence of badly written software. McCarthy suggested that we are facing the same situation when we "complain about having to replace poorly written software now" if UTC is redefined. Seaman disagreed, clarifying that compliance with the existing standard does not make software poorly written. McCarthy offered that he was "sure that they [the drug company] wrote to their standards too," but McCarthy's main point would be that "any software that is more than ten years old should be changed anyway"—a point reportedly echoed from Kamp. After a long pause followed by disbelieving murmurs throughout the room, Seago jested that such a recommendation, if acted upon, would seem to personally benefit the software developers.

REFERENCES

[1] Wallard, A. (2007), "Consultative Committee on Time and Frequency Note on Coordinated Universal Time." CCTF 09-27, 3 September 2007.(URL http://www.bipm.org/cc/CCTF/Allowed/18/CCTF_09-27_note_on_UTC-ITU-R.pdf)

[2] Duncombe, R., P.K. Seidelmann (1977), "The New UTC Time Signals." *Navigation: Journal of the Institute of Navigation*, Vol. 24, No. 2, Summer 1977, p. 162.

Session 5:
SPACE OPERATIONS

AAS 11-673

UTC AND THE HUBBLE SPACE TELESCOPE FLIGHT SOFTWARE

David G. Simpson[*]

Many scientific spacecraft include on-board computers whose flight software implicitly assumes a correspondence between the UT1 and UTC time scales. Using the Hubble Space Telescope flight software as an example, we examine the aspects of on-board computer flight software that may make use of these time scales, and consider how the software may be impacted by allowing the two time scales to diverge.

INTRODUCTION

Since the early days of the Space Age, many artificial satellites have been equipped with on-board computers. The software that runs on spacecraft on-board computers, called *flight software*, may serve a number of functions, such as:

- Spacecraft attitude determination and control;

- Autonomous commanding to the spacecraft hardware while out of ground contact;

- Spacecraft health and safety checks; and

- Scientific instrument support.

On-board computers have the advantage of providing a great deal of flexibility to a mission, since the flight software may be re-programmed at any time to compensate for hardware failures or adapt to changing mission requirements.

The Hubble Space Telescope (HST) is an example of a modern scientific spacecraft employing on-board computers. In addition to some dedicated microprocessors, the HST spreads its computational work among *two* general-purpose on-board computers: a 486 computer (used for attitude determination and control and spacecraft health and safety monitoring), and an NSSC-I (NASA Standard Spacecraft Computer I) used to support the scientific instruments. The two computers are capable of communicating with each other so that, for example, the 486 computer can signal the NSSC-I to safe its scientific instruments if a dangerous spacecraft attitude it detected.

Among the calculations performed by HST's 486 attitude-control computer are the calculation of ephemerides for the Sun, the Moon, the HST spacecraft, and the Tracking and Data Relay Satellites (TDRS) used for communication with the ground. The 486 computer also computes a geomagnetic field model as backup to the on-board magnetometers for use in its attitude control algorithm. Such

---

[*] Science Data Processing Branch and Geospace Physics Laboratory, NASA Goddard Space Flight Center, Greenbelt, Maryland 20771, U.S.A.

calculations are typical of computations done in attitude-control computers of scientific spacecraft, and implicitly assume a close correspondence between the UT1 and UTC time scales. Here we'll examine some of the details of such calculations, and how they might be impacted by a proposed change in the definition of the UTC time scale that would eliminate leap seconds and allow UTC to drift with respect to UT1.

SPACECRAFT CLOCKS

Scientific spacecraft generally include an on-board clock driven by a crystal oscillator. The flight software can track oscillations of the on-board clock in the form of a simple counter, which serves as the basis of time calculations in the software. For example, the ground can provide a set of clock calibration coefficients to allow the flight software to convert the on-board clock counter to units of SI seconds elapsed from some specified epoch such as J2000,[*] and this calibrated clock can then be used as input to the ephemeris and geomagnetic field calculations. The on-board clock calibration coefficients need to be updated from the ground periodically to allow for drift in the crystal oscillator.

The flight software generally places the clock counter into the downlinked telemetry stream, where software on the ground can convert it to UTC using a similar set of calibration coefficients. This ground-calibrated clock is used to time-tag the received telemetry.

SOLAR EPHEMERIS

Scientific spacecraft may compute solar ephemerides for a number of reasons. A solar observatory, for example, may compute the position of the Sun at frequent intervals to ensure that the instruments stay correctly pointed directly at the Sun's disk. Many spacecraft contain solar arrays as their primary source of electric power, and these arrays must be kept pointed normal to the Sun's direction. If the solar arrays are steerable, then a knowledge of the direction of the Sun allows the on-board computer to keep the arrays properly oriented.

For an astronomical observatory like HST, the Sun is something to be avoided: the sensitive instruments can be permanently damaged if they are exposed to direct sunlight. In this case a safing test can be written to place the entire observatory into a safe condition if the instruments are pointed too close to the Sun.

Another reason for computing the position of the Sun in flight software is to correct the true position of astronomical targets for velocity aberration effects. By calculating both velocity aberration and parallax corrections, the flight software can correct the true position of an astronomical target to its *apparent* position, so that it can be located by the scientific instruments.

It is a relatively straightforward task to compute the position of the Sun using a low-precision analytical model.[1] Beginning with the time n (measured in days from epoch J2000 in the UT1 time scale),

$$n = \text{JD} - 2451545.0 \tag{1}$$

where JD is the Julian day, we compute the mean longitude L and mean anomaly g of the Sun from

$$L = 280°.460 + 0°.9856474\, n \tag{2}$$

$$g = 357°.528 + 0°.9856003\, n. \tag{3}$$

---

[*] J2000 is the instant of January 1, 2000, at 12:00:00 TDB.

We then find the ecliptic longitude λ and ecliptic latitude β, taking the latter to be $0°$ since the Sun lies in the ecliptic plane:

$$\lambda = L + 1°.915 \sin g + 0°.020 \sin 2g \tag{4}$$
$$\beta = 0°. \tag{5}$$

The Earth-Sun distance in astronomical units may be found from the mean anomaly g using

$$R = 1.000\,14 - 0.016\,71 \cos g - 0.000\,14 \cos 2g. \tag{6}$$

Here λ, β, and R form a set of spherical polar coordinates of the Sun. We can convert the ecliptic coordinates to equatorial coordinates referred to the plane of the equator using

$$\tan \alpha = \tan \lambda \cos \varepsilon \tag{7}$$
$$\sin \delta = \sin \lambda \sin \varepsilon, \tag{8}$$

where α is the right ascension, δ is the declination, and ε is the obliquity of the ecliptic for epoch J2000, $\varepsilon = 23° \, 26' \, 21''.448$. Converting from spherical to cartesian coordinates gives the geocentric cartesian coordinates of the Sun:

$$x = R \cos \alpha \cos \delta \tag{9}$$
$$y = R \sin \alpha \cos \delta \tag{10}$$
$$z = R \sin \delta \tag{11}$$

which is typically what is required for on-board safety check and velocity aberration calculations. This analytical formula for the position of the Sun is accurate to about 1 minute of arc.

LUNAR EPHEMERIS

Not all spacecraft require the calculation of an on-board lunar ephemeris, but sometimes it is required. For the Hubble Space Telescope, for example, the full Moon is sufficiently bright to pose a potential danger to the sensitive scientific instruments. A lunar ephemeris is therefore required to implement a Moon-pointing safing test, which will command the instruments to a safe state if the Telescope ever points too close to the Moon. The lunar ephemeris is also required to calculate the *velocity* of the Moon, which is a (small) part of the velocity aberration corrections described earlier.

The position of the Moon is much more difficult to calculate to high accuracy than the position of the Sun. While the solar ephemeris calculation is essentially a two-body (Earth-Sun) problem to good accuracy, the lunar ephemeris calculation forms a *three*-body (Earth-Moon-Sun) problem. In calculating the position of the Moon, one must be realistic in assessing how much accuracy is required: calculating the position of the Moon to high accuracy is a difficult problem, and may consume too much of the on-board computer's resources. For a safing test, an accuracy on the order of the lunar diameter ($\sim 0.5°$) is usually adequate, and any errors in the calculation of the position may be absorbed in an extra "buffer zone" around the Moon's position.

A lunar ephemeris calculation typically used in flight software is an analytical model[1] based on Brown's lunar theory.[2,3] One begins by computing the time T in Julian centuries elapsed from epoch J2000 (with fractions of a day being on the UT1 time scale):

$$T = \frac{\text{JD} - 2451545.0}{36\,525}, \tag{12}$$

where JD is the Julian day. One then computes the ecliptic coordinates of the Moon (λ, β) and the horizontal parallax π directly from

$$\begin{aligned}\lambda = {} & 218°\!.32 + 481\,267°\!.881\,T \\ & + 6°\!.29\sin(135°\!.0 + 477\,198°\!.87\,T) - 1°\!.27\sin(259°\!.3 - 413\,335°\!.36\,T) \\ & + 0°\!.66\sin(235°\!.7 + 890\,534°\!.22\,T) + 0°\!.21\sin(269°\!.9 + 954\,397°\!.74\,T) \\ & - 0°\!.19\sin(357°\!.5 + 35\,999°\!.05\,T) - 0°\!.11\sin(186°\!.5 + 966\,404°\!.03\,T) \end{aligned} \quad (13)$$

$$\begin{aligned}\beta = {} & +5°\!.13\sin(93°\!.3 + 483\,202°\!.02\,T) + 0°\!.28\sin(228°\!.2 + 960\,400°\!.89\,T) \\ & - 0°\!.28\sin(318°\!.3 + 6003°\!.15\,T) - 0°\!.17\sin(217°\!.6 - 407\,332°\!.21\,T)\end{aligned} \quad (14)$$

$$\begin{aligned}\pi = {} & +0°\!.9508 \\ & + 0°\!.0518\cos(135°\!.0 + 477\,198°\!.87\,T) + 0°\!.0095\cos(259°\!.3 - 413\,335°\!.36\,T) \\ & + 0°\!.0078\cos(235°\!.7 + 890\,534°\!.22\,T) + 0°\!.0028\cos(269°\!.9 + 954\,397°\!.74\,T)\end{aligned} \quad (15)$$

The horizontal parallax π is directly related to the Earth-Moon distance R:

$$R = \frac{R_\oplus}{\sin\pi}, \quad (16)$$

where $R_\oplus$ is the radius of the Earth. Knowing the ecliptic coordinates and Earth-Moon distance, one then computes the geocentric cartesian coordinates of the Moon using Eqs. (7–11). The error in the Moon's position with this model is about $0°\!.4$, which is comparable to the diameter of the lunar disk.

SPACECRAFT EPHEMERIS

Many spacecraft flight software systems require a computation of the position of the spacecraft at any given time. This may be necessary for computing a parallax correction to the apparent position of an astronomical target, or for computing a position vector to a TDRS communications satellite. A number of methods have been used to compute a spacecraft ephemeris in flight software; a typical method is the two-body Keplerian orbit propagator described here, which is used for the Hubble Space Telescope.

In a two-body ephemeris calculation,[4] one calculates the position of the spacecraft from a known set of orbital elements, which are stored on-board in the flight software. New values of these elements must be uplinked periodically: since the Earth is not a perfect point mass and the spacecraft is subject to perturbations due to the Sun and Moon, the orbital elements change will change with time.

Beginning with a time t measured in days from some epoch time T_0, we begin by finding the mean anomaly M of the spacecraft at time t:

$$M = M_0 + 2\pi N(t - T_0), \quad (17)$$

where M_0 is the mean anomaly at the epoch time T_0, and N is the mean daily motion in rev/day. Knowing the mean anomaly M and orbit eccentricity e, one then solves Kepler's equation to find the eccentric anomaly E:

$$M = E - e\sin E. \quad (18)$$

Elaborate iterative methods for solving Kepler's equation may not be appropriate, since the computer time available for computing a spacecraft position will be limited in flight software. Instead, a direct method (such as a truncated series expansion via the equation of the center[5]) may give a quick solution to sufficient accuracy, especially for spacecraft in near-circular orbits.

From the eccentric anomaly E, the next step is to compute the true anomaly f of the spacecraft:

$$\tan\left(\frac{f}{2}\right) = \left(\frac{1+e}{1-e}\right)^{1/2} \tan\left(\frac{E}{2}\right), \tag{19}$$

and the Earth-spacecraft radial distance r:

$$r = a(1 - e\cos E). \tag{20}$$

The coordinates (r, f) form the plane polar coordinates of the spacecraft position in the plane of the orbit; the remaining calculations convert these coordinates to geocentric cartesian coordinates. To accomplish this, we begin by updating the longitude of ascending node Ω and argument of perigee ω to correct for perturbations in the orbit:

$$\Omega = \Omega_0 + \dot{\Omega}(t - T_0) \tag{21}$$
$$\omega = \omega_0 + \dot{\omega}(t - T_0). \tag{22}$$

We then compute the argument of latitude u,

$$u = \omega + f, \tag{23}$$

and finally compute the geocentric cartesian coordinates of the spacecraft from

$$x = r\cos u \cos\Omega - r\sin u \sin\Omega \cos i \tag{24}$$
$$y = r\cos u \sin\Omega + r\sin u \cos\Omega \cos i \tag{25}$$
$$z = r\sin u \sin i. \tag{26}$$

An ephemeris for the TDRS communications satellite may be computed in the same fashion, or a simplified model may be used that takes advantage of the fact that the TDRS spacecraft will always be stationed at the same longitude and directly over the equator.

GEOMAGNETIC FIELD MODEL

On many spacecraft, a set of electromagnets called *magnetic torquer bars* is placed on the spacecraft body which can generate magnetic fields that interact with the Earth's magnetic field. When the spacecraft's reaction wheels have spun up to their limiting rate, they may be slowed down by applying a torque to them while simultaneously using the torquer bars to "push" against the Earth's magnetic field to keep the spacecraft stationary. In effect, excess angular momentum in the reaction wheels is thus transferred to the Earth. Doing this requires a knowledge of the geomagnetic field vector at the spacecraft position, which can be found from magnetometer data and the spacecraft's orientation. But if the magnetometer data is not available, the flight software may incorporate a geomagnetic field model as a backup.

The geomagnetic field vector is traditionally found by expanding the geomagnetic scalar potential V into a spherical harmonic series:[6]

$$V(r, \theta, \lambda) = a \sum_{n=1}^{k} \left(\frac{a}{r}\right)^{n+1} \sum_{m=0}^{n} [g_n^m \cos m\lambda + h_n^m \sin m\lambda] P_n^m(\cos\theta), \tag{27}$$

where a is the radius of the Earth, r is the distance of the spacecraft from the center of the Earth, λ is the longitude of the spacecraft subsatellite point, θ is the co-latitude of the subsatellite point, and P_n^m are Schmidt-normalized associated Legendre polynomials of the first kind. The coefficients g_n^m and h_n^m are published as the International Geomagnetic Reference Field (IGRF), and are updated every five years to correct for time variations in the field.

The geomagnetic field vector **B** at the spacecraft position is found from the gradient of the scalar potential:

$$\mathbf{B}(r, \theta, \lambda) = -\nabla V. \tag{28}$$

This gives the geomagnetic field vector at the spacecraft position, knowing the spacecraft radial distance r and the longitude λ and co-latitude θ of the sub-satellite point. But the on-board ephemeris gives the spacecraft position in the inertial geocentric cartesian frame, so we need to know where the Earth is in its rotation to convert between the inertial frame and the frame rotating with the Earth. The angle between the prime meridian and the vernal equinox is called the *sidereal time at Greenwich* (GST), and is given by the formula[7]

$$\begin{aligned} \text{GST} = &\ 280°.460\,618\,37 + 360°.985\,647\,366\,29\,(\text{JD} - 2451\,545.0) \\ &+ 0.000\,387\,933\,T^2 - T^3/38\,710\,000, \end{aligned} \tag{29}$$

where JD is the Julian day (including any fractional day, in UT1) and T is given by Eq. (12).

DISCUSSION

Currently, leap seconds are introduced into the UTC time scale in order to keep UTC to within 0.9 seconds of UT1, a time scale based on Earth rotation. The proposed change in the definition of the UTC time scale would eliminate leap seconds, causing the UTC and UT1 time scales to diverge beginning in 2018.

We can estimate the magnitude of this drift using historical data. Stephenson and Morrison[8] have found that the difference between terrestrial time (TT) and universal time (UT1) over several centuries can be fit to a mean parabola given by

$$\text{TT} - \text{UT1} \approx 0.003086 y^2 - 11.23 y + 10199 \quad \text{sec}, \tag{30}$$

where y is the year. To find the predicted difference between the "new" UTC and UT1, note that terrestrial time (TT) is related to atomic time (TAI) through

$$\text{TT} = \text{TAI} + 32.184 \text{ sec}. \tag{31}$$

If we estimate roughly four leap seconds will occur between 2011 and 2018, then beginning in 2018 UTC and TAI would be related by

$$\text{TAI} \approx \text{UTC} + 38 \quad \text{sec}. \tag{32}$$

Combining Eqs. (30) through (32) gives a relation for the estimated long-term difference between UTC and UT1, if the proposed change to UTC is implemented in 2018:

$$\Delta \equiv \text{UTC} - \text{UT1} \approx 0.003086 y^2 - 11.23 y + 10129 \quad \text{sec}. \tag{33}$$

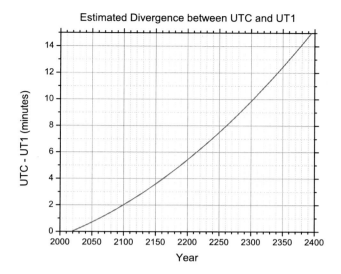

Figure 1. Predicted divergence of UTC and UT1 time scales, if the proposed change to UTC is adopted beginning in 2018. The predicted rate of change is based on the mean parabola fit to historical data given by Reference 8.

The *rate* at which the two time scales would diverge is therefore given approximately by

$$\frac{d\Delta}{dy} = 0.006173y - 11.23 \quad \text{sec/year.} \tag{34}$$

The predicted divergence of the UTC and UT1 time scales based on the mean parabola is shown in Figure 1.

Assuming these rates of divergence of the two time scales, it is possible to estimate the corresponding errors in the calculation of solar and lunar ephemerides *if UTC continues to be used in place of UT1*. From the rates of apparent motion of the Sun, Moon, and HST spacecraft relative to the Earth, we find the results shown in Figure 2.

As shown in the figure, the solar ephemeris would be least affected by an error in using a re-defined UTC time scale instead of UT1, with the error approaching the model error of $0.\!^{\circ}01$ after roughly 400 years. The error in the lunar ephemeris accumulates more rapidly, but the lunar ephemeris model also has a higher inherent error; if a re-defined UTC time scale were used in place of UT1 in the lunar ephemeris, the resulting accumulated error would roughly equal the model error in about the same time, 400 years.

A more significant issue is the spacecraft ephemeris. In this case the position of the spacecraft is computed from an algorithm that uses the number of seconds elapsed since an epoch time. The epoch time is one of several orbital elements that are derived from observations of the spacecraft position. If the time scale used in computing the orbital elements is the same time scale used onboard to compute the spacecraft ephemeris, then there should be no difficulty with the spacecraft ephemeris calculation. But if the on-board ephemeris calculation and the derivation of the orbital elements are computed using two *different* time scales (i.e. one with UT1 and the other with re-defined UTC), then there is the potential for errors to accumulate rapidly—perhaps approaching the

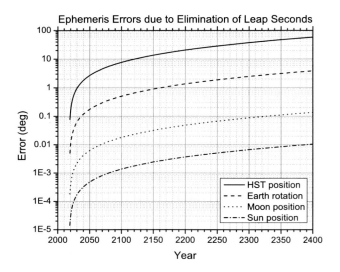

Figure 2. Predicted errors in ephemeris calculations, if UTC the proposed UTC change is adopted and ephemeris calculations continue to use UTC as their time basis. The errors shown for the HST spacecraft are the expected errors if the spacecraft orbital elements are not derived from the same time scale as is being used for the on-board ephemeris. The "Earth rotation" curve applies to error in calculating the position of a geosynchronous communications satellite (e.g. TDRS).

model error in less than five years (Figure 2). Since the orbital elements are typically produced by a different group than the flight software team, there is potential for confusion over which "universal" time scale is being used.

Some spacecraft flight software systems (such as HST's) maintain ephemerides for several geosynchronous communications satellites, in order to properly point the high-gain antennas. Since geosynchronous satellites rotate with the Earth, it is necessary to maintain either a UT1 clock or a calculation of Greenwich Sidereal Time on-board to properly track the Earth's rotation. If a UTC clock is used in place of a UT1 clock, errors would accumulate at the "Earth rotation" rate shown in Figure 2. This would lead to ephemeris errors on the order of the model error in a few decades.

For the geomagnetic field model, the error in the magnetic field vector depends in a complex way on the error in Earth rotation, but the latter should amount to less than $1°$ for several decades.

CONCLUSIONS

Spacecraft flight software systems making use of the solar, lunar, and spacecraft ephemeris models and geomagnetic field model described here implicitly assume the use of the UT1 time scale, although the on-board clock is typically calibrated to UTC. As seen in Figure 2, if UTC without leap seconds is used in place of UT1, the error in the calculated solar position due to the omission of leap seconds will approach the error in the solar ephemeris model ($0°.01$) after roughly 400 years. For the Moon, the error accumulates more quickly, but the model error is also larger ($0°.4$); the error in time would approach the modeling error at around the same time.

Spacecraft ephemeris calculation errors would grow much more rapidly, approaching the model error in much less time, from five years or less for the ephemeris of a low-Earth orbiting spacecraft like HST to a few decades for a geosynchronous spacecraft. But even if the accumulated errors do not approach the model error during the expected lifetime of the mission, one must consider the issue of *software reuse*: many flight software systems use software adapted from earlier missions to help lower software development costs. Failing to replace UTC with UT1 where appropriate in today's flight software systems could then cause an unexpected accumulation of errors in future missions that inherit today's software.

Another issue to consider is the time-tagging of the telemetry stream. Some missions (like HST) insert time tags into the telemetry stream by simply downlinking the on-board clock counter. In this case, the ground software could continue to convert the clock counter to UTC as before. But if an on-board UTC clock calculation is required (as in the Extreme Ultraviolet Explorer flight software), a flight software change may be required to allow the flight software to maintain separate UT1 and UTC clocks on-board.

Of course, the simplest scenario is to maintain the *status quo*, where the UTC time scale continues to be maintained in synchronization with UT1 via the insertion of leaps seconds. A change in the definition of UTC would require a mission-by-mission consideration of the possible impacts on the flight software systems, such as have been described here.

REFERENCES

[1] *The Astronomical Almanac for the Year 2012*. Washington, DC: U.S. Government Printing Office, 2012.
[2] E. W. Brown, *An Introductory Treatise on the Lunar Theory*. Cambridge: Cambridge University Press, 1896.
[3] B. Emerson, *Nautical Almanac Office Technical Note No. 48: Approximate Lunar Coordinates*. Greenwich, UK: Science and Engineering Research Council, Royal Greenwich Observatory, June 1979.
[4] S. W. McCuskey, *Introduction to Celestial Mechanics*. Reading, Mass.: Addison-Wesley, 1963.
[5] C. D. Murray and S. F. Dermott, *Solar System Dynamics*. Cambridge: Cambridge Univesity Press, 1999.
[6] J. A. Jacobs, *Geomagnetism (Volume 1)*. London: Academic Press, 1987.
[7] J. Meeus, *Astronomical Algorithms*. Richmond: Willmann-Bell, 2nd ed., 1998.
[8] F. R. Stephenson and L. V. Morrison, "Long-term fluctuations in the Earth's rotation: 700 BC to AD 1990," *Phil. Trans. R. Soc. Lond. A*, Vol. 351, 1995, pp. 165–202.

DISCUSSION CONCLUDING AAS 11-673

Neil deGrasse Tyson asked if it were possible to update space-borne ephemerides such that redefined UTC might become the explicit variable. David Simpson said that this could be accomplished by calculating a new series with UTC as the independent variable and uploading the new series onto the spacecraft. Tyson noted that the process of changing the independent variable seemed like it could be computer automated, thereby the cost of change in this situation seems pretty minor compared to other situations affected by UTC redefinition. Simpson clarified that without leap seconds series expansion itself should need to be re-generated. Frank Reed said that the recreation of a new series would be "child's play" and he could do it tomorrow. Simpson wondered if a lunar-model reformulation would be as simple as suggested.

Ken Seidelmann noted that there ought to be two other timing requirements for Hubble. One would be the antenna pointing for space-to-ground communications, which would seem to require UT1, and another for tracking solar system objects, which would seem to require a dynamical or coordinate time scale. Simpson clarified that the issue of ground communications is mentioned in the written paper but was omitted from the presentation. In addition to the lunar and solar ephemerides, a spacecraft ephemeris is also maintained onboard. Additionally, there is a simplified two-body ephemeris maintained onboard for the TDRS geostationary communications satellites which uses UT1 as its independent variable. This is used to compute the direction from the spacecraft to the TDRS satellites for proper antenna pointing. Seidelmann said the accuracy requirements on those ephemerides should be more critical than the lunar and solar ephemerides. Rots said that he performed a calculation at 500 km altitude assuming a one-second error, and that should introduce a maximum error of about 3 arc-minutes with a geostationary object.

With regard to tracking solar system objects, Simpson said that an explicit ephemeris is not used. Instead, a bias is introduced into the spacecraft attitude-control law which allows the telescope to slowly drift in order to follow a moving object. Seidelmann wondered if timekeeping issues affected the anticipated fields of view which are quite restricted for some instruments. Simpson replied that could be handled on the ground as part of the bias calculation, but agreed that this would still need to be handled properly within the ground calculations. Rots and McCarthy asked if star trackers were used for orientation. For Hubble, Simpson said that wide-field star trackers are used in conjunction with the Fine Guidance Sensors. McCarthy responded, "When you do that, this always falls away and you are just using your internal calibration and your platform."

McCarthy said that he did not understand how a lunar ephemeris could be produced relative to UT1. Kaplan said that was a good question, that he was aware of the formulae being cited, and that he would investigate as soon as possible, because the creation of lunar and solar ephemerides in terms of UT1 seemed unnatural. Kaplan said that if UT1 was the independent variable, then someone had already applied a model for ΔT, because normally such ephemerides are a function of a more uniform time scale like TT. Simpson noted that the independent variable used throughout the *Astronomical Almanac* labeled "UT" he would interpret to mean "UT1". Kaplan noted that UT is used primarily in the *Almanac* wherever Earth orientation is of primary interest, but not for solar-system ephemerides. Seidelmann wondered if these expressions originated from the navigation departments, which tend to reference UT. Kaplan said that ΔT model would still have been necessary for UT to have been used as the independent variable. McCarthy said that he would be interested in knowing what was used for ΔT. Kaplan said the model for ΔT could be years out of date, and certainly the people who formulated these equations probably never anticipated that a mission like the Hubble Space Telescope would be using them. Simpson said that the model was not used blindly; the series were compared against the DE200 lunar ephemerides and the errors were observed to be well within allowable tolerances. Tyson commented that this level of testing seems to affirm that a ΔT model was in fact used. Simpson reiterated that the independent variable in the analytical lunar model is time T in Julian centuries elapsed from epoch J2000 on the UT1 time scale. Reed noted that since the accuracy of the lunar ephemeris was measured in terms of fraction of a degree, the precise definition of the independent variable was not consequential to that case, to which Simpson agreed.

Seaman asked a rhetorical question about the degree to which "flight software" has been investigated regarding a possible change in the definition of UTC. Simpson agreed that this issue "tends to be low on people's radar" and speculated that many people have not heard of this issue, and therefore concluded that there has likely been little to no study. In order to look into the ramifications, someone at the management level would first have to decide to spend money to execute an impact study, and Simpson said that he was not aware of anyone within NASA who had done this.

McCarthy said that he imagined that the flight instrumentation is calibrated to an onboard clock and that onboard clock could be calibrated to whatever time scale is preferred. Seaman responded that in relation to a process, a phrase like "I imagine" would indicate that planning should have happened that did not. In response to McCarthy, Terrett offered that there is the potential for confusion as to what that time scale ought to be if UTC is redefined. It might be still be labeled or thought of as UTC, but if onboard clock oscillations were tied to something other than redefined UTC, then that could result in data-processing complications. Seidelmann said that it is an issue between the time-tagging of the instrument data, the spacecraft, and the telemetry data, and the meaning of those times relative to other events. Simpson suggested that the few people at NASA who might actually know about this issue may be waiting to see the result of the ITU-R vote before spending any money toward this.

Rots said that he did not see a significant issue here, and that time tagging relative to TT seems to make more sense anyway. He said that his spacecraft telemetry data are received tagged relative to UTC; the first task is to convert these tags to TT with systems that must be aware of leap seconds. The only confusing aspect from this conversion is there is a minority of users that do not pay sufficient attention and presume that these data are still tagged to UTC, but that is a different issue.

Seaman asked how UTC gets into the telemetry stream. Simpson said that in his experience, the telemetry is actually tagged according to the onboard clock counter and it is the task of the ground crews to correlate that count to whatever time scale is required. Tyson said that the analysis of astrophysical phenomena in the time domain is a burgeoning frontier, so the precise synchronization of clock times between different instruments tied to different epochs will only grow in importance. Simpson said that for time-critical applications (*e.g.*, gamma-ray bursts) one would probably embed some kind of precision time within the science telemetry stream itself. On further reflection of his earlier statement, Tyson offered that data transformations are routinely performed all the time, citing the calculation of stellar coordinates as an example. Rots said that a star tracker which is calibrated against guide-star coordinates from the ICRF catalog will inherently provide results relative to the ICRF; in that case no transformation is involved.

Reed asked how the simple orbital elements of the TDRS satellites were handled, as he was curious how often the elements needed to be updated due to the simplicity of the dynamical model. Simpson said that all of the ephemeris calculations were handled by the same code in the original DF-224 onboard computer, but when the HST was upgraded to an i486 onboard computer, an optionally simplified model was provided that took advantage of the fact that the TDRS spacecraft is always kept over the same spot on Earth. Called "TDRS-on-a-stick", the model assumes that the TDRS spacecraft are physically attached to the equator of the Earth and rotate with it at a given longitude. Seidelmann noted that in this case, onboard knowledge of UT1 is seemingly required because the TDRS model is explicitly tied to Earth rotation. Rots reiterated that a one-second orientation error should cause about 3 arc-minutes of pointing error.

AAS 11-674

COMPUTATION ERRORS IN LOOK ANGLE AND RANGE DUE TO REDEFINITION OF UTC[*]

Mark F. Storz[†]

With the decision on whether or not to discontinue leap seconds scheduled for January 2012, it is important to develop tools to evaluate the error that can be expected in operational software that tracks space objects from the ground or ground objects from space. These tools focus on the error that would occur in software that uses Coordinated Universal Time (UTC) as an approximation to Universal Time 1 (UT1). These error evaluation tools input the difference (in seconds) between UT1 and redefined UTC and a user-specified altitude for space objects. From these inputs, one of the tools plots a grid of look angles in polar coordinates thus generating a "sky plot" as seen from a particular ground location. This tool shows the true position in the sky when one uses UT1 to compute Earth orientation and the biased position in the sky when one uses redefined UTC. The two positions are connected by an arc from the true position to the biased position. This arc is the path the biased position would take as one gradually increases the separation between UT1 and redefined UTC. The color of the arc changes according to the bias in range. This tool also outputs the true and biased values for range, azimuth angle and elevation angle at all grid points. This, and related tools provide the user a sense of the adverse operational impacts as the biased position deviates more and more from the true position.

INTRODUCTION

The time scale that forms the basis for civil time is known as Coordinated Universal Time (UTC). UTC is typically shifted by an integer number of hours to produce local civil time for various time zones worldwide. UTC is tied to the orientation of Earth about its axis relative to the sun. This orientation is defined in terms of the angle measured eastward from the point on the equator opposite the sun's mean position to the Greenwich meridian. When measured in hours of arc (1 hour = 15°), this angle serves as a "time" known today as Universal Time 1 (UT1).

Strictly speaking, UT1 is not a true time scale because it slows down in an irregular fashion as the tidal forces on Earth slow Earth's rotation rate. On average, Earth's slowing rotation rate causes the solar day to increase by about 1.4 milliseconds per century. The irregular slowdown is due primarily to the changing mass distribution within Earth and its atmosphere and oceans. Despite this variability, Universal Time has been used for centuries as the basis for civil time.

---

[*] This paper is declared a work of the U.S. Government and is not subject to copyright protection in the U.S. DoD Distribution A. Approved for public release; distribution unlimited.

[†] Chief, Force Enhancement Analysis Branch, HQ Air Force Space Command, Analyses and Assessments Division, HQ AFSPC/A9AE, 250 S. Peterson AFB, Colorado 80914-3090, U.S.A.

Because UT1 is primarily an Earth orientation angle, it is used to compute the position of space objects (e.g. satellites and celestial objects) relative to observation points on Earth, as well as the position of Earth objects relative to points in space. To compute Earth's orientation in an inertial coordinate frame, UT1 is converted from an angle relative to the *mean sun*[*] to an angle measured eastward from the vernal equinox point, known as *right ascension*. The right ascension of the Greenwich meridian θ_G in degrees is often given by the polynomial:

$$\theta_G = 280.46061837 + 360.98564736629 \times d + \left[\frac{d}{1854436}\right]^2 - \left[\frac{d}{12355622}\right]^3 \qquad (1)$$

where d is UT1 measured in days elapsed since January 1st in the year 2000 at 12:00, an epoch referred to as J2000.0. Figure 1 shows the geometry associated with this formula at a time when Earth has an orientation of 30°. The θ_G angle is essential for computing the inertial coordinates of space objects relative to a rotating Earth as well as the Earth-fixed coordinates of objects on Earth relative to an inertial coordinate frame in space.

Unlike UT1, atomic time is a true time in the strictest sense. It does not vary with Earth's slowing rotation rate. It is based on a solar day with exactly 86,400 System International (SI) seconds. The SI second is defined as, "the duration of 9,192,631,770 periods of the radiation corresponding to the transition between the two hyperfine levels of the ground state of the cesium 133 atom."[1] This number of periods was chosen to be as close as possible to one second of Ephemeris Time (ET), now known as Terrestrial Time (TT). The ET second was defined as 1/31,556,925.9747 the length of the tropical year for the year 1900.[2] Today, the TT second is defined to be identical to the atomic SI second. There are three major atomic times in common use today, each differing by a constant offset in seconds: Terrestrial Time (TT), International Atomic Time (TAI) and GPS Time (GPST). These three atomic times progress at a steady rate and never gain on each other. On the other hand, centuries ago, Universal Time (UT1) was significantly faster than atomic time, but today is generally slower.

Aesop's "tortoise and the hare" fable is used in Figure 2 to illustrate the relationship between the steady atomic-time "tortoises" and the erratic universal time "hare." As the centuries progress, the different times advance upward as shown on the left-hand-side of the figure. When referenced to the progression of the atomic time "tortoises", as shown on the right-hand-side of the figure, the universal time "hare" catches up and surpasses the atomic times and then falls behind. The progress of UT1 and the three atomic times over the last four centuries is shown in Figure 3, adapted from Figure 1 in Nelson *et al.*[2]

Coordinated Universal Time (UTC) is a "hybrid" time scale, having characteristics of both UT1 and atomic time. It was established 1 January 1972 as the new "civil time" replacing Greenwich Mean Time (GMT). UTC is kept within ±0.9 second of UT1 through periodic leap second adjustments. This way, UTC is always an integer number of seconds off from International Atomic Time (TAI). Figure 4 was adapted from a download off the Earth Orientation Center web page.[†] It shows the relationship between UT1 and UTC since 1972. UTC was 10 seconds behind TAI in 1972 and today is 34 seconds behind. GPS Time is a constant 19 seconds behind TAI. Since UTC is kept within ±0.9 second of UT1, UTC may be used to approximate

---

[*] The term *mean sun* is no longer used in current realizations of the motion of Earth around the sun, but is still a useful concept for heuristic purposes.
[†] http:/hpiers.obspm.fr/eop-pc/earthor/uts/leapsecond.html

Earth's orientation within ±13.5 arc seconds which is equivalent to ±418 meters of error on Earth's equator at the surface. Many space tracking applications can tolerate these errors and use this approximation.

Despite UTC's usefulness as both a time and as an approximate Earth orientation angle, several commercial systems needing precise time have difficulties introducing leap seconds whenever they occur. Consequently, a group of international stakeholders has been advocating discontinuance of leap second adjustments. However, they do not want to eliminate leap seconds from only their own internal system time, but want the whole world to transition to a "civil atomic time" so they can remain consistent with a standard worldwide time scale.

The International Telecommunication Union - Radiocommunication Sector (ITU-R) has been sponsoring a series of international meetings to decide whether or not leap seconds should be discontinued with a decision expected at the upcoming ITU Radiocommunication Assembly in January 2012. If the delegates vote to discontinue leap seconds, discontinuance could go into effect as early as 1 January 2018, resulting in a civil time no longer tied to Earth's orientation. The Assistant Secretary of Defense for Networks and Information Integration issued a policy letter (29 Jun 09) to the US State Department supporting the discontinuance of leap seconds, but asked that leap seconds be discontinued no earlier than 1 January 2019 to "allow sufficient time for necessary system modifications to be accomplished." The "necessary system modifications" are likely to be upgrades to software and firmware used to track space objects from the ground or to track ground objects from space. The source of potential software problems lies in the fact that the difference between UT1 and a new "civil atomic time" without leap seconds will likely grow indefinitely, a situation that some space tracking software has not been designed to handle, as discussed by Seago and Storz.[3]

ERROR DISPLAY TOOLS

This section describes two tools used to display the error in look angle and range one can expect from space tracking software that uses UTC as an approximation for UT1. Both of these tools were written in MATLAB®[*] and were developed at Headquarters Air Force Space Command. Both tools make use of standard coordinate transformations as described in P.R. Escobal.[4]

DUT1 Sky Plot Error Tool

This tool (*DUT1-Skyplot*) was developed to investigate the slow degradation with time (after leap second discontinuance) in space tracking algorithms that use UTC to approximate UT1. This tool plots the error in elevation angle, azimuth angle, and range as a function of direction in the sky, latitude of the ground site, altitude of the space object, and the value for DUT1 (= UT1-UTC). The length of the arc represents error in look angle (both elevation and azimuth). The color of the arc represents the range error. Deep red indicates computed ranges more than 10 km greater than truth; deep blue indicates computed ranges more than 10 km less than truth. In between is a rainbow color scheme with yellow-green indicating a zero range error.

Figure 5, Figure 6, and Figure 7 show the error in look angle and range due to an unapplied DUT1 value of 30 seconds for space objects at an altitude of 100 km. Notice that at both the equator and at 60° N latitude, the arcs are aligned along great circles spanning from the eastern horizon to the western horizon. The look angle error is largest at the equator as shown in Figure 5

---

[*] MATLAB® is a high-level language and interactive environment that enables one to perform computationally intensive tasks. MATLAB was developed by The MathWorks, Inc.

and is about 8° at the zenith. For applications that use UTC to approximate UT1, this is the kind of look angle error one can expect 30 to 50 years after leap seconds are discontinued. Since the error is proportional to the cosine of the latitude, it falls to about 4° at 60° N latitude at the zenith (Figure 6), and is zero at the north pole (Figure 7). The very small look angle arcs in Figure 7 are actually parallel to lines of constant elevation angle. This is revealed in Figure 8 where the DUT1 was artificially increased to 900 seconds for visual effect. Notice in Figure 8 that the arcs appear to spin around the zenith which corresponds to the celestial ephemeris pole (CEP) at this latitude.

Figure 9, Figure 10, and Figure 11 reveal how the error arcs transition from an east-west orientation to an orientation circling the celestial ephemeris pole as the altitude of the space objects increase. A fixed latitude of 60° N was chosen to demonstrate this. Notice in Figure 6 that, for space objects at an altitude of 100 km, the orientation of the arcs is almost purely in an east-west direction and each arc is nearly aligned with a great circle reaching from the east cardinal point to the west cardinal point on the horizon. The arcs do not appear to spin around the celestial ephemeris pole. This is due to the proximity of objects at 100 km altitude. They are so close to the observer that they behave like the corners of ceiling tiles in a building as an observer walks a few steps toward the east. However, as one increases the altitude of the space objects, there is a gradual transition to the geometry where the arcs appear to spin around the celestial ephemeris pole. Figure 9 shows the behavior of the error arcs for space objects at an altitude of 1000 km. This is the altitude where the arcs begin to spin around the north cardinal direction, thus appearing to be dots in that direction. Figure 10 shows the behavior for objects at an altitude of 10,000 km. Notice that the arcs are spinning around a "false" celestial pole that has a lower elevation angle than the true celestial ephemeris pole. Figure 11 shows the behavior for objects at an altitude of 1,000,000 km. At this altitude, the space objects behave like stars with look angle error arcs spinning around the celestial ephemeris pole. The value of DUT1 had to be increased with altitude for visual effect; otherwise the arcs would be too small to display.

DUT1 Cardinal Direction Error Tool

This tool (*DUT1-Cardinal*) was also developed to investigate the degradation in accuracy of space tracking algorithms that use UTC to approximate UT1. The tool plots the error in elevation angle, azimuth angle and range as a function of satellite altitude in km, and the value for DUT1 in seconds. The plots in Figures 12 through 16 display the error in 5 cardinal directions.

Figure 12 shows how the elevation angle error varies with space object altitude when an observer on the equator is looking toward the zenith. The zenith direction at the equator produces the largest look angle errors. The seven curves in Figure 12 correspond to a DUT1 of 0, 5, 10, 15, 20, 25 and 30 sec. Notice that for a DUT1 of 30 seconds, the elevation angle error is over 8°. This would take the computed look angle outside the field of view for many sensors.

Figure 13 and Figure 14 show how the azimuth angle error varies with space object altitude when an observer on the equator is looking north or south (respectively). The seven curves in Figures 13 and 14 also correspond to a DUT1 of 0, 5, 10, 15, 20, 25 and 30 sec. Notice that Figure 13 is the mirror image of Figure 14.

Figure 15 and Figure 16 show the range error when an observer on the equator looks east or west (respectively). The seven curves in Figure 15 and Figure 16 also correspond to a DUT1 of 0, 5, 10, 15, 20, 25 and 30 sec. Notice that the range error does not vary with altitude.

Figure 17 though 21 are analogous to the previous five figures. The only difference is that these plots were generated for latitude 60° N instead of the equator. Since the errors vary with the cosine of the latitude, they are roughly half the error values at the equator. However, the north and south cardinal points on the horizon exhibit a marked asymmetry. Notice that the curves in Figure 18 for the north cardinal point converge near 1000 km altitude. This is related to the transition of the error arcs from an east-west orientation to an orientation circling the celestial ephemeris pole as the altitude increases. In particular, this convergence point occurs at the altitude where the "false celestial pole" is at the north cardinal point as was shown in Figure 9. The altitude where these curves converge varies with latitude. At high latitudes, this point occurs at a lower altitude than at low latitudes. For the southern hemisphere, Figure 19 (south cardinal direction) would exhibit this convergence point instead of Figure 18 (north cardinal direction). Figures 20 and 21 look very much like Figure 15 and Figure 16 (respectively), but the magnitude of the error is about half as much due to the dependence on cosine of the latitude.

CONCLUSION

Discontinuing leap seconds would fundamentally change the way civil time is defined. Civil time would no longer be kept close to (within 1 second of) UT1, it would no longer be tied to the orientation of Earth relative to the sun, and it would resemble another atomic time. The Space Community needs to develop a plan for upgrading operational software in case leap second discontinuance goes into effect. These error evaluation tools are important for assessing the potential operational impacts to space tracking software. These tools could also play a part in upgrading operational software, especially if the upgrades are made at the back end of legacy algorithms, after erroneous look angles and ranges have already been computed.

ACKNOWLEDGMENTS

I would like to acknowledge Mr. Bill Guilfolye and 1Lt Lacey Castaneda, both with Air Force Space Command, Analyses and Assessments Division for help in coding the MATLAB tools.

APPENDIX: FIGURES

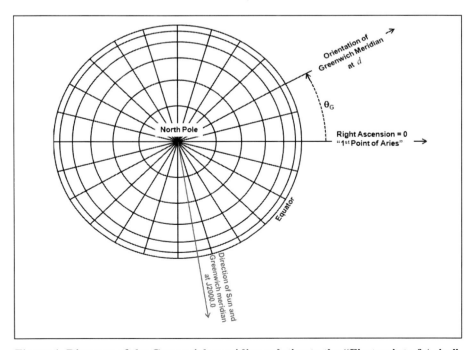

Figure 1. Diagram of the Greenwich meridian relative to the "First point of Aries".

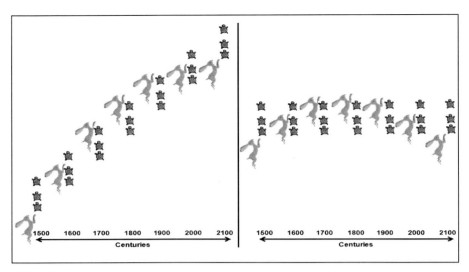

Figure 2. The "tortoise and the hare" analogy for atomic times and universal time.

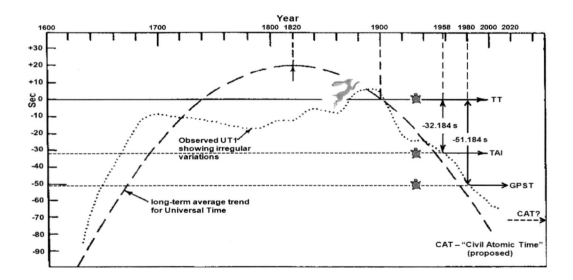

Figure 3. UT1 and atomic times over the last four centuries.

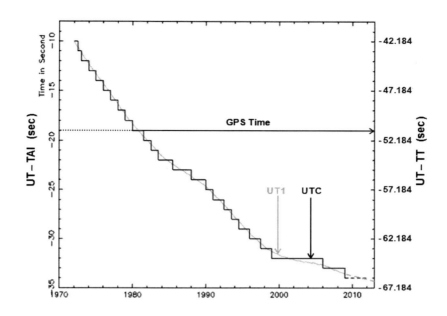

Figure 4. UT1 and UTC since 1972.

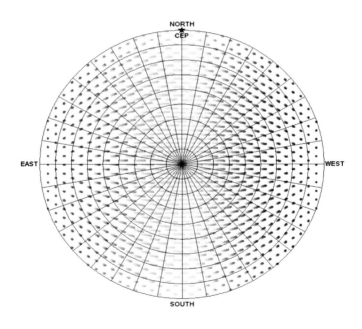

Figure 5. Sky Plot at Equator for space objects at 100 km altitude (DUT1 = 30 sec)

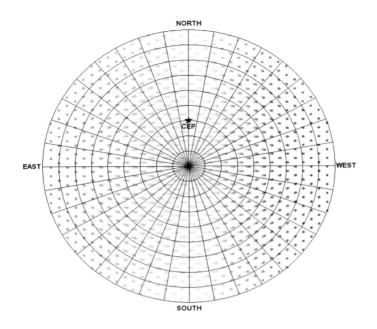

Figure 6. Sky Plot at 60° N latitude for space objects at 100 km altitude (DUT1 = 30 sec)

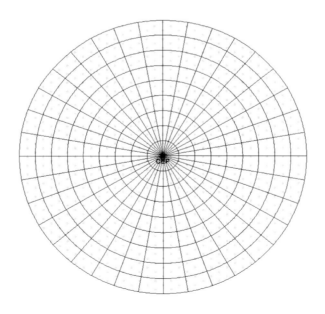

Figure 7. Sky Plot at 90° N latitude for space objects at 100 km altitude (DUT1 = 30 sec)

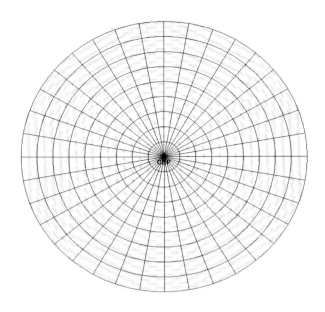

Figure 8. Sky Plot at 90° N latitude for space objects at 100 km altitude (DUT1 = 900 sec)

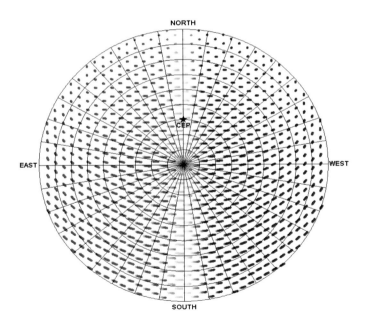

Figure 9. Sky Plot at 60° N latitude for space objects at 1000 km altitude (DUT1 = 300 sec)

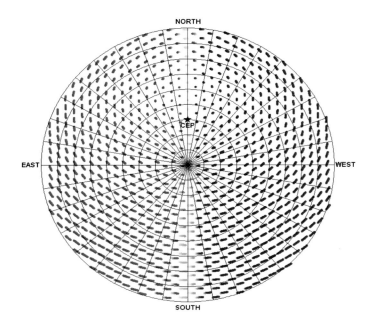

Figure 10. Sky Plot at 60° N latitude for space objects at 10,000 km altitude (DUT1 = 600 sec)

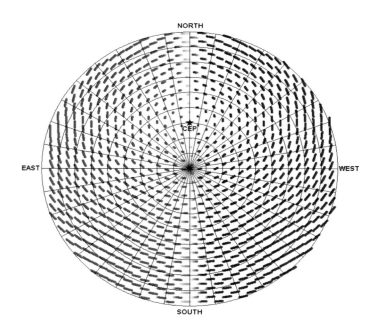

Figure 11. Sky Plot at 60° N latitude for space objects at 1,000,000 km altitude (DUT1 = 900 sec)

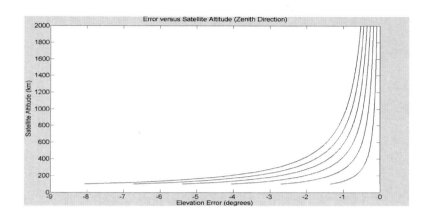

Figure 12. Elevation angle error at the equator when looking toward zenith

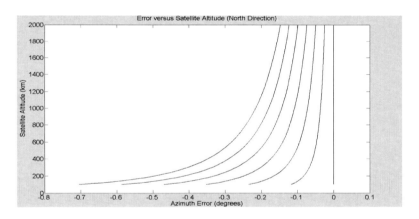

Figure 13. Azimuth angle error at the equator when looking north

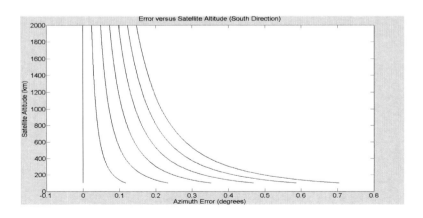

Figure 14. Azimuth angle error at the equator when looking south

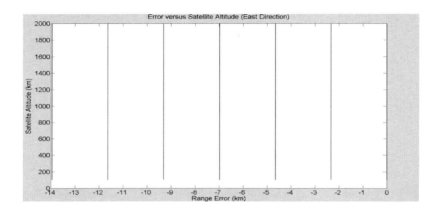

Figure 15. Range error at equator when looking east

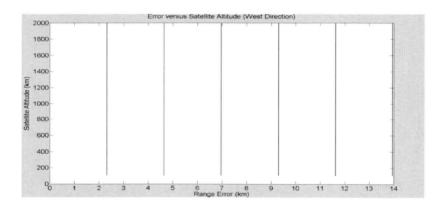

Figure 16. Range error at equator when looking west

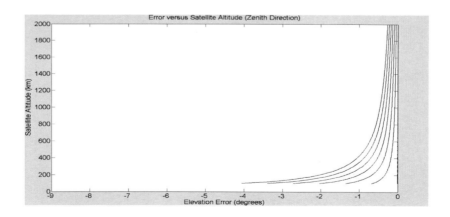

Figure 17. Elevation angle error at 60° N latitude when looking toward zenith

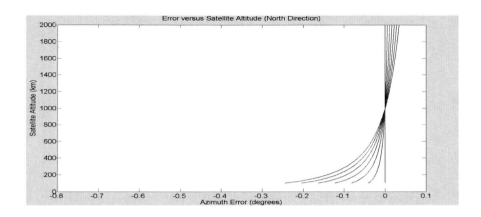

Figure 18. Azimuth angle error at 60° N latitude when looking north

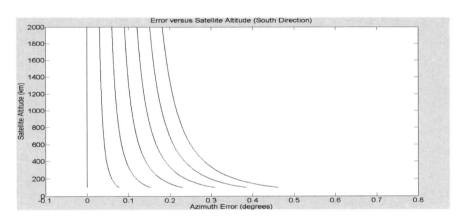

Figure 19. Azimuth angle error at 60° N latitude when looking south

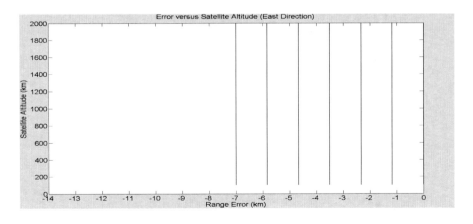

Figure 20. Range error at 60° N latitude when looking east

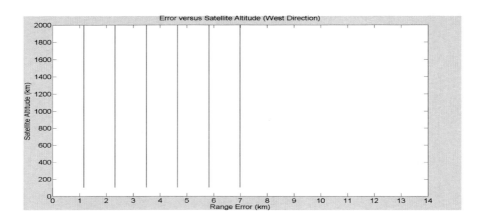

Figure 21. Range error at 60° N latitude when looking west

REFERENCES

[1] Finkleman, David; Seago, John H. and Seidelmann, P. Kenneth "The Debate over UTC and Leap Seconds." AIAA 2010-8391, AIAA Astrodynamics Specialist Conference, Toronto, Ontario, Canada, 2-5 August 2010.

[2] Nelson, D.D. McCarthy, S. Malys, J. Levine, B. Guinot, H. F. Fliegel, R. L. Beard and T. R. Bartholomew, "The leap second: its history and possible future." *Metrologia*, 38, pp 509-529, 2001.

[3] Seago, John H. and Storz, Mark F., "UTC Redefinition and Space and Satellite-Tracking Systems," in: *Proceedings of the ITU-R SRG Colloquium on the UTC Timescale,* IEN Galileo Ferraris, Torino, Italy, 28-29 May 2003

[4] Escobal, P. R.; *Methods of Orbit Determination*, Appendix I, "A compendium of thirty-six basic coordinate transformations," R.E. Krieger Publishing Co., 1976.

DISCUSSION CONCLUDING AAS 11-674

Dennis McCarthy remarked that a certain level of permissible inaccuracy exists whenever UTC is used as a surrogate for UT1. He asked if there was a minimum threshold for accuracy established for the systems familiar to Mark Storz. Storz responded that error tolerances would likely be system-dependent. He noted that the field of view constraints on space surveillance sensors might provide one example: for target acquisition, a sensor with a smaller field of view requires more accurate pointing. McCarthy asked if a numerical specification could be cited; Storz explained by citing the US Space Surveillance Network (SSN) as an example. He noted that these varied tracking systems make recurring observations of orbiting space objects because the orbital behavior cannot be predicted perfectly. These systems rely on the Simplified General Perturbations Theory #4 (SGP4). This theory, and the entire network, assumes that the wall clock time is a measure of Earth rotation in order to predict where objects will appear within the field of regard of each sensor. Another process likely to be adversely affected by discontinuing leap seconds is *observation association*; this is the task of tagging observations of space objects taken relative to the terrestrial frame to the predicted ephemerides from known orbits relative to a celestial frame. If the relationship between the terrestrial frame and celestial frame becomes too imprecise, there is a risk of tagging the observation to the wrong object, or not tagging the observation at all, thus treating the object as an uncorrelated target.

McCarthy replied that "one second seems to be okay," so what level of discrepancy would cause operational issues or other adverse reactions? Storz speculated that higher frequency radars such as X-band might start having issues as soon as five years out, but also noted that the issue has not been studied thus far and would need to be investigated. He also noted that optical trackers such as Ground-based Electro-Optical Deep Space Surveillance (GEODSS) telescopes have relatively narrow fields of view. GEODSS telescopes use reference stars to accurately calibrate the field of view in right ascension and declination. However, if the Earth orientation angle is computed incorrectly or inaccurately, then the computed right ascension and declination of the space object will be erroneous and potentially outside the telescope's field of view. Based on Storz's comments, McCarthy deduced that space surveillance systems may start having operational difficulties by about 2030. Storz clarified that some operational failures are to be anticipated much sooner because of software that is constrained to assume the |UT1-UTC| is limited to 0.9 seconds.

McCarthy asked if some space-based sensors might self-calibrate their orientation relative to space using star sensors. Storz said that star calibration may be used to calibrate the orientation of gimbal-tracking encoders relative to an inertial frame, but the predicted surface target positioning will still be incorrect unless the longitude of the target is adjusted for UT1-UTC. Seaman wondered if large time differences could ever induce non-linear feedback issues like gimbal lock or problems with servo loops in tracking systems. Storz said that the first task will be to review and revise system-wide requirements to specify that DUT1 is now required where it has never been required before, and that unbounded growth of DUT1 must not cause adverse system behaviors. Terrett said that Earth-resource satellites must account for UT1-UTC for imaging resolutions better than one kilometer, but lower-resolution imaging with older spacecraft haven't had to worry about it. He clarified that space-borne software and some older systems simply cannot be updated.

AAS 11-675

PROPOSAL FOR THE REDEFINITION OF UTC: INFLUENCE ON NGA EARTH ORIENTATION PREDICTIONS AND GPS OPERATIONS[*]

Stephen Malys[†]

The Earth-Centered, Earth-Fixed World Geodetic System 1984 (WGS 84) reference frame is realized by the adoption of a self-consistent set of highly-accurate coordinates for the Department of Defense (DoD) Global Positioning System (GPS) Monitor Stations. Over the past several decades, procedures, Interface Control Documents, and operational, configuration-controlled software have been designed for UT1-UTC to be less than 1.0 second. In some cases, automated 'limit checks' have been established on this parameter. The proposed discontinuation of leap seconds and redefinition of UTC will impact these operations. A significant amount of time, effort, and funding will be required for NGA and other organizations to identify and assess all operational software impacted by the change. While the proposal may benefit other communities, a redefinition of UTC and the elimination of leap seconds offer no benefits or improvements to National Geospatial-Intelligence Agency (NGA) or GPS operations.

INTRODUCTION

Over the past several years, an International Telecommunication Union – Radiocommunication Sector (ITU-R) Study Group has been investigating the following three questions:

1. What are the requirements for globally-accepted time scales for use both in navigation/telecommunication systems, and for civil time keeping?

2. What are the present and future requirements for the tolerance limit between UTC and UT1?

3. Does the current leap second procedure satisfy user needs or should an alternative procedure be developed?

The goal of this effort is to make recommendations to the greater ITU and ultimately to the international community as a whole. This paper briefly assesses the effects of changing the definition of UTC on NGA's mission.

---

[*] Approved for Public Release Case 11-469.

[†] Senior Scientist for Geodesy and Geophysics, InnoVision Basic and Applied Research Office, National Geospatial-Intelligence Agency, 7500 GEOINT Drive, Springfield, Virginia 22150, U.S.A.

NGA is responsible for maintaining the WGS 84 global geodetic reference frame and other related geophysical models. The WGS 84 Earth-Centered, Earth-Fixed reference frame is realized by the adoption of a self-consistent set of highly-accurate coordinates for the DoD GPS Monitor Station antennas located around the world (Figure 1). These monitor station antennas are attached to the crust of the rotating Earth. The GPS time scale used for all GPS orbit processing is directly tied to UTC (USNO) through well-established, operational processes. The orbit determination process requires transformation between the Earth-Centered, Earth-Fixed reference frame and the Earth-Centered Inertial reference frame and therefore knowledge and accurate prediction of small changes to the Earth's rotation angle (UT1). The quantity UT1-UTC is used in these transformations and will remain an essential bit of information for GPS and other satellite missions requiring high-accuracy orbit determination.

Over the past several decades, procedures, Interface Control Documents (ICD), and operational, configuration-controlled software have been designed to deal with Leap Seconds. In some cases, automated 'limit checks' have been established that test for the UT1-UTC quantify to be less than 1.0 second. The ICD-GPS-211, the ICD between NGA and the GPS Operational Control Segment (OCS), for example, describes such limit checks.

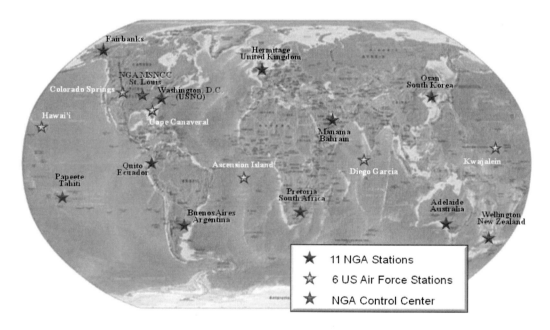

Figure 1. DoD GPS Monitor Station Network.

NGA OPERATIONAL SUPPORT

The NGA GPS Network Control Center supplies 24/7 real-time support to the U.S. Air Force 2nd Space Operations Squadron's (USAF 2SOP's) GPS operations. This support is a collaborative effort between NGA and the USAF that includes the sharing of data, anomaly resolution, and collaboration on GPS performance monitoring. NGA's GPS tracking data directly improve GPS broadcast accuracies and integrity monitoring. NGA also routinely generates its own precise (post-fit) GPS ephemerides in support of world-wide geodetic surveying and quality control for GPS operations.

Daily Earth Orientation Parameter (EOP) predictions and post-fit estimates, generated using USNO's daily EOP solutions, are used in earth-fixed to inertial reference frame transformations. The required accuracy of these EOP predictions are documented in the Interface Control Document ICD-GPS-211 and shown in Table 1. Currently, ICD-GPS-211 between NGA and the USAF 2SOPS indicates that the UT1-UTC parameter is to be maintained to less than 1.0 second. This ICD will soon be replaced by ICD-GPS-811. This ICD defines the requirements for data transfer between NGA and the USAF's Next Generation Operational Control Segment known as 'OCX'. This new interface treats the UT1-UTC parameter in a similar fashion. Because this parameter is now limited to 1.0 second or less the automated 'Limit Checks' have been established in code to assure compliance with current and future ICDs. Therefore, the handling of the leap second has been 'institutionalized' within procedures, software, and documentation.

Table 1. ICD-GPS-211D Accuracy Requirements for EOP predictions*

Predictions (days)	Polar Motion X & Y (mas)	UT1-UTC (ms)
1	3	3
7	7	5
14	10	7

*One standard deviation measured over any 1-year period

EFFECT ON GPS OPERATIONS

As indicated earlier, the ICDs state requirements for the transfer of data between NGA, GPS OCS, and other satellite operations. Replacing the current definition of UTC with a continuous time scale will require the modification of these documents. It will also require the identification, modification, testing, and validation of operational software to allow UT1-UTC to grow beyond the current 1.0 second limit.

While eliminating this 'sanity check' may seem trivial, changes to the code, documentation and execution of thorough testing will require resources and in some cases, contract modifications. The costs and time needed for the initial investigation and subsequently to make the required changes to the operational software are unknown at this time, but they are expected to be significant. Furthermore, considering that this possible redefinition of UTC and the elimination of the leap second offer no benefits to NGA GPS operations and GPS users, pursuit of such a fundamental change appears to be an inefficient use of limited resources.

CONCLUSION

DoD GPS Monitor Stations, the 'starting point' for all GPS Positioning, Navigation and Timing (PNT) are located on the rotating Earth and will be for the foreseeable future. Therefore, a regular source of UT1-UTC predictions will continue to be needed for NGA, GPS, and other DoD satellite operations.

The proposed discontinuation of leap seconds and redefinition of UTC will impact the operational software and automated transfer of Earth Orientation Prediction Parameters between NGA, the GPS OCS, and other DoD organizations. A significant amount of time, effort, and funding will be required for NGA and other organizations to identify and assess all operational software that references, tests for, or applies the UT1-UTC parameter in high-accuracy orbit determination processes. The costs and time needed for the required changes to the operational software and ICDs are unknown at this time, but they are expected to be significant.

While the proposal to re-define UTC may offer benefits to other communities, a redefinition of UTC and the resulting elimination of leap seconds offer no benefit or improvement to NGA or GPS operations. Our recommendation, therefore, would be to maintain the current definition of UTC.

ACKNOWLEDGMENTS

I would like to thank Dr. Thomas Johnson for his assistance in preparing this manuscript.

DISCUSSION CONCLUDING AAS 11-675

Ken Seidelmann noted that arguments favoring the abolition of leap seconds are often posed as somehow benefiting the GPS program, which Steve Malys' presentation seemed to deny. Dennis McCarthy countered that he has heard no such arguments. John Seago noted that Seidelmann's perception may be accurate, yet based on the fact that some GPS receiver technology may benefit from changes to UTC, rather than the GPS program itself. McCarthy agreed that this is possible because GPS time receivers have a spotty record in their handling of leap seconds. His first-hand experience with such receivers suggests that some manufacturers neglect the details of the GPS interface control document (ICD). Malys agreed that many manufacturers fail to honor the GPS ICD specifications.

Rob Seaman reported that he attended a GPS birds-of-a-feather (BOF) meeting at JavaOne many years ago, and the discussions by the attendees—many of whom write embedded applications for GPS receivers—lacked a depth of understanding compared to the discussions within the colloquium. McCarthy guessed that perhaps 10% to 20% of GPS timing receivers do not handle leap seconds well; the proper treatment of leap seconds may now be overlooked due their declining frequency in recent years. Malys clarified that these issues do not affect positional accuracy, and it was agreed that the current discussion was limited to GPS time receivers. Seidelmann noted that one receiver manufacturer failed to account for the leap years properly.

Seaman noted that the issue of leap seconds is a two-edged sword. While developers like Poul Henning-Kamp have testified that there are computer science issues surrounding the introduction of leaps seconds, Seaman's own experience "as a computer-science type of guy" suggested that the abolition of leap seconds will certainly make his tasks more difficult rather than less difficult. Either option—keeping leap seconds or abolishing them—would have consequences. The previous three presentations affirmed that the issues remain mostly unstudied up to now, and decisions should not be made until the issues are studied carefully and the trade-offs are thoroughly understood.

Reed opined that a changeover might be hard for legacy software, but for future software it should be "definitely easier". Seaman countered by way of example. He is on the Large Synoptic Survey Telescope (LSST) data management list; he sees many LSST messages and has even attended LSST meetings focused on timekeeping, and it seemed that timekeeping issues were not "on their radar." One should expect legacy software to be recycled without guidance on these issues, and the current process as a whole has not provided guidance. LSST is expected to deploy about the time that leap second would go away, but that project hasn't been giving the issue any attention.

Based on Malys' presentation, John Seago was impressed that it might not only be expensive to retire intercalary adjustments, but it seemed almost impossible to reintroduce them once they were taken away. Any sort of an alternative proposal involving a less-frequent adjustment, such as something like a "leap minute", reduces the visibility of the issue with system designers and does not motivate designers to accommodate them. Seago then referred back to McCarthy's earlier point that the current rarity of leap seconds already causes some people to neglect them now;

loosening the current degree of coupling would foster a situation whereby the re-alignment of the background time scale for civil timekeeping to Earth rotation becomes impractical.

David Simpson commented that there seems to be a prevailing attitude that it will take money to investigate the impact of making a change, yet no one wants to make that investment until a decision has been made. However, once the decision has been made, it then becomes too late for cost estimates to affect the decision.

Session 6:
GROUND OPERATIONS

AAS 11-676

UTC AT THE HARVARD-SMITHSONIAN CENTER FOR ASTROPHYSICS (CFA) AND ENVIRONS

Arnold H. Rots[*]

The Smithsonian Astrophysical Observatory is involved in the operation of observatories across the entire electromagnetic spectrum and associated with several other data providers. Although the reliance on UTC and constraints on the value of DUT1 vary, there is considerable apprehension about changing the definition. In particular there is a sense that it may cause considerable confusion and misunderstanding in the context of the Virtual Observatory.

STATUS AT SAO

At the CfA, the Smithsonian Astrophysical Observatory (SAO) operates (or is a partner in the operations of) observatories across the entire spectrum. Some of these "run on" UTC, but there are differences when it comes to whether the observatories depend on $|DUT1| < 0.9$ s. I will provide an overview of how the observatories and related activities would be affected by the abolition of leap seconds in UTC.

Sub-Millimeter Array

The Sub-Millimeter Array (SMA, on Mauna Kea, HI) synchronizes its clock to GPS and keeps track of leap seconds and *DUT1* as published by International Earth Rotation and Reference Systems Service (IERS). As long as IERS continues to publish *DUT1* there will be no impact.

Ground-based Optical Observatories

The optical ground-based observatories are not terribly sensitive to UTC definition changes, but it would help them if a timescale were maintained that is related to earth rotation. This may sound cryptic, but it does express a preference to leave things as they are – because one never knows.

Chandra X-Ray Observatory

The Chandra X-Ray Observatory is space-based, where UTC is an irrelevant concept. Chandra's clock is calibrated in UTC, but the first thing we do is to convert it to Terrestrial Time (TT) for science data. Not all space missions manage time in the same way on board. Chandra's onboard clock essentially runs on linear time. But since commanding is done in UTC, it needs to be made aware of what UTC is doing. As a result, each leap second creates a problem, since it is never inserted at the right moment, but rather when internal ephemerides are (necessarily) up-

---

[*] Archive astrophysicist, Chandra X-ray Center, Smithsonian Astrophysical Observatory, 60 Garden Street – MS 67, Cambridge, Massachusetts 02138, U.S.A.

dated to include that second. Specifically, it is inserted while no scientific observing is going on and there is always something that goes wrong; at the last leap second insertion orbit ephemeris files needed to be recreated due to a 1-second error that was found several weeks after the fact. So, in a sense, Chandra would be better off without any leap seconds. On the other hand, the idea that one would use UTC on board a spacecraft (rather than, say, TT or TAI) is a bit absurd by itself. Even for ground-to-space communications a 1-second error in the on-board clock amounts to no more than a 3-arcminute offset, which is well within the spacecraft antenna's beam size.

VERITAS

VERITAS (Mt Hopkins, AZ) observes at the very high energy end. It uses GPS clocks and is in the same position as the optical observatories: as long as IERS continues to publish $DUT1$ or UT1 – UTC, they will not be affected by any changes. VERITAS commonly works at 10 μs precision.

Minor Planet Center

The Minor Planet Center (MPC) converts all its times to TT, but prefers UTC to stay close to earth rotation, since it is close to UT1 and thereby reduces chances of confusion.

BEYOND SAO

There is strong involvement in the Virtual Observatory (VO) at CfA. Although the VO can in principle ensure, through its metadata standards, that time is handled correctly, one needs to be extremely careful. Data in the VO comes from many places all over the world and, even though that may be regrettable, there is no guarantee that all of them will get it right.

CfA has always had a close relationship with the American Association of Variable Star Observers (AAVSO). Arne Henden, its director, did not think the disappearance of leap seconds would cause great problems for AAVSO members, but summarized his position as: "Don't fix it if it ain't broke" – meaning: keep UTC as it is.

CONCLUDING REMARKS

There is a consideration regarding ground-based moderately-high accuracy timing data. Right now we know that, with the observatory's ITRS coordinates and UTC, one can get timing accurate to $1.5 \times \cos(latitude)$ μs. That will not hold true anymore if $|DUT1|$ is not constrained to be less than 0.9 s.

Staff at the SAO is involved in the formulation of two standards in astronomy that touch on time issues. The Flexible Image Transport System (FITS) World Coordinate System (WCS) standard for Time is finally getting close to becoming reality. The International Virtual Observatory Alliance (IVOA) has had a Space-Time Coordinate metadata standard since 2007. Both provide guidance for the proper usage of time scales and the completeness of metadata. However, both are also agnostic with respect to any constraints on $DUT1$.

The bottom line that emerges from polling users across the spectrum is that people are apprehensive about changing something that has worked for a very long time. It may not break everybody's software, but it is guaranteed to cause confusion. This is particularly true for the Virtual Observatory where we have little control over how data providers and users choose to interpret the standards and where they may or may not be aware of any implicit assumptions (*e.g.*, about constraints on the value of $DUT1$).

The inescapable conclusion appears to be that, if time services prefer to broadcast a linear timescale, it would be preferable that they choose a proper one (TAI would be an excellent choice), continue the leap seconds in UTC, and include TAI − UTC, as well as the traditional *DUT1* or UT1 − UTC, in the distribution.

ACKNOWLEDGEMENTS

The Chandra X-ray Center is operated by SAO under contact NAS 8-03060 with NASA.

DISCUSSION CONCLUDING AAS 11-676

The need to broadcast DTAI (TAI-UTC) was questioned because it is such a slowly changing quantity. Arnold Rots responded that the broadcast of DTAI was needed because it addresses a software management issue: if DTAI is included in future timing broadcasts, then any software that uses the information will automatically pick up the leap seconds. If leap seconds have to be manually inserted into software systems, a software-change request must be submitted every time a leap second is introduced because files need to be updated and tested.

George Kaplan asked about the astronomical communities that perform studies involving precise radial velocity (*e.g.*, exo-planet studies) and whether they might need to know the Earth rotation rate very accurately. Terrett suggested that these studies may be spectroscopic in nature and therefore may not be overly sensitive to UT1-UTC. Allen noted that barycentric corrections must be applied in the analysis of exo-planetary data but the accuracy requirements are on the order of 5 to 30 seconds to keep the barycentric correction from introducing unwanted noise. After some discussion of relative versus absolute measurements, Rots clarified that the point is that an observer's velocity vector along the line of sight is not going to change very much in one second. It is assumed that observatories that do these analyses are aware of DTAI whether or not leap seconds are being introduced.

Rob Seaman noted that effectively Rots had conducted another survey, internal to SAO. With this understanding, Seaman asked if Rots had any sense of the degree of attention his respondents were giving to the issue, given the fact that many respondents were apprehensive to change and that many of Rots' reported responses were seemingly terse. Rots confirmed that it was his impression that, on the whole, there is not much awareness of this issue, a typical response perhaps being "Oh yeah, there is something like that going on; we aren't worried much about it but it would be nice if you didn't change it [UTC]." Nevertheless, because of the way the vast majority of SAO systems tend to work, he supposed that changes to the definition of UTC will not be much of an issue either way. What seemed to be of much more concern was the possible confusion that would result; the one thing that is generally understood by people with any sensibility of timekeeping and time scales is that UTC is approximately UT1.

Reed commented that it was his impression that there is a conversion of American Association of Variable Star Observers (AAVSO) data to Terrestrial Time (TT) or the equivalent and in that sort of situation the cessation of leap seconds should provide reduced complexity. Rots replied that coping with leap seconds has not seemed to be a problem in those situations; also, there are probably observers in such programs who might become more confused by the disappearance of leap seconds in UTC. John Seago asked if more or less confusion might result if the terms "UTC" and "Coordinated Universal Time" were discontinued along with the leap second. Rots replied that the absence of something called UTC might be even more confusing to some observers.

AAS 11-677

AN INVENTORY OF UTC DEPENDENCIES FOR IRAF

Rob Seaman[*]

The Image Reduction and Analysis Facility is a scientific image processing package widely used throughout the astronomical community. IRAF has been developed and distributed by the National Optical Astronomy Observatory in Tucson, Arizona since the early 1980's. Other observatories and projects have written many dozens of layered external application packages. More than ten thousand journal articles acknowledge the use of IRAF and thousands of professional astronomers rely on it. As with many other classes of astronomical software, IRAF depends on Universal Time (UT) in many modules throughout its codebase. The author was the Y2K lead for IRAF in the late 1990's. A conservative underestimate of the initial inventory of UTC "hits" in IRAF (*e.g.*, from search terms like "UT", "GMT" and "MJD") contains several times as many files as the corresponding Y2K ("millennium bug") inventory did in the 1990's. We will discuss dependencies of IRAF upon Coordinated Universal Time, and implications of these for the broader astronomical community.

INTRODUCTION

The practice of astronomy is very heavily dependent on software, precisely because it is the most observationally oriented of the sciences; experiments in astrophysics are typically conducted with computer models of one sort or another. Astronomical software falls into several classes and is developed and maintained by numerous institutions; much is open source, but proprietary packages are also prevalent. The users of these applications and systems are diverse, from unprecedented professional research projects to "citizen science" with the general public. Systems layered on astronomical software are deployed in some of the most extreme environments on Earth, from remote mountaintops to beneath the ice of the South Pole, from spacecraft and airplanes and balloons to submarine sensor arrays. Astronomical software systems may be networked between several continents or assets on-orbit – or they may have to operate untended in isolation for months or years.

One class of astronomical software is the desktop data reduction and analysis package. The Image Reduction and Analysis Facility[†] (IRAF), developed by the National Optical Astronomy Observatory in Tucson, Arizona, has been the premier such package for three decades. The installed base is several thousand machines on all continents. More than 10,000 refereed journal articles have cited the use of IRAF.[1] IRAF is also used in a variety of pipeline processing and workflow applications.

---

[*] Senior Software Systems Engineer, National Optical Astronomy Observatory, 950 N Cherry Ave., Tucson, Arizona 85719, U.S.A.

[†] http://iraf.net.

A contributing factor to the longevity of IRAF is its extreme portability. The exact same applications code runs on numerous flavors of both BSD and System V Unix, but also in the past under VAX/VMS and Data General's AOS, for instance. It achieves this portability via an architecture of a rich Virtual Operating System (VOS) layered on top of an isolated kernel, but also via a controlled programming environment including its own SPP language and library APIs.

The very broad usage and long lifetime of IRAF has resulted in the creation of numerous external packages layered on the core capabilities and application tasks of the system. For all these reasons, IRAF provides a useful test case for collecting remediation requirements should fundamental astronomical standards and protocols be redefined.

COORDINATED UNIVERSAL TIME IN ASTRONOMICAL SOFTWARE

Time has always been important within astronomy, and in turn astronomy has performed important duties in timekeeping.[2] Astronomers make intensive use of many different time scales, including Coordinated Universal Time (UTC). While usage varies, the general case is that UTC is presumed to approximate Universal Time (UT), that is, mean solar time in Greenwich, England.[*] The use of UT permits simple deterministic algorithms to determine Earth orientation for many purposes, including pointing telescopes. UTC also provides broadly understood timestamps as with any other community or class of software. And UTC is typically an intermediate step in recovering either International Atomic Time (TAI) or predictions of UT1.[3]

If UTC were to be redefined to no longer provide a representation of Universal Time, there would be a variety of impacts on astronomical software and software systems for related communities. Since one second of time is 15 seconds of arc—a large angular distance for most astronomical purposes—this change would need to be mitigated immediately in the astronomy community. Mitigation requirements would include the following.

Software and systems assuming UTC ≈ UT would need to follow one of at least three paths:

1. be rewritten to explicitly distinguish between these two newly separated meanings of "Universal Time",
2. be isolated to receive a vetted UT1 input, or
3. be retired and/or replaced.

Software and systems currently accommodating a DUT1 correction would have to take both of these steps:

4. be vetted for proper operation under values of DUT1 > ±0.9s, and
5. be isolated to receive a vetted DUT1 input, likely from a new source.

Some software modules would have to accommodate all of the above. Mitigation of astronomical software and systems to accommodate a redefined UTC standard would require performing an inventory of all potentially affected modules, and building a plan for remediating each file. Some files would be rewritten and others retired. New library and application code would be needed. New infrastructure for delivering new (and currently undefined) UT1 and/or DUT1 time signals would have to be designed, deployed, and maintained indefinitely.

---

[*] Per CCIR Recommendation 460-4, "GMT may be regarded as the general equivalent of UT."

CLASSES OF ASTRONOMICAL SOFTWARE

We cannot hope here to cover the range of astronomical software, let alone the much greater varieties of software pertaining to other communities. IRAF is representative of the class of desktop data-reduction software packages used by optical and infrared observational astronomers. Astronomy is itself a broad and often compartmentalized discipline. To give a sense for the scale of the problem, here is a non-exhaustive list of some other types of astronomical software, in particular related to nighttime ground-based optical/infrared observational projects:

Observing preparation tools

Astronomers no longer peer through the eyepieces of telescopes. Rather, they collect photons using complex cameras and plan their observing sessions in advance in detail. This includes contingencies based on such factors as the weather and the whirling celestial sphere – that is, taking into account Earth orientation. A particular night's worth of data will often form just a small part of a larger "campaign" of observations, for instance, perhaps a survey of vast areas of the sky. Access to the facilities will often involve writing a detailed proposal for how the camera and telescope will be used. Often more than one telescope will be used in a coordinated fashion to combine target acquisition with spectrographic follow-up (for instance).

Innumerable software tools are implicit to the previous paragraph. These include "phase 1" and "phase 2" planning – roughly divided between issues generic to the objects being studied and issues specific to the facilities that will be used to study them. Other tools include classes of tailored calculator applications to estimate parameters such as how long the camera's shutter must remain open to adequately expose a previously unobserved celestial object. Focal-plane masks must be prepared for multi-object spectrographs – masks that include calculated effects such as differential atmospheric refraction, thus depending on the Earth orientation via the topocentric coordinates of the targets.

Often the astronomer does not perform the observations herself, but rather a staff observer will construct a nightly queue schedule combining targets from several different programs. This will have to adapt to changing conditions as the Earth rotates.

Astrometry & catalogs

Before a celestial object is observed, its location must be cataloged. The history of astronomy is written in a succession of laboriously constructed catalogs of different types of objects. Each target in the sky is first discovered in some type of coherently designed survey, and then its location is computed using the esoteric techniques of astrometry such that it can be located again for further study by other astronomers or using other instruments. Celestial objects vary in both intrinsic ways (e.g., pulsating variable stars) and extrinsic ways (e.g., eclipsing variables). They move themselves and the Earth moves underneath them. All of these effects must be measured and cataloged using tools and data formats that are very dependent on the definition of UTC.

Telescope control software

Given an observing plan and coordinate information, a telescope must acquire and track the objects. This requires a real-time clock and innumerable calculations back-and-forth from one coordinate system to another. The concept of Universal Time is embedded throughout these systems, for example, in the pointing model that links the hardware servos and mechanical stresses on the trusses of the telescope structure to the rotating celestial coordinates.

Telescopes are very widely divergent mechanisms. Two telescopes may resemble each other less than a digital clock does a sundial. Telescopes that are steered via hour angle and declination rely on a direct connection between their operation and Earth orientation. Many modern telescopes point using altitude and azimuth, and convert to celestial coordinates in software. And some telescopes have no moving parts at all, modeling the swirling universe entirely in the computer. Telescopes of all types may have to accommodate observations of sources that are not fixed on the celestial sphere in the first place, thus complicating the pointing further.

Modern telescopes are complex systems of systems in which timekeeping signals are exchanged via messaging protocols, whose telemetry is logged and persisted for maintenance purposes over many years, whose status is provided to diverse users for different technical and scientific purposes, and that often engage in coordinated activities with other telescopes. Pertinent time scales may be very brief (fractions of a second) or very long (decades).

Instrument control software

If anything, two astronomical instruments may resemble each other even less than two telescopes. Only some can be described as cameras, but these may have CCDs and shutters like a Nikon and thus the need to attach a time stamp to the resulting exposure. Others detect electromagnetic radiation from radio to gamma rays, or non-EM phenomena such as neutrinos or gravitational waves. For these applications, space-time coordinates are often complexly intertwined.

The software for controlling the instruments is equally variable. In addition to time stamps, each observation will be tagged with innumerable bits of scientific and logistical metadata, much of which is related either directly or indirectly to timekeeping and thus UTC. Most instruments support the concept of observing sequences that may include different types of exposures (for those instruments for which "exposure" is a coherent concept). These sequences place additional constraints on timekeeping. Many instruments support advanced observing modes of one type or another that may involve coordinated observing with external facilities (and their clocks).

Data handling & transport

Observatories are located in remote locations such as distant mountaintops, orbital platforms, in the ice beneath the South Pole, under the Mediterranean Sea, in salt mines. On the other hand, the modern scientific paradigm relies on large collaborations of partners at institutions and universities around the world. Data collected from esoteric instruments on one-of-a-kind telescopes must be quickly and reliably transported from distant lands to diverse partners, if it is be useful. The data may cross many timezones each day, and each year a research project may need to accommodate several changes in distinct daylight-saving rules. UTC is embedded throughout the infrastructure and operational logistics of astronomical data handling.

Archiving

Each snapshot of the universe is different. Astronomical facilities are rare resources; the resulting data are expensive; their archival value increases with time. Both space and ground-based observatories have seen an ever-increasing trend in support of the building of publicly accessible archives. The scientific interpretation of archival content relies on timekeeping. Their permanent value depends on capturing and curating the historical meaning of UTC and other time scales.

Pipeline processing

Modern astronomical projects, just like science of all types, rely on the automated pipeline processing of large and continually growing data sets. Needless to say, this processing often cares very deeply about Earth orientation. Universal Time has often been used as the foundation for these pipelines. A pipeline is not so much written as it is commissioned. Preserving its detailed behavior under differing inputs is an explicit goal. A small change to the logistics of the pipeline may require completely re-commissioning the pipeline and perhaps having to re-ingest the entire data set. Redefining UTC is not a small change.

Virtual Observatory & astro-informatics

The future of astronomical software is some combination of trends including astro-informatics and the Virtual Observatory. As with bio-informatics and similar domains, astro-informatics is the application of semantic technologies to the coherent management of astronomical data with the goal of realizing a transition to a "fourth paradigm" of scientific enquiry.[4] As with other virtual technologies, the Virtual Observatory is a distributed service-oriented grid-enabled infrastructure tying together community assets in a scalable architecture. These activities are, however, layered on exactly the same standards as traditional astronomy. What they bring to the table is a much more complex set of dependencies on UTC.

Time domain astronomy

Recent progress in the science of astronomy has often focused on the "time domain", that is, on variability in astronomical objects and transient events on the celestial sphere. The characterization and astrophysical comprehension of these processes depends on coherently accumulating time series data, often across a broad spectral range and thus coordinating observations made using every category of software listed above. A physically coherent model of timekeeping is required to have any possibility of achieving these goals. A redefinition of a fundamental scientific time scale cannot fail to leave signatures in the data that distinguish between data taken before and after the change is made.

PERFORMING A UTC INVENTORY

Some aspects of the proposed redefinition of UTC bear a striking similarity to the "Millennium Bug", also known as the Y2K (for "Year 2000") bug. In particular, the removal of the 0.9s limit on DUT1 would embed a similar ticking time bomb in unexamined code. As a reminder, the key issue with the Y2K situation was that software data structures with two-digit year representations had an implicit underlying assumption of 20th Century dates that would be violated as the clock and calendar turned over on New Year's Day, 2000. The implications were diverse and broad; the necessary fixes were extremely widespread. The total cost of Y2K remediation activities is estimated to have been in excess of 300 billion U.S. dollars.[5]

Other aspects of the redefinition of UTC are quite different. Where the fundamental fix for Y2K was simply to replace instances of two-digit years with four-digit fields,[*] accommodating an entirely new concept of civil timekeeping (distinct from mean solar time) will require far more subtle (and thus more expensive) changes. What then is the likely cost for mitigating a redefinition of UTC?

The answer is that nobody knows. Proponents of making such a change to the meaning of UTC make two assertions: 1) the affected codebase is much smaller, and 2) there is also a cost for issuing leap seconds. Each of these is rather an argument for conducting a coherent inventory. Is the range of affected software systems smaller? Then it will be quicker and easier to complete the inventory. (Or perhaps the inventory would be larger and broader than currently imagined.) Is there a cost for leap seconds? Then this should form part of the inventory. (Or just possibly the ongoing cost for leap seconds would be found to be negligible.)

However, the most fundamental observation is that this is not a zero-sum issue. Ceasing the issuance of future leap seconds as proposed would certainly require extensive code changes in

---

[*] This itself is not a permanent fix, since the same issue will recur in the year 10,000.

astronomical and related software, but it would not remove all the costs related to past leap seconds. Redefining the fundamental civil time scale has the effect of partitioning the handling of dates into separate rules before and after the change was made. Since many timekeeping use cases are retrospective in nature as with archives and databases, knowledge of and handling for prior leap second instances will be a permanent future responsibility.

A BRIEF UTC INVENTORY OF IRAF

A UTC mitigation plan for IRAF would bear many similarities to the IRAF Y2K remediation plan.[*] The main difference, of course, is that the timing of the Y2K crisis was forced by external events whereas a UTC crisis would be created by the actions of the Radiocommunication Sector of the International Telecommunication Union. The first step in constructing such a plan is to perform an inventory of potentially affected files. This was relatively straightforward for Y2K and involved searching for strings like "century", "year", and "19". Such an inventory will be much harder to conduct for UTC – not only will be there many more possible search terms, but there can be no guarantee that a particular module is secure from implicit dependencies. Note that the Y2K remediation project for IRAF consumed a significant fraction of the effort of several group members for three calendar years, perhaps totaling 1.5-2.0 FTE-years overall at NOAO.

The search terms also will vary from software package to software package. For IRAF, an initial inventory has been performed with these terms, roughly in descending order of efficiency in generating good hits:

- UT, UTC, GMT, JD, MJD, DUT, LST
- Hour, minute, second
- Year, month, day
- Solar, sidereal
- Clock, calendar

Other terms are too general:

- Date, time

And others simply do not appear:

- Leap second
- Intercalary

A "good" hit is a file with a plausible connection to timekeeping. With Y2K the search terms often resulted in a short list of hits with a high yield of needed changes. With UTC the lists are often longer (or non-existent); it remains to be seen how efficient the corresponding yield turns out to be. The ultimate goal is to identify all files requiring mitigation, without fail. For complex scientific code this requires human review to comprehend the intent of the algorithms in each file and the data formats, data structures, interfaces, and documentation that connect the system into a unified and useful whole.

---

[*] http://iraf.noao.edu/projects/y2k/y2kplan.html

Table 1 contains a count of the number of files containing each of the search terms so far identified. The final inventory may well rely on additional search terms and a UTC inventory of other software packages will respond to different terms, e.g., ISO-8601[6] related fieldnames. There are 1312 total files that contain one or more of these search terms, about 11% of the entire IRAF codebase (excluding documentation and other non-source files).

For comparison, the IRAF Y2K inventory included roughly 124 files (including documentation), or less than 1% of the codebase. A detailed direct comparison will not be possible, of course, until the completion of a full UTC inventory, but it is clear now and has been clear for many years that an IRAF mitigation project for a redefinition of UTC will be a significantly larger, longer term, and more expensive project than Y2K remediation. Certainly not less than the Y2K workload, a more reasonable estimate would be in the 3-5 FTE-year range just for the IRAF core system explored in this inventory.

Search term	Hits
UT	250
UTC	23
GMT	38
JD	158
MJD	63
LST	67
Second	857
Minute	66
Hour	145
Day	156
Month	68
Year	100
Sidereal	20
Solar	65
Calendar	10
Clock	73
Total	1312 (*out of 11,600*)

Table 1. Number of files in the IRAF core system (including NOAO packages) that contain the word or term indicated. Each term is case-independent and must occur at the beginning of a token. The total only counts a single time files that have hits from multiple terms. The total number of source files searched is about 11,600. At least 11% of the IRAF source tree must be vetted for UTC dependencies.

Approximately two-thirds of the files surveyed required no changes during the IRAF Y2K project, but this does not mean that no work was involved for those files. Each file identified in the inventory actually represents several other files that were reviewed. For instance, a package directory with several hits would often trigger a complete code review for that package. Y2K remediation also involved writing a new ISO-8601 compliant interface that was introduced "upstream" of other files that were not themselves modified.

In the case of a redefined UTC, modules that require actual Universal Time will need to be provided with a new source of UT, either real-time or reconnected in a laborious fashion from inputs gathered throughout diverse user interfaces, parameter sets, catalog databases and so forth. Even if 90% of the files identified using the current UTC search terms were to turn out not to require direct modification, the remaining file count would still represent four times as many files as were modified for Y2K. However, a review of a sample of the files suggests that the Y2K experience is likely representative of the UTC expectation. Since there are roughly ten times as many candidate files to evaluate, we should thus expect roughly ten times as many files needing modification. Few packages or libraries were untouched by the inventory, and the exercise would also correspond to basically a complete code review of the IRAF system.

CONCLUSION

The original notion for this quick inventory was to provide a few example code fragments demonstrating a range of issues that would have to be changed to accommodate a redefinition of Coordinated Universal Time. Once the scale of the problem became clear this seemed like a vain hope short of conducting the ultimate full inventory (necessary to create a coherent plan of attack).

The issue is pervasive and may perhaps best be illustrated simply by listing some of the many reference texts that programmers have borrowed from over the years when creating the algorithms embedded in their code: 1) the Astronomical Almanac,[7] 2) the Explanatory Supplement to the Astronomical Almanac,[8] 3) Astrophysical Formulae,[9] 4) Allen's Astrophysical Quantities,[10] 5) Numerical Recipes,[11] and 6) Astronomical Algorithms.[12] Consulting the corresponding references, one will discover that each of these useful volumes has been released in multiple editions over the years. A particular programmer may have implemented subtly different algorithms depending on which edition of which reference was consulted. The precise reason the system of systems of astronomical software works at all is that each of these references (and many more) is tied to the same underlying set of standards. The standards evolve, but they do not generally mutate overnight. If a sudden change is made to a fundamental standard such as UTC, dramatic changes will be required to astronomical software, systems, procedures, logistics, documentation, training, and on and on.

REFERENCES

[1] Tody, D., "The IRAF Data Reduction and Analysis System", in *Instrumentation in astronomy VI*, p. 733, 1986.

[2] McCarthy D.D., P.K. Seidelmann (2009), *Time—from Earth Rotation to Atomic Physics*. Wiley –VCH.

[3] Kaplan, G., "Time Scales in Astronomical and Navigational Almanacs" Paper AAS 11-671, from *Decoupling Civil Timekeeping from Earth Rotation—A Colloquium Exploring Implications of Redefining UTC*. American Astronautical Society Science and Technology Series, Vol. 113, Univelt, Inc., San Diego, 2012

[4] Hey, T., Tansley, S. & Tolle, K., ed., *The Fourth Paradigm: Data-Intensive Scientific Discovery*, Microsoft Research, 2009.

[5] Mitchell, R. L., "Y2K: The good, the bad and the crazy", Computerworld, 28 December 2009.

[6] ISO 8601, *Data elements and interchange formats – Information Interchange – Representation of dates and times*, International Organization for Standardization, 2004.

[7] *Astronomical Almanac*, U.S. Government Printing Office, annual.

[8] Seidelmann, P. K., ed., *Explanatory Supplement to the Astronomical Almanac*, University Science Books, 1992.

[9] Lang, K. R., *Astrophysical Formulae, 2nd Ed.*, Springer-Verlag, 1980.

[10] Cox, A. N., ed., *Allen's Astrophysical Quantities, 4th Ed.*, Springer-Verlag, 2000.

[11] Press, W. H., Teukolsky, S. A., Vetterling, W. T., & Flannery, B. P., *Numerical Recipes: The Art of Scientific Computing, 3rd Ed.,* Cambridge University Press, 2007.

[12] Meeus, J., *Astronomical Algorithms, 2nd Ed.,* Willmann-Bell, 1999.

DISCUSSION CONCLUDING AAS 11-677

Neil deGrasse Tyson wondered if there was some value in letting errors happen in order to discover them more efficiently. Arnold Rots replied that if civil time is decoupled from Earth rotation, small errors are likely to gradually increase with time, such that by the time they are noticed, years or decades of recorded data might need to be corrected or reprocessed. David Terrett added that the concern here was that, unlike Y2K issues, there really is no catastrophic failure to give warning that a problem is occurring. Mark Storz clarified that some software should be expected to crash. As an example, Ken Seidelmann noted one organization did not make necessary corrections to predicted ephemerides caused by a change of standards caused by the 1925 redefinition of GMT in almanacs.

Steve Malys suggested that there are two very different categories of operations. Interruptions to science missions or research activities have limited impact and these programs are likely to be managed flexibly. In contrast, military missions cannot easily tolerate operational downtime and are therefore managed very rigidly. As an example, so many things are now dependent on an operational system like GPS; the ramifications of GPS going offline for even an hour are hard to image and such capability cannot be allowed to suffer failure. Therefore, in these cases it takes much more testing and analysis to ensure that problems will not happen from changes to timekeeping standards. Terrett suggested that the US DoD has more resources than, say, astronomers, to address such issues. Malys agreed but clarified that building and maintenance of DoD systems tends to be spread out amongst contractors outside the DoD itself, the management of which creates an institutionalized environment, such that the potential for flexibility and collaboration is greater amongst research astronomers compared to military operations.

Ken Seidelmann asked about the verification of software owned by individuals in need of correction. Rob Seaman agreed that this is not a minor issue, but that situation was obviously beyond his immediate influence.

Dennis McCarthy asked what Seaman would like to see from the IERS in terms of improving infrastructure for UT1 dissemination. Terrett offered that topic was the subject of his upcoming paper. Seaman noted that the modern paradigm for astronomical infrastructure is the Virtual Observatory, an internet-based service-oriented architecture with very clear standards for usage in principle. Therefore, a clear protocol and a robust infrastructure must exist to transport the messaging. The infrastructure should consider that telescopes are operated untended and without reliable network connectivity due to their remoteness, so there may be a need for something more sophisticated than synchronizing time through the USNO web site. Terrett disagreed, suggesting that the simplest possible approach should be preferred. McCarthy noted that IERS representatives were in the room and this meeting was the ideal opportunity to inform the IERS regarding user requirements. Seaman noted that the issue of UT1 dissemination is a good issue to address independent of any leap second issues.

Terrett said that he was only familiar with his own applications and accuracy requirements and was unsure of the timing requirements of other communities, and that his proposal may not meet all needs. Rots wondered if encoding of GPS signals might be an option for time broadcasts in the future. Terrett said that mostly observatories simply need to accurately synchronize their clock to UT1. Rots was not convinced that UT1 was all that an observatory would want from a time service. Seaman admitted that a service at the complexity and robustness of NTP might suffice; there might be two different kinds of time transmitted although that creates its own complications. Terrett clarified that he was interested in a service that broadcasts a correction to civil time rather than Earth-rotation time itself. Seaman said that if the correction is slowly changing (as DUT1 precise to 0.1 seconds is now), then there is no need for a continuous broadcast correction service unless higher precision is needed. The point of Seaman's paper was that there are many cases that are not necessarily attempting to "point something at something else." Terrett said his main concern was a means of providing precise real-time Earth orientation.

McCarthy said that at some point the IERS had casually discussed maintaining a time signal steered to UT1 and making that signal accessible to users who needed a precise realization of UT1 in real time. The drawback to such a service is that some may use the service without understanding what it provides. Seaman noted that if leap seconds cease, then symmetry is broken between UTC and UT1 and there will be a need to keep track of and distribute these scales separately regardless.

George Kaplan asked for clarity regarding the FITS standard for time; Allen replied that FITS refers to data formats that have been in use in astronomy for a couple of decades now, and the time standard refers to World Coordinate System Paper V, which is basically a specification of how to write metadata, such that a machine should be able interpret the scale being used.[1] Rots noted that the reason for mentioning FITS was that this standard should be able to cover whatever happens regarding the future of UTC.

REFERENCES

[1] Rots, A.H., P.S. Bunclark, M.R. Calabretta, S.L. Allen, R.N. Manchester (2009), "Representations of Time Coordinates in FITS" Astronomy & Astrophysics Manuscript No. WCSPaperV0.73, September 28, 2009. (URL http://hea-www.cfa.harvard.edu/~arots/TimeWCS/WCSPaperV0.73.pdf)

TELESCOPE SYSTEMS AT LICK OBSERVATORY AND KECK OBSERVATORY

Steven L. Allen[*]

The telescopes in active use at Lick Observatory and Keck Observatory were constructed over an interval spanning more than a century. All of the telescope systems were designed in an era when systems which provide civil time were based on the rotation of the earth. Existing software systems for the control of telescopes at Lick Observatory and Keck Observatory use UTC as a close approximation to UT1. If UTC abandons leap seconds then ongoing operation will require various strategies suitable for each different telescope.

INTRODUCTION

The 400 years of telescope history have seen huge changes in the practices and technologies. The pointing of telescopes has changed considerably. Early observers like Galileo casually aimed their small telescopes. Teams of laborers pulled ropes to hoist the framework holding their lord's large telescope and their lord himself. Iron workers produced large bearings and gears for precision equatorial mountings with clock drives that modelled the rotation of the earth. Now robotic control engineers build alt-az telescopes whose operation depends entirely on models in software. Obtaining the desired pointing results during interaction with the models in these systems involves algorithmic conventions about earth and sky.

COORDINATE REFERENCE SYSTEMS AND FRAMES

Human commerce is facilitated when all parties agree on the meanings of the words describing products and procedures. The late 1800s saw trans-continental railways and trans-Atlantic telegraph cables making new connections between communities with little previous contact. One of the efforts to standardize human endeavor was the International Meridian Conference of 1884.[1] The Prime Meridian at Greenwich was one terrestrial result of the conference, but the resolutions also prescribed conventions for the measurement of time as a subdivision of calendar days.

Technical details of the implied metrology were not specified by the diplomats at the conference. Among the practitioners of the metrology was Simon Newcomb who had the funding of the US government along with the data from the the American and European observatories. Before the end of the century he had overseen calculations to produce mathematical expressions of celestial motions[2] so impressive that all the directors of national ephemerides agreed to use them.

---

[*] Programmer/Analyst, UCO/Lick Observatory, ISB 1156 High Street, Santa Cruz, California 95064, U.S.A.

Astronomical changes over the past century

Shortly after the 1919 formation of the International Astronomical Union (IAU) the national ephemerides changed their tabulations of time to conform with the 1884 International Meridian Conference resolution that the Universal Day should be reckoned from midnight. In an effort to avoid confusion the American almanac engraved a warning paragraph on its cover.[3] To the chagrin of many, the British Admiralty insisted on keeping the same name, GMT, even though it had formerly been used for time reckoned from Greenwich mean noon. This resulted in an IAU decision to replace GMT with Universal Time (UT),[4] a change of nomenclature with no effect on telescope pointing.

Despite its non-relativistic basis, errors known at the time of adoption, and the discovery of earth rotation variations, time service bureaus used Newcomb's 1895 expression for UT across almost 90 years. Atomic chronometers, digital computers, satellite geodesy, very long baseline interferometry (VLBI), and Lunar Laser Ranging (LLR) contributed to new metrology technologies. These revealed serious deficiencies in existing models and motivated a new expression for UT1 starting in 1984, but its authors took great care to match the new definition to the old one.[5] Existing telescope systems did not need a large change in procedures for handling celestial coordinates.

At the turn of the 21s century ongoing measurements with high-precision technologies led to another, newer definition of UT1 that does not need the concept of equinox.[6] This change is accompanied by a complete change of the underlying concepts for celestial coordinates.[7] These IAU 2000 changes are essential for the precisions required with VLBI, LLR, interplanetary navigation, timing of phenomenon, etc. For the pointing requirements of optical telescopes, however, there will be little discernible difference before the end of the 21st century.

Pedagogical aspects

At the beginning of the 20th century the underlying concepts of astrometry had not changed much since Ptolemy. By the end of the 20th century the underlying concepts had been completely changed twice within the span of a productive career. A rate of change like that obsoletes procedures, software, and human expertise. It produces a strong need to review the pedagogical resources and their limits of validity.

The textbook from Smart contained the same kinds of haversine table look-ups used by navigators for the preceding century.[8] The *Explanatory Supplement*[9] encapsulated the early changes to procedures which preceded use of the FK5 system in 1984.

Texts from Murray[10] and Green[11] gave early treatments for new methods for computing and the FK5 conventions. The new *Explanatory Supplement*[12] covered the FK5 conventions in detail, but only a few years later those were replaced by the sweeping changes of the IAU 2000 conventions. Aside from the problem of old texts, some texts contain errors, and students who learned from any of these may continue to employ old concepts and algorithms for the duration of their career. This unfortunate truth is a strong argument for changes in the conventions and definitions to be as inconsequential as possible.

Calculations using the current IAU 2000/2006 framework are described by USNO Circular 179,[13] Wallace and Capitaine,[14] and in the IERS Conventions.[15] These are useful documents until such time as the next revision of *Explanatory Supplement* appears.

Astrometric software

The original complexity of astrometric calculations was within the capability of a trained human navigator with a book of mathematical tables. The complexity of the current conventions exists because of digital computers but the algorithms require expertise not likely to be found in many programmers. As a result most current computations for telescope pointing rely on libraries of algorithms used widely across the astronomical community. One early example for the FK4 and FK5 systems is the Starlink Library for Astronomy (SLALIB).[16] For the IAU 2000/2006 conventions the IAU fostered the Standards of Fundamental Astronomy (SoFA) effort.[17]* The USNO provides another implementation in its Naval Observatory Vector Astrometry Software (NOVAS)[18] which is completely free of intellectual property issues.†

TELESCOPE POINTING OPERATIONS

UCO/Lick observatory operates the telescopes on Mt. Hamilton in California and collaborates to oversee the operation of the Keck telescopes in Hawaii. Here is a quick survey of the effects we would see in the absence of leap seconds, along with some strategies for continuing operations.

Lick refractor

Figure 1. The 36-inch James Lick refractor is pointed by manual effort.

The James Lick telescope on Mt. Hamilton (Figure 1) has a 36 inch objective on an equatorial mount.[19] The Lick, other large telescopes from the 19th century, and many subsequent telescopes

---

*http://www.iausofa.org/
†http://aa.usno.navy.mil/software/novas/novas_info.php

are pointed using the physical effort of the astronomer. The absence of leap seconds would have no effect on the operation of the Lick telescope.

Shane reflector

The Donald Shane telescope on Mt. Hamilton (Figure 2) has a monolithic 3 m primary in an equatorial mount. The Shane saw first light in 1959. Its original pointing relied on mechanical systems and analog electronics.[20] In the 1970s the analog pointing systems gained digital assistance from an 8-bit 6502 microprocessor. Recent upgrades to the Shane replaced the 6502 with Unix-based "POCO" software on computing hardware with roughly 1000 times greater capability.[21]

All Shane slewing remains under direct control of a Telescope Technician (TT). When a blind slew does not bring a target into the field of the guide camera the TT makes manual corrections based on experience with the telescope. In the absence of leap seconds the TTs could continue to point the Shane for several years without significant degradation.

The source code for POCO belongs to Lick Observatory. POCO employs SLALIB for its astrometry, so within a few decades POCO will need an upgrade to conform to the IAU 2000 conventions. The absence of leap seconds would trigger a need to spend manpower resources upgrading POCO within a few years instead of a few decades. In an era of tight state budgets this is not a welcome change.

Figure 2. The 3-m Shane reflector and the 1-m Nickel reflector have human operators who compensate for pointing problems.

Nickel reflector

The Anna Nickel telescope on Mt. Hamilton (Figure 2) has a 1 m primary in an equatorial mount. The pointing for the Nickel was designed early in the era of digital control systems.[22] For its first 25 years astronomers operated the Nickel from the dome or the adjacent control room. Astronomers slewed the telescope manually, and they could correct pointing errors using the finder scopes.

Recent upgrades to the Nickel added new encoders, motors, and the POCO software. These changes allow astronomers at remote sites to operate the Nickel using the Internet.[23] The field of view of the Nickel telescope guide cameras is about 7 arcminutes, so occasions when the telescope pointing is outside the guide field are uncommon. If the pointing does fail then a local observer or technician must use the finders. These conditions indicate that the requirement to upgrade POCO for the Shane would ensure that the Nickel telescope will never experience pointing problems due to the absence of leap seconds.

Keck reflectors

The Keck telescopes on Mauna Kea (Figure 3) have segmented 10 m primaries in alt-azimuth mounts. The Kecks saw first light in the 1990s. Keck pointing systems rely on software, but all telescope slewing is under direct control of an Observing Assistant (OA). In many cases the Keck telescopes slew to within 7 or 8 arcsec of the target. Pointing may rarely be as much as 40 arcsec off target; in such cases the OA typically locates a nearby catalog star before proceeding to target. Nightlog Tickets during 2011 indicate about 1 hour of observing time lost to pointing issues.[24]

Figure 3. The 10-m Keck reflectors have human operators.

Guide cameras for Keck instruments have fields of view around 3 to 3.5 arcminutes. The absence of leap seconds would begin to affect pointing procedures within a year. The skill of the Keck OAs should be able to handle pointing for several years after that, but in that time the celestial coordinates reported with the science data would become increasingly aberrant.

The pointing source code for the telescopes belongs to Keck. It employs SLALIB for the astrometry. As with the Shane telescope, the absence of leap seconds would trigger a need for Keck to spend manpower updating the pointing code within a few years instead of a few decades.

APF reflector

The Automated Planet Finder (APF) telescope on Mt. Hamilton (Figure 4) has a monolithic 2.4 m primary in an alt-azimuth mount. APF saw first light in 2009. It is intended to perform fully robotic observation without human attendants. The APF telescope and dome were purchased as a complete system of hardware and software. The specification required pointing within 10 arcseconds. The vendor achieved this specification using a commercial GPS receiver to provide time to the telescope software. Lick Observatory does not have the source code for the APF software.

Figure 4. The 2.4-m APF reflector is robotic. Its pointing software will cease to acquire targets within around a year if leap seconds are abandoned.

The robotic operation means there will be no humans on site supplying their skill to correct the pointing. In the absence of leap seconds the APF telescope will fail to meet its specification within the first year. There is no guarantee that the vendor will be available to provide an update for the software. In the absence of leap seconds and new software, several strategies are options for continuing operation of APF.

We could corrupt the input coordinates we provide to the APF telescope software, adjusting the right ascension by the number of missed leap seconds. This has a drawback because the coordinates supplied from the telescope software to the science data files will be wrong. Lick built the science instrument for APF, so we could remedy this by a second hack to our software which receives the output coordinates. Eventually the time offset would grow large enough to displace the notion of zenith, and that would affect the pointing model, but this would not happen during the expected lifetime of the telescope. The result, however, would be confusing if astronomers attempt manual use. Users would have to be trained to perform the right ascension offset before entering targets into the GUI and to expect the coordinates visible in the GUI to be wrong.

The design of the APF software provides a more desirable alternative for handling the absence of leap seconds. This technique relies on the concept of the Ephemeris Meridian originally proposed by Sadler.[25] Although the APF telescope relies on a GPS receiver for time, the telescope pointing software does not use the geodetic coordinates from GPS. The telescope software obtains its geodetic coordinates from a configuration file, and the coordinates are not exposed in any relevant fashion. This fortuitous aspect of the software design means that in the absence of leap seconds we expect to modify the specified longitude of the telescope by the amount of drift in the Ephemeris Meridian resulting from the missing leap seconds. This option is available for APF science because we are only concerned with one coordinate system, the celestial sphere. Many telescopes from the same vendor, however, are used for satellite tracking. The same technique would probably not work for satellite applications because they rely on knowing the relations between both celestial and terrestrial coordinate systems.

CONCLUSION

In the absence of leap seconds the significant difference for the operability of telescopes at Lick and Keck observatories is not the mount nor the software. The critical difference is the role of humans in the operation of the telescopes. The telescopes which remain operated by humans will not be affected for years after cessation of leap seconds. The telescope which is entirely operated through software will be affected within a year after cessation of leap seconds.

The cost of changing software and procedures for the human-operated telescopes will be unwelcome, but straightforward to absorb as a part of routine maintenance during the years before problems arise. For the APF telescope the concept of Ephemeris Meridian allows us to exploit a trivial "hack" to the software system inputs which will not be visible in the data stream. This hack, however, relies on the particulars of assumptions made by its software system designers and on the particulars of its operational goals. It is not reasonable to generalize these cost results to other telescopes and software systems. Every telescope pointing system needs its own analysis.

REFERENCES

[1] US Department of State, ed., *International Meridian Conference*. Washington, D.C.: Gibson Bros., 1884. Protocols of the Proceedings.

[2] S. Newcomb, "Tables of the motion of the earth on its axis and around the sun," *Astronomical papers prepared for the use of the American ephemeris and nautical almanac, v. 6, pt. 1, [Washington, Bureau of Equipment, Navy Dept., 1895], 169 p. 29 cm.*, Vol. 6, 1895, pp. 2–+.

[3] *The American Ephemeris and Nautical Almanac for the year 1925.* 1925.

[4] Stratton, F. J. M., ed., *Proceedings of the 3rd General Assembly*, Vol. 3 B of *Transactions of the IAU*. Cambridge University press, 1928.

[5] S. Aoki, H. Kinoshita, B. Guinot, G. H. Kaplan, D. D. McCarthy, and P. K. Seidelmann, "The new definition of universal time," *Astronomy & Astrophysics*, Vol. 105, Jan. 1982, pp. 359–361.

[6] N. Capitaine, B. Guinot, and D. D. McCarthy, "Definition of the Celestial Ephemeris Origin and of UT1 in the International Celestial Reference Frame," *Astronomy & Astrophysics*, Vol. 355, Mar. 2000, pp. 398–405.

[7] M. Soffel, S. A. Klioner, G. Petit, P. Wolf, S. M. Kopeikin, P. Bretagnon, V. A. Brumberg, N. Capitaine, T. Damour, T. Fukushima, B. Guinot, T.-Y. Huang, L. Lindegren, C. Ma, K. Nordtvedt, J. C. Ries, P. K. Seidelmann, D. Vokrouhlický, C. M. Will, and C. Xu, "The IAU 2000 Resolutions for Astrometry, Celestial Mechanics, and Metrology in the Relativistic Framework: Explanatory Supplement," *Astronomical Journal*, Vol. 126, Dec. 2003, pp. 2687–2706, 10.1086/378162.

[8] W. M. Smart, *Text-book on spherical astronomy*. The University press, Cambridge, 1931.

[9] *Explanatory Supplement to the Astronomical Ephemeris and the American Ephemeris and Nautical Almanac.* 1961.

[10] C. A. Murray, *Vectorial Astrometry.* Techno House, Redcliffe Way, Bristol BS1 6NX: Adam Hilger Ltd, 1983.

[11] R. M. Green, *Spherical astronomy.* Cambridge University Press, 1985.

[12] Seidelmann, P. K., ed., *Explanatory Supplement to the Astronomical Almanac.* University Science Books, 1992.

[13] G. H. Kaplan, *The IAU Resolutions on Astronomical Reference Systems, Time Scales, and Earth Rotation Models: Explanation and Implementation.* Washington, D.C. 20392: US Naval Observatory, 2005. (USNO Circular 179).

[14] P. T. Wallace and N. Capitaine, "Erratum: Precession-nutation procedures consistent with IAU 2006 resolutions," *Astronomy & Astrophysics,* Vol. 464, Mar. 2007, pp. 793–793, 10.1051/0004-6361:20065897e.

[15] G. Petit and B. Luzum, eds., *IERS Conventions (2010).* Verlag des Bundesamts für Kartographie und Geodäsie, Frankfurt am Main: IERS Conventions Center, 2010. (Technical Note 36).

[16] P. T. Wallace, "The SLALIB Library," *Astronomical Data Analysis Software and Systems III* (D. R. Crabtree, R. J. Hanisch, & J. Barnes, ed.), Vol. 61 of *Astronomical Society of the Pacific Conference Series,* 1994, pp. 481–+.

[17] P. T. Wallace, "SOFA: Standards of Fundamental Astronomy," *Highlights of Astronomy,* Vol. 11, 1998, pp. 191–+.

[18] G. Kaplan, J. Bangert, J. Bartlett, W. Puatua, and A. Monet, *User's Guide to NOVAS 3.0.* Washington, D.C. 20392: US Naval Observatory, 2009. (USNO Circular 180).

[19] E. S. Holden, "VIII. Description of the astronomical instruments," *Publications of Lick Observatory,* Vol. 1, 1887, pp. 59–77.

[20] J. Osborne, "The 120-inch Telescope," Tech. Rep. 32, Lick Observatory, 1983.

[21] J. Gates, W. T. S. Deich, A. Misch, and R. I. Kibrick, "Modern computer control for Lick Observatory telescopes," *Society of Photo-Optical Instrumentation Engineers (SPIE) Conference Series,* Vol. 7019 of *Society of Photo-Optical Instrumentation Engineers (SPIE) Conference Series,* Aug. 2008, 10.1117/12.787995.

[22] J. Osborne, "A Spare Parts 40-inch Telescope," *Sky & Telescope,* Vol. 60, 1980, pp. 97–+.

[23] B. Grigsby, K. Chloros, J. Gates, W. T. S. Deich, E. Gates, and R. Kibrick, "Remote observing with the Nickel Telescope at Lick Observatory," *Society of Photo-Optical Instrumentation Engineers (SPIE) Conference Series,* Vol. 7016 of *Society of Photo-Optical Instrumentation Engineers (SPIE) Conference Series,* July 2008, 10.1117/12.789490.

[24] Keck Observatory, "Nightlog Tickets," 2011.

[25] D. H. Sadler, "Ephemeris Time," *Occasional Notes of the Royal Astronomical Society,* Vol. 3, 1954, pp. 103–113.

DISCUSSION CONCLUDING AAS 11-678

Regarding Steve Allen's suggestion of tricking a propriety telescope control system by giving it a fictitious "ephemeris longitude", McCarthy asked about the telescope's field of view requirements and how that would affect the frequency of "ephemeris longitude" updates. Steve Allen clarified that for his system, robotic operations were not yet fully functional and the operational field of view was not yet established for the new guider, but it was anticipated to be less than 3 arcminutes and that there should be no issue changing the "ephemeris longitude" with sufficient frequency. The process does pose the inconvenience of shutting down the telescope control system, updating the longitude, and restarting the telescope control system.

John Seago noted that use of an "ephemeris longitude" may not be a usable workaround for some users with similar systems, such as satellite tracking systems. Allen agreed, noting that his observatory's exact location relative to the terrestrial frame was not critical to his instruments pointing and guidance operations. Rob Seaman wondered how many systems like Allen's might be fielded.

Frank Reed felt that it was in a vendor's best interest to support their customers, and Allen replied that technical representatives had already discussed the possibility of moving away from their propriety codes and adopting NOVAS libraries. Allen went on to cite an example of an astrodynamics textbook having errors which may well have been the basis of some incorrect control-system programming. Allen noted that this is one more example of how incorrect or outdated information can persist for a long time. The solution requires large amounts of re-education and delving into code to discover who may have used an incorrect prescription of changed standards. Ken Seidelmann noted that it has been his experience within the US Department of Defense (DoD) that propriety systems often keep government customers dependent upon the vendor. Allen noted that the University of California does not have the resources of the US government and there is no financial benefit for a vendor in his situation.

Mark Storz noted that his organization has entertained the possibility of an "ephemeris longitude" approach for some applications, but this is not viable across the US DoD. Daniel Gambis wondered about some of the choices with regard to the telescope catalog database; Allen clarified that these choices were made by the vendor, and in some cases, out of unfamiliarity.

MIDDAY ROUND-TABLE DISCUSSION OF OCTOBER 6, 2011

MIDDAY ROUND-TABLE DISCUSSION OF OCTOBER 6, 2011

The draft of the proposal to be voted upon at the ITU-R Radiocommunication Assembly in January 2012 offers only two options: maintain the status quo with leap seconds, or abandon the insertion of leap seconds altogether. This discussion contemplated the viability of other possible options. Variations on the handling of time have already been tried. In most cases there have been quirks which limit the applicability of various options or which cause confusion among users needing precise time.

Neil deGrasse Tyson wondered if the main problem with UTC was not with leap seconds *per se* but rather with their lack of predictability within software. He wondered if a low (second) order model of Earth rotation rate might be a viable option, acknowledging that such would be less accurate yet usefully predictable. He suggested that a model could be used to plan leap adjustments centuries in advance, effectively becoming part of the calendar, and would therefore be useful for protecting software systems over their service lifetime.

David Terrett noted that this would be a change from the current convention, and that any change would bring about issues. Daniel Gambis noted that decadal variations can be significantly larger than the long term trend, making estimation and prediction of a trend difficult in the long-term. Steve Allen noted that this option or a variation of it was an early consideration but it did not seem acceptable to anyone at the time, and that there are more fundamental problems that such a solution does not address. Ken Seidelmann noted that the length of year is now known rather accurately and this is what makes predicted leap days in the Gregorian calendar workable. Terrett also noted that programmers still get the rules of the Gregorian calendar wrong, so having a predictable leaping rule for time of day for software does not necessarily solve the software issue. Tyson questioned whether the dismissal of a predictable rule should be based on the anticipated incompetence of some programmers, but then it was appreciated by most in the room that the proposed redefinition of UTC was already being motivated by the very same argument.[1]

Frank Reed noted that the second-order model was perhaps accurate to only two significant figures at best, and therefore was far too inaccurate to be useful as a prediction model. Dennis McCarthy said that any such approach would require loosening of the current ±0.9-second tolerance for UT1-UTC, but that in principle the proposal should meet a concern of programmers regarding the predictability of leap adjustments. McCarthy also shared that some programmers do not appreciate that accurate leap-second insertions are not predictable, and Allen noted that "future leap seconds" is a very common internet search phrase. Rob Seaman suggested that extending the predictions from six months to a few years may be helpful in some cases. Daniel Gambis said that the IERS could confidently predict leap seconds in advance of "two or three years." McCarthy said that prediction years in advance could be a good option for almanacs and other publications that are printed well in advance. At this point Steven Malys reminded the attendees that the proposal for consideration in January was that UTC should diverge from Universal Time indefinitely and without further adjustments, which McCarthy and others affirmed.

Ken Seidelmann asked Gambis to affirm whether leap seconds might really be predictable out to three years; however, Allen suggested that three years' prediction still would not be enough to satisfy software people. Seidelmann wondered if extended prediction might help some software, but Terrett said that he thought the primary software issues were really caused by programmer ignorance about the existence of leap seconds, rather than a need for extended forecasts. Allen suggested that it was more than ignorance, and his upcoming presentation would discuss the fundamental problems, and that satisfaction would not be had by any scheduling scheme. Reed said that leap seconds are just generally confusing for software people.

Mark Storz said he had mistakenly thought that the IERS might continue to announce leap seconds as a service to systems that expect them, but they would simply be ignored in future UTC broadcasts. John Seago asked if there was any risk of an independent party attempting to coordinate and publish future leap seconds announcements should this service no longer exist. Rob Seaman replied that there are commercial entities who try to make money in the time-service business through unconventional ways. Allen noted that the authority of such an announcement would be the issue, citing Gregorian calendar reform as an example where universal agreement could not be had even by papal decree. Wolfgang Dick noted that almanacs that need to predict or publish UT1-UTC in advance will effectively have this information. Storz noted that while people could calculate potential adjustments from that information, their insertion into timekeeping systems would not be coordinated. Allen noted that such a situation would result in the "proliferation of independent time systems" that the ITU-R was reportedly attempting to avoid via the cessation of leap seconds.[2] Dennis McCarthy suggested that truncation of the leading digit could lead to a simple rule as to when to insert such a correction. John Seago and Mark Storz also wondered if rounding might also lead to a suggested rule.

Kaplan said that when the Very Large Array (VLA) became operational in the 1980's, he had hoped that astronomical systems would take up its example of running on TAI as a standard. He noted that a significant complication with UTC is that the computation of precise time intervals was not algorithmic and required archived leap-second tables. Allen noted that people will always be faced with the application of tabulated corrections, because our ancestors have always taken actions that seemed stupid to later generations yet cannot be undone.

Tyson noted that Nobel-Prize winner Joe Taylor lamented the unexpected effort that was required to process pulsar measurements tagged to UTC due to leap-second insertions, citing this as an example affirming that many astronomers are likely unfamiliar with the definition of UTC. Arnold Rots gave an example from a recent IAU symposium where astronomers generically labeled data and graphs without any acknowledgement of the underlying time scale. Dennis McCarthy offered that there are IAU resolutions recommending proper time scale nomenclatures, such as endorsement of the use of Modified Julian Date (MJD). Terrett asked how the time scale should be stipulated and Allen clarified that guidance was noted in 1997 Resolution B1 of the IAU.[3] Allen also suggested that it is useful to provide users with a citation to educate and convince the reader that the presentation is the proper one.

Seago offered that the civil dating of events is traditionally done in terms of a "calendar day" and "time of day" within an astronomically based calendar, and that UTC's leap seconds are a consequence of maintaining an astronomically based calendar. Without such adjustments, the calendar becomes entirely algorithmic, and the complete detachment of a "metric" day from celestial motions potentially raises some issues mentioned by Paul Gabor. Seago suggested that an extrapolation of this line of thinking could also argue for the abolition of civil time zones and calendars, yet that level of change had yet to be advocated. Seaman noted that France attempted unconventional changes to measuring time after the French Revolution which proved unpopular.

Seidelmann noted that conventional Julian Days already exist for those who might prefer that type of representation.

Kaplan noted that many proposed solutions are seemingly worse than the problems they are trying to solve, as they could cause unnecessary complications in the distant future. One idea of particular concern might be the introduction of an NTP service of variable frequency that attempts to track UT1 (as UTC did before 1972). Kaplan felt that such a service would inevitably be used by someone inadvertently thinking that it was UTC. The tagging of data by such a service would pose significant complications in the distance future as it could be difficult to relate the time supplied by the service back to uniform atomic time. Steve Malys wondered if we might be creating unknown problems for the future with the current proposal to decouple civil timekeeping from Earth rotation, as it seems that motivations for the decoupling are not adequately justified. The major argument favoring changes now seems to be the inconvenience to software-programmers, which appears to be a weak reason for denying our progeny civil time linked to the astronomical day. Tyson agreed that this reasoning would be embarrassing to admit a century from now.

Seaman reminded the group that the colloquium topic is about the decoupling of civil timekeeping and Earth rotation, and that UTC is simply the current solution. There are vast possibilities for civil timekeeping that could be explored beyond leap seconds that could stimulate interesting discussion; however, Seaman expressed skepticism as to whether such discussion could readily turn into a simple solution, noting in particular that TAI-like time is not going to be the solution to every civil-timekeeping problem. Seaman also wondered if the status of TAI might change if UTC were redefined; if so, would it be wiser to start labeling epochs according to TT's system of dating events which is offset from TAI by 32.184 seconds. Terrett said that TAI exists by definition indefinitely as a sequence of *SI* seconds. Allen suggested that the existence of TAI presupposes that the BIPM maintains TAI. McCarthy clarified that really there is no clock maintaining TAI, but Seidelmann countered that TAI could be made available by trivially subtracting leap seconds from UTC.

Tyson wondered if there is anything that could be learned by imagining ourselves placed a century into the past to understand the challenges that were important then, and then exploring if our priorities today will seem like folly to our progeny. Terrett suggested that trying to predict requirements a century forward would likely be impossible for our generation, but Tyson noted that is part of the value of the exercise; it might reveal the kinds of issues that were made obsolete or otherwise solved over time. Seidelmann noted that such an exercise would likely reveal that we simply cannot anticipate what technological changes will occur that will have an effect on the definition of civil time scales; therefore, there is limited value in making seemingly arbitrary changes now when future technology will eventually force changes regardless. He cited variable Earth rotation and general relativity as examples of physical issues that were not well known a century ago and have resulted in great complications in our modern timekeeping. Seaman added plate tectonics to list of physical causes that were unknown a century ago.

Allen offered that changes occur not because people have a clear idea of what will be needed in the future, but they happen whenever the state of the art becomes so impractical that making *any* change outweighs the perceived risk of making the wrong decision for the future. Simpson noted that, because there seemed to be no compelling reason to abolish leap seconds right now, the current situation did not seem to meet the condition for an immediate change that Allen described. Allen replied that perceptions may differ depending on the situations of different communities.

Gambis wondered if there might be some value in thinking much longer term, such as 1000 years in advance. As an example of how technology can change with time, Seidelmann offered that the second itself may be redefined according to a more-precise optical standard in the not too distant future. Tyson asked what applications might be driving the need for this extra precision. Allen responded that a new standard would simply measure what cannot be measured now. Seidelmann agreed by noting that metrological improvements feed technological advancement.

McCarthy said that the key in trying to decide in what's going to happen in the future is to make things flexible enough right now so that we can accommodate things down the road and to not hardwire in things like leap seconds. Tyson replied that he did not get a sense from these conversations that flexibility was a primary consideration. Terrett said that a point in his talk was that openness and willingness seems to be needed already, particularly as it relates to software inertia, and that software needs to be designed with a certain level of flexibility that it has not enjoyed yet. McCarthy noted that people needed to realize that Earth rotation rate will continue to vary in ways that cannot be accurately predicted. For everyone's amusement, Tyson speculated that geo-engineering might somehow allow the Earth's rotation to be controlled far into the future.

Within the context of Tyson's suggestion of placing ourselves in the past and thinking forward to the present, Seago noted that Simon Newcomb seemed very concerned regarding the accuracy of his *Tables of the Sun*, perhaps believing that they would be in use for a long time, and it appeared that his goal was to do the best he could with the methods and data available to him at the time to minimize adverse impact on future generations.[4] Seago then noted that there are really only two proposals under consideration at the moment: atomic time without any allowable adjustments, and *status quo* UTC which allows for adjustment. With regard to McCarthy's issue of maintaining software flexibility for the future, Seago said of these two options, the decoupling of civil timekeeping from Earth rotation appeared to provide the *least* flexibility for the future, and software support of small calendrical adjustments promoted the greatest flexibility. Malys agreed with this assessment, believing that it would be very difficult to resynchronize civil time with Earth rotation if they were allowed to separate, and that the level of difficulty would grow with size of the difference.

Tyson said that much of the conversation had focused on engineering and technology, but if the issue is really about "civil time", *e.g.*, if civil time can be defined as someone wanting to catch a train, then the discussion had not approached that issue very far. Allen pointed out that a cellular telephone with a stock Android® operating system will display Global Positioning System time by default, which is presently fifteen seconds ahead of UTC, and the telephone must be hacked to display civil time. Seago repeated that, with regard to issue of the meaning of "civil time", Newcomb recognized two fundamental representations governing the dating of events: calendar day and time of calendar day, both of which are astronomically based.[5] Seago offered that astronomical concepts like *day* and *time of day* may be fundamental to what the populace expects from civil time, for without an astronomical basis, the long-term meaning of "day" becomes an added issue in the debate. If *time of day* means something that tracks the Earth, then engineering and technology would place requirements on the desired level of synchronization. Some technologies may desire closer synchronization than others, while some technologies, such as telecommunications, may not demand close synchronization to the astronomical day at all.

Malys noted that GPS serves to differentiate itself by representing system time in terms of week number, and seconds of week. Seago agreed and noted that there may be confusion as to the meaning of the very term "time scale". He thought that, to most people, a "time scale" was little more than a dating system for labeling events, and that this description is consistent with past metrological publications that he had seen.[6] Therefore, when people refer to, say, "TAI", they are not intending to refer to the paper clock via BIPM *Circular T*, but they are suggesting a method of labeling real-time atomic UTC seconds uniformly, *i.e.*, UTC + DTAI. Seago wondered if different ideas about the meaning of "time scale" might be inhibiting broader discussions about available options for uniform dating schemes.

Gabor noted that the European Galileo navigation system would create yet another timing system. Allen noted that much effort was expended to ensure that Galileo signals would not interfere with GPS signals and that the GPS and Galileo system times would be highly compatible. Reed pointed out that GLONASS provides yet another GNSS system time. Malys commented that GPS provides a realization of UTC time and Galileo should be able to provide yet another realization that would be slightly different than UTC(GPS). Allen noted that the differences of realization would likely be at the nanosecond level. McCarthy clarified that the differences between internal GPS system and internal Galileo system time would be broadcast by both systems so that future GNSS system can interoperate with both sets of signals, and that the level of offset would likely be at or below 10 ns. McCarthy also noted that the reference epoch for both GPS and Galileo time scales would be the same. Allen noted that the epoch of the Chinese Beidou navigation system was offset from the GPS and Galileo epoch by 14 seconds.

Kaplan noted that many people think of "GPS time" as UTC provided via a GPS receiver, so it is often unclear whether people are referring to the internal time scale of the GPS system or UTC as available through corrected GPS signals. McCarthy also added that some GPS receivers can toggle between UTC and GPS system time and consequently some users do not know which time is being used.

REFERENCES

[1] Kamp, P.-H. (2011), "The one-second war." *Communications of the ACM*. Vol. 54, No. 5, pp. 44-48.

[2] Beard R. (2005), "Future of the UTC time scale." 45th Civil GPS Service Interface Committee Meeting (Long Beach, CA, 13 September 2005).

[3] Resolution B1 of the IAU XXIIIrd General Assembly - Transactions of the IAU Vol. XXIII B Proceedings of the 23rd General Assembly Kyoto, Japan, August 18 - 30, 1997 Ed. J. Andersen Kluwer Academic Publishers.

[4] Newcomb, S. (1895), *The Elements of the Four Inner Planets and the Fundamental Constants of Astronomy*. Supplement to the American Ephemeris and Nautical Almanac for 1897. Government Printing Office, Washington. p. 188.

[5] Newcomb, S. (1906), *A compendium of spherical astronomy*. Macmillan Company, p. 114-7.

[6] Blair, E.B., (ed.), *Time and Frequency Fundamentals*, NBS Monograph 140, p. 4.

Session 7:
CONTINGENCY PROPOSALS

AAS 11-679

AUTOMATING RETRIEVAL OF EARTH ORIENTATION PREDICTIONS

David L. Terrett[*]

As the range of applications requiring knowledge of UT1-UTC increases, the demand for access to this information in a machine readable format over the Internet will increase. The current format, the text of the IERS Bulletin A, is not ideally suited to this purpose for a number of reasons. The exact format of the file is not defined so that anyone writing a program that extracts information from the bulletin has to guess the rules governing the format by inspecting samples. Any such program is at risk if, for any reason the format changes. This paper explores whether there are alternatives to Bulletin A that would be more suitable for ingestion by computer systems and could be implemented within the resources available to the IERS. Suitable standards-based technologies exist but must have both a long expected lifetime and be practical to implement both for the producer and the consumer. A concrete proposal based on XML standards is included.

INTRODUCTION

There are some real-time control applications that require information about the earth's orientation; most obviously, the control systems of astronomical telescopes. This information comes from the IERS predictions of UT1-UTC and polar motions as published in Bulletin A. A typical operational scenario is for the telescope operator to look up UT1-UTC for the current date and type it into a computer at the start of the night. This is a relatively minor step in the, often complex, process of getting the telescope ready for observing, carried out under time pressure and at some observatories, at high altitude. There are obviously opportunities for errors in this process, such as picking the wrong date, or, if there are technical problems demanding the operator's attention, the step may be overlooked altogether. Fortunately, the consequences are usually not particularly serious, typically just a small error in the telescope pointing and worst that can happen is that there is a short delay in getting observing under way. UT1-UTC changes sufficiently slowly that entering a new value can be neglected for days or even weeks, before the error becomes significant (provided of course that the control system retains the current value when the system is re-started).

An experienced telescope operator knows the valid range for UT1-UTC and plausible values for the polar motions (where required, many systems ignore them as they are only marginally significant for ground-based telescopes and only then for telescopes with the very best pointing accuracy achievable) so the chances of making a gross error are small. However, as observatory

---

[*] RAL Space, STFC Rutherford Appleton Laboratory, Harwell Oxford, Didcot, Oxfordshire, OX11 0QX, United Kingdom.

budgets come under every increasing pressure it is becoming more and more common for, even quite large, telescopes to no longer have a dedicated operator. The task of readying the system is then in the hands of an astronomer, often a visitor to the observatory and so working in an unfamiliar environment and who may operate a telescope only a few times a year. The chances of an error occurring clearly becomes greater and the time taken to recover becomes longer.

The opportunity for error can be all but eliminated by having the control system retrieve the current value of UT1-UTC from the IERS without operator intervention and, in the case of a robotic telescope, where there is not operator at all, it is essential. However, the current format of the IERS Bulletin A, is not ideally suited to this purpose for a number of reasons. The exact format of the file is not defined (at least, not in an easily discoverable place) so anyone writing a program that extracts information from the bulletin has to guess the rules governing the format by inspecting sample bulletins. Any such program may fail if, for any reason the format changes. As long as leap seconds keep UT1-UTC less than 0.9 seconds, the validity of the input can be checked so that most errors can be detected but if this constraint is removed, checking for errors becomes much more difficult.

As UTC (or whatever civil time is called in the future) drifts away from the rotation of the earth, the range of applications which requires knowledge of UT1-UTC will increase, and along with it, demand for access to this information in a machine readable format over the internet. Also, Bulletin A contains more than just UT1-UTC and is somewhat intimidating for the non-specialist. A programmer who is told that UT1-UTC is needed and is given the URL of the Bulleting A is likely to be a bit uncertain as to whether the column labeled UT1-UTC is, in fact, all they need or whether other information contained in the bulletin is somehow relevant. All the necessary explanatory information is available on the IERS web site but in too much detail for the non-specialist to comprehend easily.

Once UT1-UTC is being imported into a software system automatically, the consequences of the process failing in some way have to be considered. In particular, the case where the failure goes undetected and an erroneous value is inserted into the system. If a safety critical system could be disrupted by a corrupted copy of the bulletin then any new way of distributing earth orientation predictions must have the necessary safeguards built in.

Any such new way of distributing earth orientation predictions ought to be straight-forward to use and accessible to the widest possible audience using readily available tools and well known techniques. It must also be implementable with the resources available to the IERS. The obvious mechanism is a text file that can be copied with the http protocol.

FILE DESIGN

Format

There can be little argument that the most appropriate file format for this application is XML[*] because it is:

- Defined by a mature and stable internationally recognized standard.
- Widely deployed across practically all applications areas.

---

[*] http://www.w3.org/TR/REC-xml/

- A wide range of tools for creating, reading and manipulating XML are available, many of them free, and on all commonly used operating systems.
- It can be interpreted by humans as well as by computer programs.
- It is unlikely to be superseded in the foreseeable future.
- The family of XML standards includes a specification for digital signing

XML is, of course, not perfect. It can be wasteful of space and expensive to parse in comparison with some other formats and it is clumsy to edit by hand. It also looks fairly ugly. However, none of these criticisms are particularly relevant for the application being proposed here.

XML is extremely flexible and distinguishing between a good design and a bad design is not easy. The principles adopted here are to keep things as simple as possible and to favor ease of parsing by a computer over readability. The latter implies using element properties rather than text elements for the data.

Content

The following example is valid XML and contains the about minimum necessary to achieve the stated purpose without being overly cryptic:

```
<earth_rotation_prediction_table>
  <earth_rotation date="2011-07-02" x="0.1714" y="0.4109" UT1-UTC="-0.30187"/>
  <earth_rotation date="2011-07-03" x="0.1727" y="0.4099" UT1-UTC="-0.30234"/>
  <earth_rotation date="2011-07-04" x="0.1739" y="0.4088" UT1-UTC="-0.30289"/>
  <earth_rotation date="2011-07-05" x="0.1750" y="0.4077" UT1-UTC="-0.30351"/>
  <earth_rotation date="2011-07-06" x="0.1759" y="0.4065" UT1-UTC="-0.30417"/>
</earth_rotation_prediction_table>
```

However, for a document that is going to be distributed outside a single organization some additional content is desirable such as:

- An XML declaration
- An indication of where the information it contains originates
- A reference to a source of explanatory information about the file contents
- A reference to an XML schema that formally defines the structure

Other improvements that can be considered are to make the units of UT1-UTC and the polar motions explicit, to encode the dates as Modified Julian Dates in addition to calendar dates and to specify the range of dates covered by the table. Finally, the element and property names should be placed in a namespace in case they conflict with names in other XML documents it might be merged with. This expands the example above to (the names chosen for the namespace and schema location are for illustration only):

```
<?xml version="1.0" standalone="yes"?>
<earth_rotation_prediction
        start_date="2011-09-09"
        end_date="2011-09-13"
        xmlns ="http://www.iers.org/xbulletins"
        xmlns:xsi="http://www.w3.org/2001/XMLSchema-instance"
        xsi:schemaLocation=
              "http://www.iers.org/xbulletins Earth_Rotation_Prediction.xsd">
   <source
        url="http://data.iers.org/products/6/14858/orig/bulletina-xxiv-036.txt">
IERS Bulletin A Vol. XXIV No. 036
```

```xml
        </source>
        <reference url="http://maia.usno.navy.mil/bullainfo.html"/>
        <reference url="http://hpiers.obspm.fr/iers/bul/bulb/explanatory.html"/>
        <earth_rotation_prediction_table>
            <earth_rotation date="2011-09-09" MJD="55813"
                x_arcsec="0.1714" y_arcsec="0.4109" UT1-UTC_sec="-0.30187"/>
            <earth_rotation date="2011-09-10" MJD="55814"
                x_arcsec="0.1727" y_arcsec="0.4099" UT1-UTC_sec="-0.30234"/>
            <earth_rotation date = "2011-09-11" MJD ="55815"
                x_arcsec="0.1739" y_arcsec="0.4088" UT1-UTC_sec="-0.30289"/>
            <earth_rotation date="2011-09-12" MJD="55816"
               x_arcsec="0.1750" y_arcsec="0.4077" UT1-UTC_sec="-0.30351"/>
            <earth_rotation date="2011-09-13" MJD ="55817"
                x_arcsec="0.1759" y_arcsec="0.4065" UT1-UTC_sec="-0.30417"/>
        </earth_rotation_prediction_table>
</earth_rotation_prediction>
```

The impact on the readability of adding more material is plain to see and, although it is easy to think of more that could be added, a balance has to be struck between what might be relevant to someone reading the file and obscuring the original purpose.

This format can be formally specified by an XML schema such as:

```xml
<?xml version="1.0"?>
<xs:schema elementFormDefault="qualified"
           xmlns:xs=http://www.w3.org/2001/XMLSchema
           targetNamespace="http://www.iers.org/xbulletins">
    <xs:element name="earth_rotation_prediction">
        <xs:complexType>
            <xs:sequence>
                <xs:element ref="source"/>
                <xs:element ref="reference" maxOccurs="unbounded"/>
                <xs:element ref="earth_rotation_prediction_table"/>
            </xs:sequence>
            <xs:attribute name="start_date" type="xs:date"
                          use="required"/>
            <xs:attribute name="end_date" type="xs:date"
                          use="required"/>
        </xs:complexType>
    </xs:element>
    <xs:element name="earth_rotation">
        <xs:complexType>
            <xs:attribute name="date" type="xs:string"
                          use="required"/>
            <xs:attribute name="MJD" type="xs:string"
                          use="required"/>
            <xs:attribute name="x_arcsec" type="xs:string"
                          use="required"/>
            <xs:attribute name="y_arcsec" type="xs:string"
                          use="required"/>
            <xs:attribute name="UT1-UTC_sec" type="xs:string"
                          use="required"/>
        </xs:complexType>
    </xs:element>
    <xs:element name="earth_rotation_prediction_table">
        <xs:complexType>
            <xs:sequence>
                <xs:element ref="earth_rotation"
                            maxOccurs="unbounded"/>
            </xs:sequence>
        </xs:complexType>
    </xs:element>
```

```
            <xs:element name="reference">
               <xs:complexType>
                   <xs:attribute name="url" type="xs:anyURI"
                                 use="required"/>
               </xs:complexType>
            </xs:element>
            <xs:element name="source">
            <xs:complexType>
            <xs:simpleContent>
                <xs:extension base="xs:string">
                    <xs:attribute name="url" type="xs:anyURI"
                                  use="required"/>
                </xs:extension>
            </xs:simpleContent>
         </xs:complexType>
      </xs:element>
</xs:schema>
```

This is more prescriptive than is strictly necessary; it forbids any other content and constrains the order of the element. With more work, a specification that guarantees the presence and format of the earth rotation prediction table element but gives more flexibility for the rest of the file could be developed.

DIGITAL SIGNING

If earth orientation information is going to be used by safety critical systems, a mechanism for assuring the integrity of this information must be implemented. The technology for doing this is digital signing using a public/private key infrastructure. The standard for signing XML documents is defined by the WC3 recommendation "XML-Signature Syntax and Processing".[*] Only a small number of organizations will be interested in checking the integrity of earth rotation predictions so the ease with which a signature can be verified is not particularly important; the few who need to verify it can be assumed to be knowledgeable about the processes involved. However, signing a document has to be sufficiently straight-forward to be practical for the IERS and the result should be as unobtrusive as possible for those not interested in the signature.

The WC3 recommendation describes several alternative structures for a signed XML document.

- The signature can be contained within (be a child of) the material being signed.

- The signature can contain the material to be signed (the signature is the parent of the signed material).

- The signature can be alongside (a sibling of) the signed material

- The signature can reside in another document entirely

Putting the signature in a separate document is the least intrusive option as the only addition to the format shown above would be a reference to the location of the signature, but it does require more administration; there is an additional file to manage and if files are moved additional steps are need to maintain the link from the document to the signature. Embedding the document in the signature alters the steps needed to access the table and makes it harder to strip out the signature if it is not required. Both of the remaining options are equally suitable.

---

[*] http://www.w3.org/TR/2002/REC-xmldsig-core-20020212/

Free software is available for both signing and verifying signatures (for example, the Java software development kit from Oracle) but some programming is required. A certificate issued by a recognized certificate authority is also need; ideally one issued by an authority who's root certificate is installed by operating system manufacturers so that if, in the future, software tools verify signatures by default, warnings about missing certificates will not be generated.

ACKNOWLEDGMENTS

The preparation of this paper and the author's attendance at the symposium was funded by the Large Binocular Observatory.

DISCUSSION CONCLUDING AAS 11-679

Wolfgang Dick noted that, in his experience, there seemed to be little demand for data provided via XML. Rob Seaman suggested that this perception may be due to a still-limited market within the user community, and David Terrett noted that demand should increase as people upgrade their control systems over time. Steve Allen quipped that many systems might be upgraded only when it is clear that they will no longer function (a possible example being overflow of format fields for files used by Mark-III VLBI correlators).

Arnold Rots agreed that XML digital signatures will be expected into the future so it would be best to introduce them from the outset, and he also suggested additional data such as numbers of leap seconds. Terrett replied that his presentation focused on the needs of his application, and that he welcomed consideration from others regarding what data should be maintained for operational purposes. However, he also signaled that it is very easy to add content to XML; such ease can create files with much unutilized information. Rots followed that consideration might be given to putting all time-relevant information into a single file. Terrett agreed, but cautioned that if files are laden with too much information, the average programmer might become discouraged. As an example, he quoted *IERS Bulletin A* which says time "T is the date in Besselian years."[*]

George Kaplan was curious if XLM parsers were standardized; Steve Allen and Terrett affirmed that parsers are easily available for almost all modern programming and scripting languages. Kaplan also queried about the meaning of a "standalone" file; Terrett said that by "standalone" there should be no need to look outside the data file to be able to parse the data file. David Simpson asked about the risk of transmission error and possible checksums; Terrett clarified that the digital signature serves this purpose. Seaman and Terrett also noted that there were advantages to having some peer review of the schema for "sanity checking" but there is not much reason for committee deliberations of schema designs for such a relatively straightforward application.

Seaman asked about proposed methods of distribution. Terrett said that distribution should be as simple as possible, suggesting file access via http. Seaman noted that may not suffice for all applications; Terrett clarified that such should be minimally available via http, and Seaman agreed http was a good starting point.

---

[*] http://maia.usno.navy.mil/ser7/ser7.dat

AAS 11-680

DISSEMINATION OF UT1-UTC THROUGH THE USE OF VIRTUAL OBSERVATORY

Florent Deleflie,[*] Daniel Gambis,[†] Christophe Barache[‡] and Jérôme Berthier[§]

Information concerning UT1-UTC and the occurrence of the leap seconds are currently made available via IERS bulletins (Bulletin D and C) sent to users in ASCII format. However, this old-fashioned procedure does not satisfy automatic systems. We have investigated the way to develop a new service based on the concept of Virtual Observatory (VO). This concept, provided by the International Virtual Observatory Alliance (IVOA), allows scientists and the public to access and retrieve UT1-UTC information using on-line distributed computational resources. We describe here how we derived the concept, using the XML-based VOTable format to build this UT1-UTC dedicated new service.

INTRODUCTION

As mentioned on the *International Virtual Observatory Alliance* (IVOA) website, the IVOA was formed in June 2002 with a mission to "facilitate the international coordination and collaboration necessary for the development and deployment of the tools, systems and organizational structures necessary to enable the international utilization of astronomical archives as an integrated and interoperating virtual observatory."[**] The work of the IVOA focuses on the development of standards. Working Groups are constituted with cross-project membership in those areas where key interoperability standards and technologies have to be defined and agreed upon.[1] The Working Groups develop standards using a process modeled after the World Wide Web Consortium. Recommendations are ultimately endorsed by the Virtual Observatory (VO) Working Group of Commission 5 (Astronomical Data) of the International Astronomical Union (IAU).

Several independent Java-based tools have been developed in the VO framework that can be used in geodesy and more generally in Earth's Science.[2] These tools can either be downloaded and set up in PCs as Java applications, or used through a web browser as Java applets. In particular, some of the tools we started to use with IERS data are "VOPlot" and "TopCat", which are devoted to astronomical data and time series plotting.

---

[*] Institut de Mécanique Céleste et de Calcul des Ephémérides, Observatoire de Paris / GRGS, 77 Avenue Denfert Rochereau F-75014 Paris, France.
[†] Observatoire de Paris / SYRTE / GRGS, 61 Avenue de l'Observatoire, F-75014 Paris, France.
[‡] Observatoire de Paris / SYRTE / GRGS, 61 Avenue de l'Observatoire, F-75014 Paris, France.
[§] Institut de Mécanique Céleste et de Calcul des Ephémérides, Observatoire de Paris / GRG , 77 Avenue Denfert Rochereau F-75014 Paris, France.
[**] http://www.ivoa.net/pub/info/

BASIC PRINCIPLES OF THE VIRTUAL OBSERVATORY

The VO-Table Data Exchange Format

VOTable is the XML-based format for representing astronomical data, recommended by IVOA (e.g. catalogues, as tables of the properties of celestial objects, celestial coordinates, brightness etc.). The VO-Table format has been defined in terms of XML in order to take advantage of computer-industry standards and to utilize standard software and tools. At the same time it is important not to lose the previous investment in astronomy-specific standards, such as the table variants of the *Flexible Image Transport System* (FITS) format. Also, astronomical tables are rich in *metadata*, which in this context means annotation, interpretable by either computers or humans, both of the tables and the individual columns that they contain. It is important to note that these metadata should be preserved with the table and the VO-Table has features to permit this.

Moreover, it is crucial to point out the fact adopting the VO-Table format does not mean giving up of its own data format: the VO-Table format can encapsulate existing files and simply supplies metadata to understand its content and facilitate data exchanges.

Why Choose the Virtual Observatory

There exists several software packages to treat metadata files, but the so-called "Virtual Observatory", as an ensemble of VO-Table-based software packages, is now widely used within the astronomical community, by several thousand users worldwide. The Virtual Observatory takes advantage on the notion of *Unified Content Descriptors* (UCD) to be inserted into metadata files to describe the data, following the self-descriptive format VO-Table based on these standards and XML. As a consequence, many tools already exist to manage, plot or analyze data supplied in VO-Table format. Converting ones own data in VO-Table format means benefiting of all existing tools, some of them providing a conversion from unformatted data files into the VO-Table format, as well. As a consequence, data will be described non-ambiguously, ensuring further exchange and better understanding between different scientific communities. By the way, the Virtual Observatory provides an easy access to all VO-Table data: they can be registered to a "registry", that means that any user can locate on the web, get details of, and make use of, any resource located anywhere in the IVOA space.

Using the Virtual Observatory in Geodesy?

VO standards have been developed for Earth-centered or body-centered reference frames in order to extend the VO to Earth and planetary sciences. Nevertheless, some improvements are to be made. Two years ago, our group proposed to the IVOA to adopt standards relevant to the Earth orientation data (polar motion, UT1-UTC, nutation, etc.) and to space geodesy. These were accepted for official use in all files relevant to VO.

SPECIFIC PROJECTS TO MAKE UT1-UTC AVAILABLE

Web application

An execute application is currently running. It makes available EOP and in particular UT1-UTC using the web service (Figure 1, Figure 2).

Figure 1. Webservice webpage at Paris Observatory: http://hpiers.obspm.fr/eop-pc/.

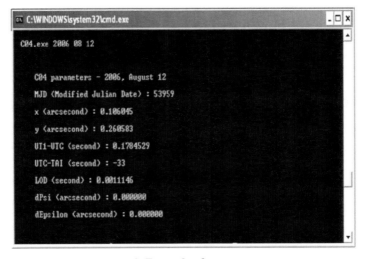

Figure 2. Example of a request

Application based on VO tools

The tool extracts and makes available EOP (Figure 4) and in particular UT1-UTC (Figure 5) over a period chosen by the user. The output is given through ASCII or VO-Table formats (XML file containing UCD). This tool is easy to be used, and is compatible with Internet Explorer, Firefox, in particular. It is made up of independent sub-programs, and made secure. Results are obtained very quickly and data do not need to be duplicated.

This process facilitates links between various user and scientific communities. The VO-Table format ensures the compatibility between external software.

The reference IERS EOP file, containing the official EOP C04 solution through the c04.62-now file is automatically retrieved of and converted into VO(XML) format compatible with VO software packages. Initially EOP parameters have to be declared accordingly to specific Unified Content Descriptors (Figure 3) for instance "time.epoch" standing for dates in MJD or "pos.eop.UT1mUTC" standing for UT1-UTC.

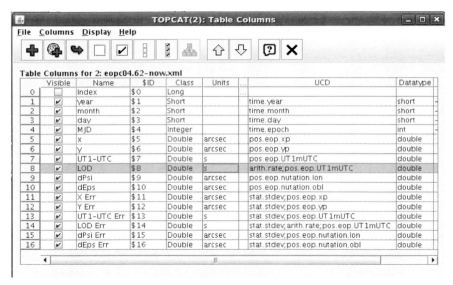

Figure 3. Unified Content Descriptors relevant to Earth rotation.

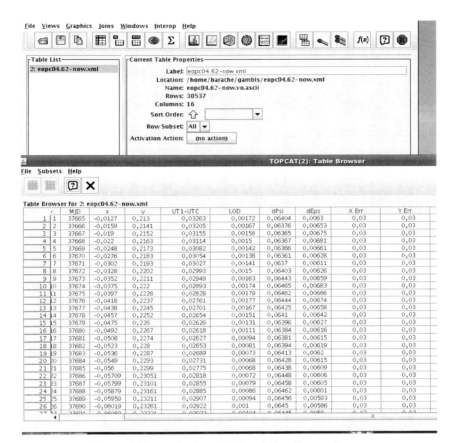

Figure 4. Example of an Earth rotation time series.

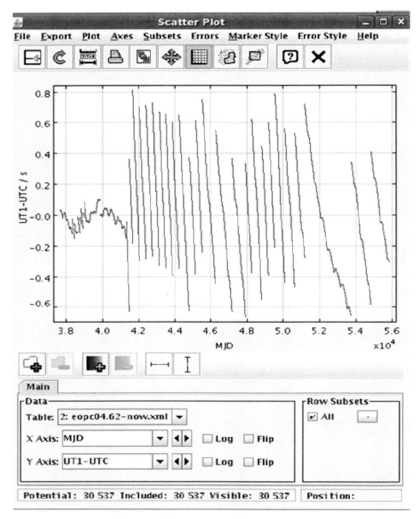

Figure 5. Time series of UT1-UTC.

CONCLUSIONS

The scientific community working in different field and requiring in particular UT1-UTC on a regular and reliable basis can benefit from the VO concept mainly through two points. On the one hand, the concept of metadata allows to gather up a single file with a description supposed to be exhaustive.

In our opinion, such tools can give an extraordinary visibility of all data and in particular earth orientation parameters and UT1-UTC derived by the IERS.

REFERENCES

[1] Souchay, J.; Andrei, A. H.; Barache, C.; Bouquillon, S.; Gontier, A.-M.; Lambert, S. B.; Le Poncin-Lafitte, C.; Taris, F.; Arias, E. F.; Suchet, D.; Baudin, M., "The construction of the large quasar astrometric catalogue (LQAC).", *Astronomy and Astrophysics*, Volume 494, Issue 2, pp.799-815, 2009.

[2] Lambert, S., F., Deleflie; A.-M., Gontier; P., Berio; C. Barache, "The Astronomical Virtual Observatory and Application to Earth's Sciences." IVS 2008 General Meeting Proc., 2008.

DISCUSSION CONCLUDING AAS 11-680

Paul Gabor asked regarding the degree of relationship of the Virtual Observatory (VO) project with the Microsoft WorldWide Telescope project. He noted that the Microsoft WorldWide Telescope has been in development for almost two decades and is free, but is not platform independent. Rob Seaman responded that VO has been collaborating with Microsoft WorldWide Telescope, citing its "knowledge" of VO tables as an example, and that the main difference between Virtual Observatory and WorldWide Telescope is implementation details. Compatibility should be anticipated.

Neil deGrasse Tyson noted that Virtual Observatory project has been discussed since early 1990's; Tyson was therefore curious about its rate of progress, understanding that new data types are always forthcoming. Daniel Gambis believed the major issues in progress with Virtual Observatory project are related to system integrity, such as how to ensure that the files are correct. Tyson noted that when the Sloan Digital Sky survey was planned, the amount of photometric data created was expected to be greater than all of the other data up to that date, so careful thought went into setting a new photometric data standard; he queried whether VO has an occasion to foresee the expanding data needs of its community going forward. Gambis could not answer the question with any certainty, but Seaman followed that VO is a mature technology and interoperation is a key aspect of VO; nevertheless it still was not clear how digital signatures should work with VO tables.

AAS 11-681

TIMEKEEPING SYSTEM IMPLEMENTATIONS: OPTIONS FOR THE *PONTIFEX MAXIMUS*

Steven L. Allen[*]

A representation for the meaning of time and the rules for handling it is built into many operational systems – civil, legal, hardware, software, etc. Many of these systems have avoided implementing the complexity required to handle leap seconds, yet some demand their existence. A plausible change to the scheme of UTC must be compatible with existing systems and should be easy to implement. I propose a small change to the representation of leap seconds which allows the tz code to describe them in a way that alleviates the underlying problems with information processing systems. It preserves the traditional meaning of civil time as earth rotation. It allows for trivial testing of the effects of leap seconds on software and hardware systems. It is a compromise that gives easy access to all forms of time information. It is not without consequences that will have to be handled.

WHO IS THE *PONTIFEX MAXIMUS*?

The calendar of the Roman Republic had months which had already abandoned any basis in astronomical observations of the moon. Months, and thus the Roman calendar dates and year, were decided by the *Pontifex Maximus*. Julius Caesar employed astronomer Sosigenes of Alexandria to re-conform the calendar with the sun. Augustus remedied a problem with the leap year implementation and produced a predictable progression of civil dates which closely tracked the sun for over 1000 years. Pope Gregory XIII employed astronomer Christopher Clavius to bring the calendar even closer to the sun, but the costs were discontinuity of 10 days and lack of international consensus for centuries after.

Figure 1. Civil Authorities: Augustus Caesar, Pope Gregory XII, and the 1884 International Meridian Conference changed the rules of the calendar

---

[*] Programmer/Analyst, UCO/Lick Observatory, ISB 1156 High Street, Santa Cruz, California 95064, U.S.A.

Figure 2. Astronomers: Sosigenes (Hume Cronyn), Christopher Clavius, and Simon Newcomb provided the mathematics for the calendar

The Lords Commissioners of the Admiralty consulted the Royal Astronomical Society and decreed that the Nautical Almanac would tabulate mean solar time instead of apparent.[1] The 1884 International Meridian Conference resolved the basis by which the day is related to the sun,[2] and Simon Newcomb produced the mathematical details for subdividing the day.[3] Astronomers, horologists, national metrology institutes and broadcast engineers provided time signals, the Bureau International de l'Heure (BIH) analyzed those, and the Consultative Committee on International Radio (CCIR) documented their best efforts in Recommendations.[*] Physicists produced atomic chronometers,[4] broadcast engineers immediately employed them, astronomers raced to calibrate them,[5] and the CCIR struggled to find an acceptable balance between technologies.[6,7,8]

All of these actors have played the role of *Pontifex Maximus*. In the thespian slang, most of these plays have been *two-handers* – one actor with political power, and one with technical ability. The role of the rest of humanity has been audience. The details of precision timekeeping, and thus the decree that a new day had started, were only available to a few.

TIME IMPLEMENTATIONS IN COMPUTING

At the time the CCIR recommended that radio broadcast time signals should have leap seconds there were few devices which could keep a continuous count of seconds. Most time keeping devices continued the tradition of approximate subdivision of days. Today most humans have routine encounters with a device that counts seconds; many of us wear them throughout the day. This does not mean that the control over time has been democratized, but the count of seconds, and failure to handle that count, is more immediately apparent than the count of days.

Among the many computer implementations of time, Unix prevails. By the 1980s the "open" systems had converged on a system clock counting seconds since 1970-01-01T00:00:00, and this was incorporated into the first Portable Operating System Interface for Unix (POSIX).[9] Unfortunately, CCIR Recommendation 460 was not openly available[†] and the committees who produced both the

---

[*] Early CCIR Recs. on broadcast time were 70 (1951), 122 (1953), 179 (1956), and 319 (1959).
[†] Recent versions of ITU-R TF recommendations became openly available online in 2010-12 http://www.itu.

ANSI C[10] and POSIX[11] standards incorporated a nonexistent concept. The "double leap second" in `<time.h>` allowed 62 seconds in a minute numbered 0 to 61.

The double leap second error was corrected in C99[12] and POSIX 2004,[13] but programmers remain confused about the implementation of leap seconds. Section 4.15 of the current POSIX[14]* specifies that "Seconds Since the Epoch" approximates elapsed seconds of Coordinated Universal Time and requires that "each and every day shall be accounted for by exactly 86400 seconds." The rationale section A.4.15 admits that UTC has leap seconds and says "POSIX time is therefore not necessarily UTC". It describes the lack of consensus, the impossibility for POSIX to mandate that a system clock matches any official clock, and allows that POSIX seconds may not all have the same length.

Independent of POSIX efforts, Dr. David Mills produced a method for synchronizing computer clocks.[15] Network Time Protocol (NTP)[16]† is routinely distributed along with computer operating systems. Many machines on the Internet, including some at national metrology institutes, provide accurate time via NTP. The RFC explains "The Coordinated Universal Time (UTC) timescale represents mean solar time as disseminated by national standards laboratories." However, the relation between seconds counted by the NTP protocol and seconds of UTC is specified not to count leap seconds. With both POSIX and NTP not counting them, every leap second produces observable time deviations as different operating systems and implementations adjust.

In contrast to the Unix model of time, IBM S/390 contained the concept of Leap Second Offset (LSO) that could optionally be set to the number of leap seconds which have occurred since 1972-01-01.[17] The document has a lengthy description of the finicky manual details required to implement the LSO. The document now applies to the newer IBM System z, or z/OS.‡ Along with the system specifics there is also a rationale which attempts to explain GMT, UT1, UTC and leap seconds. The text seems to originate from the CCIR/ITU-R documents and it includes 'UTC is the official replacement for (and generally equivalent to) the better-known "Greenwich Mean Time" (GMT).' Online discussions among IBM sysadmins indicate that setting the LSO is rarely practiced.

Microsoft operating systems have run on machines from many different vendors with wide variations in the capabilities of the hardware clock. This means that Microsoft Windows has not needed to consider support for leap seconds because of limitations in the hardware. A Microsoft Support boundary describes their version of NTP: "The W32Time service cannot reliably maintain sync time to the range of 1 to 2 seconds. Such tolerances are outside the design specification of the W32Time service."§

Insufficient information, unavailable or unclear standards, and lack of consensus mean that various vendors and researchers continue to try different schemes for handling leap seconds. Michel Hack and a team from IBM tried handling the 2008 leap second by hacking Linux kernels in a fashion similar to z/OS.[18] The results were not POSIX compliant. Site reliability engineer Christopher Pascoe described how Google handled the 2008 leap second.¶ Their result was POSIX-conformant, but they changed the length of seconds in way unsuitable for real-time control processes.

---

int/rec/R-REC-TF/en
  *http://www.unix.org/
  †http://www.ntp.org/
  ‡http://www.ibm.com/systems/z/os/zos/
  §http://support.microsoft.com/kb/939322
  ¶http://googleblog.blogspot.com/2011/09/time-technology-and-leaping-seconds.html

TIME ZONES AND DAYLIGHT TIME

There is another form of *Pontifex Maximus* well known to about half the population of earth. Civil authorities routinely exercise control of time by moving zone boundaries and changing the dates of transitions between standard time and daylight (or summer) time.

In the 1960s and 1970s some counties in Indiana ignored federal and state regulations specifying their time zone rules. In 1999 several Australian states changed their daylight rules for the 2000 Olympics with less than a year of advance notice. In 2006 the rules were changed again for the Commonwealth Games. In 2005 the governor of Indiana enacted a law requiring all counties to observe daylight time starting less than a year hence. In 2007 president Hugo Chávez announced that Venezuela would shift from GMT-04:00 to GMT-04:30 the next week (other authorities persuaded him to wait until year's end).

Figure 3. Indiana governor Mitch Daniels and Venezuela president Hugo Chávez insisted that their constituencies adopt new rules for the start time of each day.

All of these changes in time zones and rules, and many others, are documented in the `tz` database.* Arthur David Olson of NIH instituted the `tz` database and has coordinated a community which performs ongoing maintenance. Although the `tz` database is not authoritative, most operating systems use some form of it to convert between system time and civil time. The vendors of most systems provide updates to the `tz` database as part of routine patches.

The `tz` database consists of source code and data. For POSIX systems the time conversion is performed by the `tzcode` using the `tzdata`; the kernel has no role. The kernel code ostensibly keeps UTC, and civil time conversions happen in user code. This removes any need for the kernel to be updated or know about changes.

---

*http://www.twinsun.com/tz/tz-link.htm

The `tzdata` distribution contains a file `leapseconds`. This file is intended to be used by the `tzcode` when one of the "`right`" timezones is selected by the sysadmin or a user. As provided the `leapseconds` file contains a list of all leap seconds which have been inserted into the broadcast time scale. Use of the "`right`" timezones, however, is not conformant with POSIX because it produces days with 86401 seconds. The "`right`" timezones are also not compatible with NTP, for they presume that the system clock value of `time_t` is a count including all leap seconds.

The recent revision of the iCalendar[19] data format highlighted the connection between calendar and clock for scheduling events. To facilitate the worldwide updating of iCalendar schedules the Calendaring and Scheduling Consortium (CALCONNECT) has tasked its TIMEZONE Technical Committee to create a timezone registry and API for a timezone service.[*] This also produced an Internet Draft proposing that the IANA should maintain the `tz` database.[†] The committee is considering how to describe and serve the `leapseconds` file along with the rest of `tzdata`. This might provide a robust and machine-readable means of distributing leap second announcements.

REINTERPRETING POSIX

What does POSIX really want for a kernel? The standard and rationale make enough apologies that the answer is not immediately clear.

POSIX does not want to know about astrometry or geophysics. As far as POSIX is concerned these are equivalent to the whims of politicians changing timezones and daylight rules.

POSIX wants to be conformant with the needs of "real-time" systems. The self-inconsistent words of the POSIX standard when it mentions leap seconds and UTC make this goal unreachable with the status quo.

POSIX does demand 86400 seconds in a day. This fact is built into far too much code. Changing this would be prohibitively expensive. Knowing nothing of astrometry or geophysics, however, POSIX also is oblivious to the kind of "day" that has 86400 seconds.

POSIX standard is mistaken when it says it wants UTC. The use of the term UTC was merely an update to the original notion of GMT.[‡] The practical result of the evolution of systems and hardware means that POSIX really wants the time scale which is internationally approved for use in radio broadcasts. POSIX systems do not care what name humans use for that time scale, and a change in the name of that time scale cannot affect the operation of the kernel.

POSIX requires that the zoneinfo mechanism be able to handle offsets expressed in hours, minutes, and *seconds* between `time_t` and the local timezone.[§] This requirement is the key to a possible compromise.

A POSIX CONFORMANT WAY TO RETAIN LEAP SECONDS IN UTC

The time scale used by the GPS satellites was equal to UTC on the inception date of the system, 1980-01-06. Several vendors supply NTP time servers which rely on signals from the GPS satellites to maintain correct time. The normal configuration of an NTP time server uses the information in the GPS signals to convert GPS time into UTC.

---

[*]`http://calconnect.org/tc-timezone.shtml`
[†]`http://tools.ietf.org/html/draft-lear-iana-timezone-database-04`
[‡]see the Definition of "Epoch"
[§]see System Interfaces for `tzset()` and Environment Variables for TZ

Owners of time servers from Meinberg and Symmetricom can configure them to provide NTP service as GPS time instead of UTC. Combining that with some small changes to the tz database produces a scenario I call "right+GPS". Figure 4 shows the leapseconds file where leaps before the GPS epoch are deleted and one test leap is added at the end. Figure 5 shows the shell script which produces the output in Figure 6.

# Leap	YEAR	MONTH	DAY	HH:MM:SS	CORR	R/S
Leap	1981	Jun	30	23:59:60	+	S
Leap	1982	Jun	30	23:59:60	+	S
Leap	1983	Jun	30	23:59:60	+	S
Leap	1985	Jun	30	23:59:60	+	S
Leap	1987	Dec	31	23:59:60	+	S
Leap	1989	Dec	31	23:59:60	+	S
Leap	1990	Dec	31	23:59:60	+	S
Leap	1992	Jun	30	23:59:60	+	S
Leap	1993	Jun	30	23:59:60	+	S
Leap	1994	Jun	30	23:59:60	+	S
Leap	1995	Dec	31	23:59:60	+	S
Leap	1997	Jun	30	23:59:60	+	S
Leap	1998	Dec	31	23:59:60	+	S
Leap	2005	Dec	31	23:59:60	+	S
Leap	2008	Dec	31	23:59:60	+	S
Leap	2011	Oct	3	20:51:60	+	S

Figure 4. the leapseconds file from tzdata hacked to demonstrate right+GPS zoneinfo

```
#! /bin/sh

then='empty'
isofmt='+%Y-%m-%dT%H:%M:%S'
MYTZ=$HOME/tzdir2011k+gps/etc/zoneinfo-leaps/US/Pacific
while true; do
  now=`date "$isofmt"`
  if [ x"$now" != x"$then" ]; then
    right=`TZ=:$MYTZ date "$isofmt"`
    time_t=`date +%s`
    echo "$time_t POSIX $now right+GPS $right"
    then=$now
  fi
  usleep 500000 2>/dev/null || sleep 0.5
done
```

Figure 5. shell script demonstrates right+GPS zoneinfo

The output shows that the time_t of the system clock incremented uniformly, and the interpretation of the clock was POSIX-conformant, but the time presented to the users included the leap second. This strategy produces a POSIX-conformant system by redefining the notion of "POSIX day" as 86400 seconds of atomic time while allowing ongoing leap seconds in the civil day. This strategy can only work if the ITU-R changes the name of the broadcast time scale along with omitting leap seconds from the broadcasts (which was the advice given to the ITU-R WP7A SRG at the 2003 colloquium they held in Torino).

This strategy requires a non-conformant NTP server and maintenance of the hacked tzdata distribution. With the current form of UTC only a few sites can afford the manpower requirements for

```
1317675120  POSIX 2011-10-03T13:52:00  right+GPS 2011-10-03T13:51:45
1317675121  POSIX 2011-10-03T13:52:01  right+GPS 2011-10-03T13:51:46
1317675122  POSIX 2011-10-03T13:52:02  right+GPS 2011-10-03T13:51:47
1317675123  POSIX 2011-10-03T13:52:03  right+GPS 2011-10-03T13:51:48
1317675124  POSIX 2011-10-03T13:52:04  right+GPS 2011-10-03T13:51:49
1317675125  POSIX 2011-10-03T13:52:05  right+GPS 2011-10-03T13:51:50
1317675126  POSIX 2011-10-03T13:52:06  right+GPS 2011-10-03T13:51:51
1317675127  POSIX 2011-10-03T13:52:07  right+GPS 2011-10-03T13:51:52
1317675128  POSIX 2011-10-03T13:52:08  right+GPS 2011-10-03T13:51:53
1317675129  POSIX 2011-10-03T13:52:09  right+GPS 2011-10-03T13:51:54
1317675130  POSIX 2011-10-03T13:52:10  right+GPS 2011-10-03T13:51:55
1317675131  POSIX 2011-10-03T13:52:11  right+GPS 2011-10-03T13:51:56
1317675132  POSIX 2011-10-03T13:52:12  right+GPS 2011-10-03T13:51:57
1317675133  POSIX 2011-10-03T13:52:13  right+GPS 2011-10-03T13:51:58
1317675134  POSIX 2011-10-03T13:52:14  right+GPS 2011-10-03T13:51:59
1317675135  POSIX 2011-10-03T13:52:15  right+GPS 2011-10-03T13:51:60  < leap
1317675136  POSIX 2011-10-03T13:52:16  right+GPS 2011-10-03T13:52:00
1317675137  POSIX 2011-10-03T13:52:17  right+GPS 2011-10-03T13:52:01
1317675138  POSIX 2011-10-03T13:52:18  right+GPS 2011-10-03T13:52:02
1317675139  POSIX 2011-10-03T13:52:19  right+GPS 2011-10-03T13:52:03
1317675140  POSIX 2011-10-03T13:52:20  right+GPS 2011-10-03T13:52:04
1317675141  POSIX 2011-10-03T13:52:21  right+GPS 2011-10-03T13:52:05
```

Figure 6. A POSIX-conformant leap second by using right+GPS. Note the shift of GPS – UTC from 15 to 16 s.

maintaining such an aberrant configuration. If the ITU-R were to change the name of the broadcast time scale then this strategy could be adapted as the default used by all systems.

The code here demonstrates that leap seconds can be handled by code that is already tested, widely-distributed, and in use by POSIX-conformant systems. A shift in the representation of time, and the nomenclature used by POSIX systems, allows a compromise which preserves the traditional meaning of civil time while enabling technologies that require a new meaning for broadcast time signals.

Furthermore, as demonstrated here, on a POSIX system this strategy allows any user to test any software at any time. No special hardware is needed to simulate the effects of a leap second. In this scheme leap seconds and UTC could become a timezone. This is a form of "localization", the term used to describe computer outputs which can be formatted differently according to cultural preferences.

CONCLUSION

Discussions on the details of UTC and leap seconds often turn into flame wars. There has been little consensus. Notions of the rules for keeping time are built into many systems. The subject is broad and esoteric. Many people have preconceptions based on outdated information resources, and misconceptions based on wrong resources. Good pedagogy is lacking.

If the ITU-R abandons leap seconds but retains the name UTC the tz database still allows jurisdictions to declare mean solar time as their legal civil time. Any authority who decides to continue inserting leap seconds can use this code and data to insert them. If such a time scale were named "Global Mean Time" or "Greenwich Meridian Time" then the currently synonymous terms UTC and GMT would gain notably distinct meanings.

Explaining the subject of UTC, GMT, and leap seconds is difficult. When talking with a journalist a good metaphor is the story of the blind men and the elephant. A journalist is likely to receive a different description from each different pundit, and there may be more blind men who never contribute their knowledge of the elephant. Getting a comprehensible description of the entire elephant takes a lot of work.

Figure 7. "Blind monks examining an elephant" by Hanabusa Itchō

REFERENCES

[1] Royal Astronomical Society, "Report of the Council to the Eleventh Annual General Meeting of the Society," *Monthly Notices of the Royal Astronomical Society*, Vol. 2, Jan. 1831, pp. 11–12.

[2] US Department of State, ed., *International Meridian Conference*. Washington, D.C.: Gibson Bros., 1884. Protocols of the Proceedings.

[3] S. Newcomb, "Tables of the motion of the earth on its axis and around the sun," *Astronomical papers prepared for the use of the American ephemeris and nautical almanac, v. 6, pt. 1, [Washington, Bureau of Equipment, Navy Dept., 1895], 169 p. 29 cm.*, Vol. 6, 1895, pp. 2–+.

[4] L. Essen and J. V. L. Parry, "An Atomic Standard of Frequency and Time Interval: A Cæsium Resonator," *Nature*, Vol. 176, Aug. 1955, pp. 280–282, 10.1038/176280a0.

[5] W. Markowitz, R. G. Hall, L. Essen, and J. V. Parry, "Frequency of Cesium in Terms of Ephemeris Time," *Physical Review Letters*, Vol. 1, Aug. 1958, pp. 105–107, 10.1103/PhysRevLett.1.105.

[6] CCIR, "Recommendation 374, Standard-Frequency and Time-Signal Emissions," *Documents of the Xth Plenary Assembly*, Vol. III, 1963, p. 193.

[7] CCIR, "Recommendation 374-1, Standard-Frequency and Time-Signal Emissions," *Documents of the XIth Plenary Assembly*, Vol. III, 1966, p. 281.

[8] CCIR, "Avis 460, Émissions de Fréquences Étalon et de Signaux Horaires," *XIIe Assemblée Plénière*, Vol. III, 1970, p. 227.

[9] Institute of Electrical and Electronics Engineers, *IEEE Standard Portable Operating System Interface for Computer Environments, Std 1003.1-1988*. 1988.

[10] American National Standards Institute, *American National Standard for Information Systems, Programming language C, ANSI X3.159-1989*. 1989.

[11] The Austin Common Standards Revision Group, *Single UNIX Specification, Version 2*. The Open Group, 1997.

[12] International Organization for Standardization, *Programming languages – C, ISO/IEC 9899:1999*. 1999.

[13] The Austin Common Standards Revision Group, *Single UNIX Specification, Version 3, 2004 Edition*. The Open Group, 2004.

[14] The Austin Common Standards Revision Group, *Single UNIX Specification, Version 4, 2010 Edition*. The Open Group, 2010.

[15] D. Mills, *Computer network time synchronization : the Network Time Protocol*. Boca Raton, FL: CRC/Taylor & Francis, 2006.

[16] D. Mills, J. Martin (ed.), J. Burbank, and W. Kasch, *RFC 5905: Network Time Protocol Version 4: Protocol and Algorithms Specification*. Internet Engineering Task Force, 2010.

[17] Ken Trowell and Marg Beal and Noshir Dhondy and Helen Howard and Greg Hutchison, *S/390 Time Management and IBM 9037 Sysplex Timer*. IBM Corporation, 1999.

[18] M. Hack, X. Meng, S. Froehlich, and L. Zhang, "Leap Second support in computers," *2010 International IEEE Symposium on Precision Clock Synchronization for Measurement Control and Communication (ISPCS)*, 2010, pp. 91–96.

[19] B. Desruisseaux, ed., *RFC 5545: Internet Calendaring and Scheduling Core Object Specification (iCalendar)*. Internet Engineering Task Force, Sept. 2009.

DISCUSSION CONCLUDING AAS 11-681

Frank Reed noted that Steve Allen's idea allows for the ability to maintain all sorts of unusual "minority" time scales. Allen noted that the capability is compatible with most Unix-like operating systems, and Microsoft has its own alternative methods of implementing such information. Neil deGrasse Tyson noted that if this option works smoothly and bug-free, then it seemingly allows any municipality to operate whatever kind of time scale offset the might want; he asked if this option opens the whole world up to going back to local mean time. Allen responded that some areas may already be there, citing Winterhaven, California as an example; it keeps Arizona time due to its proximity to the larger community of Yuma. Terrett noted that airlines need consistent local time tables for the development of schedules. Allen said that the International Air Transport Association maintains official tables correlating airports with time zones. Terrett wondered whether people would resort to using whatever the airlines used for time out of convenience, in much the same way municipalities adopted the zone times of railroads in the 19th century. Tyson noted that the airlines would only care about the declared time of the airports and not the time kept by adjacent communities. Terrett clarified that if the situation became chaotic for communities then they would likely fall back to some convention like the zone time of the local airport. Allen noted that 130 years ago St. Louis had six times because six railroads operated through it, each tied to a different meridian.

Reed noted that when standard time was adopted across the US it was major news which some local municipalities resented for philosophical or religious reasons. George Kaplan recommended a book by O'Malley "Keeping Watch—A History of American Time" that records some of the resistance to time zones, with daylight-saving time (DST) being another contentious issue.[1] McCarthy noted that astronomical observatories were funded by railroads for a long time to observe sidereal time and convert it to solar time. Reed commented that astrologers were interested in cataloging all the different local DST rules before US standardization. Kaplan remarked that he would like to know about any old records or catalogs related to DST rules because the US Naval Observatory gets many requests about how local time was maintained in specific locations.

REFERENCES

[1] O'Malley, M. (1990), *Keeping Watch; A History of American Time*. Viking Penguin.

CONCLUDING ROUND-TABLE DISCUSSION

Concluding Round-Table

CONCLUDING ROUND-TABLE DISCUSSION

>In the concluding discussion of *Decoupling Civil Timekeeping from Earth Rotation*, each attendee expressed parting thoughts regarding the topics raised during the presentations and earlier discussions. Issues of timekeeping terminology, standards, infrastructure, and public perceptions were raised. Short- and long-term planning, especially for software and broadcast systems, were discussed.

John Seago noted that there were no preconceived plans on how to finalize this colloquium, but the co-chairs were in agreement that each participant should be provided an opportunity to express any final thoughts, perceptions, and impressions based on the presentations and discussions of the meeting.

Dennis McCarthy offered some notes that he had written down that he had yet to share. He first wanted to clarify the nomenclature surrounding the term "Coordinated Universal Time." He felt that many people think that term Coordinated Universal Time refers to the fact that Universal Time is somehow coordinated with respect to the Earth and somehow that is the source of the name Coordinated Universal Time, which is incorrect. Coordinated Universal Time is a historical term and has more to do with ephemerides than actual time. When the USA and UK decided to begin to put their ephemeris work together, in the late 1950's Bill Markowitz and Humphry Smith decided that they would begin "coordinating the changes in time" which at that time were done by optical astronomical observations. These observations were used to make physical adjustments to the observatory clocks because Earth rotation was more accurate than the clocks of the time. They decided at that point, because the almanacs had *coordinated* their work, that the USA and UK would also coordinate their changes to clocks.[1] McCarthy hypothesized that these changes were announced by teletype in that era. So for these reasons, McCarthy said they called the time "Coordinated Universal Time" because Universal Time was the astronomical time that they were trying to maintain with astronomical observations. The practice of coordination grew with time and the terminology had nothing to do with the fact that the time scale is somehow coordinated with the rotation of the Earth.

Seago said that when Essen first proposed the idea of the leap second, he wrote that "Time scales have traditionally provided the time of day and the season of the year, as well as time interval, and if it is to be of universal use the atomic scale must be *coordinated* with astronomical scales."[2] Seago agreed with McCarthy that the term "Coordinated Universal Time" had been used prior to leap seconds and that the word "coordinated" was introduced into that title for some of the reasons McCarthy noted; yet, Seago added that it would not be a misnomer to claim that "Coordinated Universal Time" is a time scale intending to be "coordinated with Universal Time" because the noted purpose of UTC with leap seconds was to be "coordinated" with Universal Time, and the word "coordinated" was the language of choice by those making the proposal. Neil deGrasse Tyson noted that the uses of the term are different; the original usage signified geopolitical coordination while the latter signified scientific coordination. Rob Seaman added that

"Universal Time" still means astronomical measurement of time regardless; McCarthy agreed, adding that astronomical time was the only basis for civil timekeeping in the 1950's.

McCarthy also offered a reminder that, while this colloquium has focused on time as being important to many applications, a more important consideration of many users is frequency. Many users rely on clocks as an accurate source of frequency, as frequency is the inverse of time interval. As an example, GPS is often used to synchronize communication systems. McCarthy closed his remarks by saying that the colloquium was "terribly instructive" and quite useful.

Arnold Rots admitted that he was curious about the mixed responses to an earlier question of whether time services should be distributing TAI, or DTAI. Rots asked, "So why not distribute TAI?" Steve Allen offered that by the time the ITU-R decided that DTAI should be distributed, it was already too late for real-time POSIX-compliant applications. While POSIX can be hacked to allow real-time applications by using alternative, less-well-known APIs, that solution is also imperfect. Coupled with the fact that the abolition of leap seconds has been on the table for the past decade, there is zero incentive to develop uncertain capabilities within operating systems. Seaman asked why the POSIX compatibility should veto every possible alternative proposal: will POSIX matter in 100 years, in 400 years, in 10,000 years…? Allen noted that weapons systems, manufacturing systems, transportation systems, all require POSIX compatibility today. Seaman clarified that POSIX shouldn't be ignored, but rather the utility of every proposal shouldn't be based solely on whether POSIX allows it.

Allen sees the POSIX issue as a study of misfortune; this unfortunate compilation of a series of decisions made over the past 30 years "means that we're stuck." Seaman asked, "Who's *we*?" as he has never had any professional concerns over POSIX. Tyson suggested that individual concerns may be irrelevant if fundamental elements of society rely on it. Agreements as to what people will generally use makes seemingly arbitrary decisions non-arbitrary and leads to the definition of a society. That becomes the force of society, rather than the individual. Tyson said if weapons systems, manufacturing systems, transportation systems are standardized to a particular convention, that convention cannot be disregarded. Seaman offered that if there was ever a more obscure societal topic than the definition of UTC, it is POSIX.

David Terrett noted that POSIX mattered only because everybody's computer runs according to some type of standard. George Kaplan said that such an argument holds only for a while, as eventually some disruptive technology will come along that will not be running POSIX and a new standard will become prevalent. Allen said that until that time the POSIX-compliant kernel does not want to see a leap second, and the reason that the time zone database exists is to keep concepts separated. The kernel's job should be easy and well defined and the complexity should be managed outside the kernel.

Because POSIX doesn't technically keep UTC anyway, but rather an underlying count of integer seconds, Rots asked why not refer those seconds to TAI? Allen said that the installed base expects clocks to be tied to the "broadcast time" and that cannot be undone either. Rots countered that the broadcast time expected by POSIX doesn't include leap seconds! Allen said that is the essential problem, and that this accidentally happened is why "we're stuck." Seago wondered if we should be trying so hard to accommodate POSIX if the standard is so confused. Allen noted that DTAI didn't exist when POSIX needed such, it doesn't have it now, and it therefore can't tolerate it.

Rots noted that somehow POSIX seems to work notwithstanding that there are leap seconds in UTC, so what is the point in doing away with leap seconds? Allen said that some real-time systems do have problems when leap seconds happen. Seaman countered that Allen's presentation proved that POSIX can't "be real time" so POSIX should have nothing to do with those real time systems. Allen replied that there are many activities that are designated as being both real time and POSIX compliant. Seaman reiterated that it is easier to change the specification than to change the program: POSIX should change. Ken Seidelmann asked whether Allen's presentation showed that the POSIX behavior could be fixed. Allen clarified that his proposed software solution operated outside the kernel and POSIX doesn't care what happens outside the kernel. Interesting quirks might show up but these would not be fatal as might happen with a problem inside the kernel. Allen also repeated that his approach requires no special hardware and it is easy to test.

Seidelmann asked how POSIX had survived leap seconds all this time. Allen admitted that most systems—at least the ones that people are willing to talk about—don't really care about leap seconds. Allen speculated that there may be systems that no one will admit has problems that are in peril. Seidelmann asked why the entire world should change because of some hypothesized group of hidden users; if they exist, "why can't they just fix *their* problem?" Seaman suggested that "*their* problem" might be that they are required to conform to POSIX when it doesn't work for their application, which seemed a very silly issue. Allen said that it is difficult to guess about what people haven't admitted to talking about, but he noticed that the hard push to abolish leap seconds came right after the POSIX committee said in 1997 that the POSIX standard needed to be fixed to work in real-time applications.

Knowing that Rots had a pressing travel schedule, Seago interrupted the discussion to ask Rots if his question "why not distribute TAI?" was reasonably addressed by the discussion taking place. Rots said that he enjoyed the discussion, but he still preferred that the explicit broadcast standard be TAI or else broadcast DTAI. Allen offered that GPS provides TAI-like time broadcasts for those that need or prefer it. McCarthy said that the cessation of leap seconds would essentially conform to Rots requests, but Seaman clarified that the name would not change; it would still be called Coordinated Universal Time. Seidelmann also reminded that there is an offset between UTC without leap seconds and TAI. Rots clarified that his preference was to broadcast TAI rather than UTC without leap seconds. Allen said that TAI broadcast would never happen; Rots concurred that he didn't expect it. Rots also noted that civil clocks could continue to display something offset to UTC as currently defined; his preference was to simply have TAI as a background or basis time exposed. Seago noted that if TAI were the broadcast time scale, this would satisfy Allen's proposal to have a uniform broadcast time scale for POSIX. Allen said in his proposal he assumed that broadcast GPS time was available as a purchasable option. Rots said that he would gladly exchange his preference for TAI time by simply offsetting the GPS epoch if GPS time were a basis for time broadcasts.

Daniel Gambis said that it was very instructive hearing the interesting points of view coming from different domains. He wondered if it would be possible to have a summary of the colloquium's main points beyond the listing of abstracts. Seaman asked if Gambis had any specific main points that he thought should be clarified; Gambis said he would contemplate this. (Following the concluding roundtable, Gambis reported that his main perception from the colloquium was that there appears to be a lack of formal study regarding the broad ramifications of redefining UTC; thus, an official decision may need to be delayed.)

David Terrett said that it was a most-interesting couple of days. He wondered what happens now. What practical outcomes can result from the discussions, understanding that we have no influence with regard to the ITU-R process? Seago asked if Terrett had ideas about what *should* happen now. Terrett affirmed that there must be some means of ensuring practical distribution of

UT1-UTC into the future. McCarthy said that it would be wise to influence the IERS Directing Board to act on that issue.

Rob Seaman also asked what the action items might be going forward. He was skeptical that this or other meetings could influence delegations to the Radiocommuncation Assembly in January 2012, but he clarified that the chairmen were planning on making proceedings available to potentially inform other meetings of the future, possibly including the upcoming Royal Society meeting on November 3-4, 2011, to which a few attendees from this colloquium would also be attending. If UTC is redefined, other meetings would likely happen, perhaps having narrower focus on more topical issues (astronomy, astrodynamics, etc.).

Steven Slojkowski was thankful to be able to attend as an observer on behalf of the NASA Goddard Space Flight Center (GSFC) Flight Dynamics Facility (FDF). He said the proceedings helped to expand his otherwise "myopic view" that the FDF would be able to handle whatever comes along. He now realizes that his organization should have greater concern about other groups providing data to the FDF, as it is unknown how well they will respond to a redefinition in UTC and whether this will affect the data products that FDF receives from them. It is also unclear now how recipients of FDF data products may handle a possible redefinition; a general lack of understanding by the FDF's user base may demand unanticipated support from the FDF.

Neil deGrasse Tyson was thankful to actively participate in the discussions. His impression was that, if there was no such thing as a computer and people only relied on wrist watches, then a decoupling of civil timekeeping from Earth rotation could happen with much more blunt corrections over time. But practically everyone relies on computers today, which are synchronized to some time-source, and this level of coordination goes far beyond the scheduling of one's daily routine. Evidenced by all of the energy that was invested in the conversations regarding all of the other elements of life, it is clear that we are all connected now in fundamental ways and we must count other systems—such as transportation and military applications—as users of civil timekeeping. Tyson doubted that changes to civil timekeeping could be applied to one sector and not another without unpredictable consequences, because of our connectivity through machines. Tyson could offer no "silver bullet" but appreciated all of the arguments heard and resonated with each one, as they all seemed to make sense. Unfortunately, this resulted in a lack of clear convergence. Tyson noted his respect for the significant gray areas, and would be happy to champion whatever outcome, not having a "horse in this race."

Tyson sent a "tweet into the Twitter-verse" to gauge people's interest in timekeeping issues. His tweet was *"FYI: Android phones use GPS time, omitting 15 leap secs added since 1980. So their clocks are wrong—15 secs ahead of iPhones."* Tyson's motivation was that he hoped people reading the message might compare clocks on different devices and thereby gain a better appreciation of the subtle complexities of civil timekeeping. Tyson asked if the language was correct (enough), and some arcane discussion of the technicalities ensued before Tyson sent the message. Allen noted that the quality of synchronization of iPhones could vary depending on many factors, including which carrier service was being used, but Reed noted that he had been following how well cellular telephones kept time for years, and affirmed that cellular phones will stay to within a second of the time supplied by the cellular network. He noted that his cellular telephone back in 2002 kept time to within 1/10 of a second relative to WWV signals; however, because people don't use cellular telephones for precise timekeeping there is little commercial benefit for modern phones to maintain time that accurately. Seaman pointed out the irony that proposed changes to UTC are to supposedly benefit a telecommunications industry that can't get cellular phones to provide time to within 15 seconds!

Tyson concluded that he finds the public to be enchanted by leap seconds and other adjustments to the calendar. People love thinking about leap days whether or not they fully understand them; they almost celebrate their introduction with anticipation. He noted that in the famous opera *The Pirates of Penzance*, a plot element hinges on how to determine the age of a character born on a February 29th. Public commentaries on time zones and daylight saving time provide evidence that calendar adjustments are entertaining to people. Tyson opined that the public will not worry over finer calendar adjustments; they will simply accept them and have fun doing it.

George Kaplan said that the colloquium was "an amazing couple of days;" it was hard to imagine that 17 fairly intelligent people could gather around a table and talk about how to count seconds for two days—some of them traveling thousands of miles to participate. It struck him that we are now talking about a very fundamental change to civil time. In his opinion the title of the colloquium was correct: we are indeed discussing the "decoupling of civil timekeeping from Earth rotation" for the first time in human history. Understanding this, one would think that a phone-book sized volume of pros and cons should already exist exploring the motivations, impacts, and repercussions that one might normally expect out of a decision-making process. Studies by the federal government generate piles of documentation for much less consequential questions, yet study documentation didn't seem to exist for this question. Rather, it appeared that the decision would be left with a group that was presumably ignorant of many of the issues; that is, the voting memberships of the ITU-R are not experts in these issues but rather work for departments of State. It also seemed wrong that consideration appeared to be restricted to just the International Telecommunication Union; this question is larger than radio-communications, internet, and wireless protocols because it involves what we essentially mean by *time*. The narrow focus of the ITU-R process appeared "disturbing"; even within that very narrow focus of technical issues the documentation appeared very sparse. Kaplan admitted that he could go with either the status quo or the cessation of leap seconds, understanding that certain simplifications might result from predicting where things are in the sky for year to come, for example. Yet the process by which all this has taken place had been disturbing.

Allen noted that the expectations of the Royal Society meeting taking place in November 2011 were unclear, but he speculated that the context might be related to the recent *Metrologia* special issue, where Terry Quinn has proposed that the ITU-R should turn over authority for UTC definition to the CGPM (the directing body of the BIPM).[3] Seidelmann suggested that the expectations of the Royal Society meeting might not be clear to anyone at this colloquium, but McCarthy offered some contextual perspective. McCarthy said that many years ago he brought the issue of UTC definition before a meeting of the Consultative Committee for Time and Frequency (CCTF), a technical advisory group to the BIPM. McCarthy testified that he was "essentially thrown out of the room" because the issue of UTC definition was felt to be beyond the scope of the International Bureau of Weights and Measures and was the responsibility of the ITU-R. Seaman noted that Quinn (former Director of the BIPM) seemed to think that UTC definition is a BIPM issue. McCarthy said he helped Quinn draft his 1999 letter which advocated the use of TAI wherever a uniform time scale was desired, which was the extent to which the BIPM involved itself.[4] Seidelmann clarified that the discussion now centered on Quinn's recent *Metrologia* article.[2] McCarthy noted that Quinn doesn't run the BIPM now; thus, he is unsure how the CCTF would react to the proposal coming from Quinn as an Emeritus Fellow. Seidelmann offered that this will presumably be Quinn's position at the Royal Society meeting and may be a topic of the discussions. McCarthy replied that the UK has been steadfastly against the proposal to cease leap seconds, and there was a feeling that there was no real technical expertise applied to that position within the UK; rather, it was a political decision to avoid the loss of GMT. McCarthy said that Quinn felt strongly that there was a need to organize some group discussion of the issue within the UK, and Quinn had been working on getting some type of meeting together for the past two years. McCarthy said that the meeting of the Royal Society originally intended to serve that purpose but it appears to have changed its goal in recent months and he was unsure what the meeting would turn out to be. McCarthy thought that the audience originally intended to be the UK members of the Royal Society, but the invitation list had been expanded.

Wolfgang Dick said that this was one of the most interesting and entertaining meetings he had ever attended, and that he learned much. As a representative of the Central Bureau of the IERS he had not expected to hear so many thoughts expressed about the work of the IERS. For the benefit of the IERS Directing Board, he would attempt to summarize some of the various thoughts expressed regarding potential improvements of IERS products by those in attendance, understanding that the IERS is discussing changes to its products and many modernization efforts will be taking place within the IERS within the next year independent of the issue of UTC redefinition. Seaman said he felt that there was a certain fondness for the IERS and its products and he particularly enjoyed receiving *Bulletin D*. Dick said that in his opinion some IERS products seemed old-fashioned. Seaman said that he and Allen work on protocols for Celestial Transient Alerts with the Central Bureau for Astronomical Telegrams; these have message formats similar to IERS Bulletins but for different types of astronomical information. Dick said that the IERS Directing Board would meet in December and he would distribute his summary to some of this colloquium's attendees for their commentary before that. Seago remarked that if Dick's summary were generated soon, the colloquium chairmen might be able to leverage his summary efforts for the colloquium proceedings.

Paul Gabor offered thanks to the colloquium chairmen for organizing "such a wonderful venue at such a wonderfully low price." Seago acknowledged that the host organization, Analytical Graphics Inc. (AGI), was to be thanked primarily for that. Gabor noted that the colloquium was both exhilarating yet slightly frustrating because he was not really sure that he understood all of the main points. Questions such as "are we going to be able to explain to future generations why this is happening?" and "what are the motivations?" have not really been answered. One of his reasons for attending was to hopefully get a clearer understanding of the motivations behind the UTC redefinition issue, as this appears to be the first time in human history that civil timekeeping will be decoupled from Earth rotation *by design*. Although things may have went wrong or "haywire" through history, there was willingness to affect an eventual correction, as timekeeping has been principally perceived as an exercise in remaining faithful to what the heavens were doing. This is no longer going to happen if the proposed decoupling occurs. Gabor noted that even if this happens, pressure may be eventually brought about from this "principle of astronomical conformity" that will lead to a desire to recouple sometime in the future. Our question should be: can we somehow facilitate that recoupling now? How can we help future generations prepare for that? Evidence presented in the Thursday morning sessions seems to indicate that the inertia of developed software is a difficult burden to overcome.

Tyson commented that in the context of modern society, the development of software is not unlike agriculture of earlier cultures: it is an aspect of civilization forced annually for the sake of survival. Seaman quipped that, like agriculture, somebody still has to deal with the manure. Seago noted an interesting juxtaposition as it relates to software: whereas McCarthy noted that some computer technologists feel software should be replaced every decade, there does seem to be software inertia. Seago thought that the computer-science position makes sense if software languages are expected to rapidly evolve with time. McCarthy said that flexibility is critical so that software does not hold us back; many changes need to be made but inflexible software becomes prohibitively expensive to modify. Forward thinking is required in software design, including anticipating how people in the future might contemplate changing things.

Seaman noted that the Image Reduction and Analysis Facility (IRAF) project is older than 30 years and is very portable. He foresees it as software that that might still be operational a century from now. It was implemented according to the standards that were in affect at the time and is not broken. Changing the standard out from underneath software does not break software; rather it changes the way software is interpreted. McCarthy said that he has run into a number of situations where software changes were so prohibitively expensive that improvements could not be implemented. Tyson remarked this expression is nothing new; in nature, genetic codes contain many seemingly unused sequences that might have been useful at one time. McCarthy quipped that while humans can adapt, software can't. Gabor wondered if software inertia might be overcome if software design (or redesign) were managed by the computers themselves; because software development and management could become such a huge task, it may become too difficult to accomplish manually. Reed noted that Seaman's software searches for UTC usage seemed to demand such automation. Seidelmann said it would be amusing to tell a military General that we will now trust the computers to solve their own problems.

Steve Malys understood that the main topic of discussion relates to very fundamental changes. We may need to remind ourselves that computers work for people; people don't work for computers. Software inertia can be overcome; it is a solvable problem that just takes resources, and the codes we have now work. Our clocks work for people too; they were invented by humanity for specific reasons. When we are trying to explain to people a century from now why we made such a fundamental change to human timekeeping, it seems that there is no justifiable reason. It is mysterious who is actually pushing for this change based on the process in place, yet they have been pushing hard enough to be able to get the ITU-R to generate votes on the matter. There is no visibility into the process and Malys agreed with Kaplan that we should expect documentation on pros and cons and feedback from the affected communities. It is not obvious that some communities have had any input.

Malys continued by saying that within the US government, some informal surveys have been conducted by technical people who know something about the subject, but apparently there has not been a coordinated process even within the US government. This appeared to be a topic that should have required a significant level of coordination before a decision was made on behalf of the USA. Malys said that the National Executive Committee for Space-Based Positioning, Navigation, and Timing Executive Committee (PNT ExCom) seemed like the kind of place where a coordinated US position could have been developed, but to the best of Malys knowledge, this topic has never been raised within the PNT ExCom.

Seaman commented that the issue is never described the way the colloquium title presents the issue; rather, the issue is usually described as "let's cease leap seconds" rather than "let's redefine timekeeping." Malys agreed that if the issue were described as being a fundamental change to the way we keep time such that it was no longer tied to the Earth, more intense discussion would take place and different reactions would occur compared to a discussion limited to leap seconds. This aspect is also missing from available documentation. Gabor asked about UNESCO as a possible venue for further discussions, but Malys commented that even within the US government opinions are divided. By analogy, Seaman noted that we could not simply get rid of February 29th, because people who were born on that day would object. Malys agreed with some of the concerns of other contributors describing the process thus far as "disturbing" but he appreciated the opportunity to participate in an organized meeting attempting to raise some of the many issues. He hoped that there would be ways for colloquium participants to stay engaged until the ITU-R vote in January, and thought that it was not too late to influence opinions given the number of outstanding issues that already seem to be coming out of this small colloquium, particularly regarding possible impacts within the US DoD which may be unknown to voting US delegates.

Seago appreciated Malys optimism but commented that the topic has been discussed for over a decade and apparently hasn't received due attention. Malys suspected that this is because the issue is being (mis-)represented as simply a matter of whether or not to continue leap seconds. Seago concurred by deferring to a point made much earlier by McCarthy, namely that phrasings can be manipulated to elicit a desired response. McCarthy commented that the "effect of decoupling from the rotation of the Earth is not anything you're going to see for the next few hundred years." Seaman clarified that depends on the application; the decoupling will certainly affect astronomical software, which may be the "canary in the coalmine" for other software issues. McCarthy clarified that, for "civil timekeeping", the general man-on-the-street is not going to notice the decoupling. Seaman replied that technical applications are tied to UTC as the present basis of civil timekeeping, and the complicating issues regarding its redefinition have not been investigated.

Ken Seidelmann found the colloquium educational and "probably the most interesting discussion of this issue" that he has heard, and much was learned from having a number of varied stakeholder presentations from outside the very limited groups representing ITU-R interests. In practice, Seidelmann believes that an even broader group should be brought together to discuss this issue because there are still many more stakeholders (users) unrepresented at this colloquium, and consideration is still needed regarding *real* options. TAI, GPS time, and UTC (with and without leap seconds) have all been mentioned at some point, yet there seems to be advantages and disadvantages to all of these suggestions which have not been addressed if these are to be offered as realistic options.

Based on the discussions of the attending group, who had demonstrated a fair amount of knowledge on this subject, Seidelmann concluded that ITU-R delegations voting at the Radiocommunication Assembly (RA) in January appear to be in no position to make an informed decision on this issue. Whatever the outcome, it will be effectively "the result of a lottery" not based on intelligence. Tyson remarked that action by the ITU-R at this time would therefore seem unwise, based on the potential impact of their decision. If the proposal is approved in 2012, Seaman offered that there may be an opportunity for discussing a reversal because the proposal does not go into effect until five years after adoption (after 2017). Allen was skeptical that this would happen, citing as an example the controversial demotion of Pluto as a planet, which has not been, and is not expected to be, reconsidered by the IAU. Seidelmann also expressed doubts, noting that there is hesitancy for people to admit mistakes in judgment after the fact. Seaman said suppression of earlier ITU decisions is not unusual, based on available records. Seago said that once the decision had been made, then cost assessments would necessarily occur and the reported expense might influence a campaign to repeal the decision. Seaman added that ITU-R approval will create complications, and that will generate an extended discussion.

Tyson asked if anyone had questioned the utility of leap seconds any earlier in their forty- year existence: why is it an issue only now? Allen suggested that now we carry devices that want to count every second whereas in the past we did not. Seidelmann offered that the issue seemed to reside more with a lack of knowledge within software engineering. Allen apologized for software engineers by saying that they lacked access to necessary information. Seaman said that problems were exacerbated because ITU-R recommendations were not freely available for decades. Seago offered that philosophical preferences may also be influencing both sides of the issue; while it is often recognized that civil timekeeping tied to Earth rotation is desirous based on traditional grounds, those who advocate pure atomic time may also perceive that civil time linked to Earth rotation "messes up" otherwise "perfect" timekeeping, and that a few purely philosophical arguments against Earth-rotation time can be found as far back as the 1970s.[5] McCarthy said that increasing automation and growth of synchronizing networks is likely motivating some desires to cease leap seconds, as interruptions as small as one second can now have great repercussions.

Seaman suggested that the fundamental turning point was when the *SI* second was defined to be approximately related to time of day. There is no particular requirement that the *SI* unit of duration should be defined close to $1/86400$ of a day, and in fact this could be addressed by doubling or halving the current *SI* unit of duration to emphasize that it has no relation to time of day. Tyson acknowledged that the apparent definitional bias to fix the *SI* second close to the mean solar second was an interesting viewpoint. Terrett said that such a redefinition should not change the situation, because telecommunications is foremost interested in the uniformity of the unit of duration, a uniform time scale will diverge from Earth rotation regardless of its name or the duration of its unit. Seaman replied that no one is against the availability of uniform frequency, and time of day could be a data structure overlaid on the source of uniform frequency. Gabor said he perceived Terrett's concern simply as being with the Earth's lack of uniform rotation, to which Seaman replied that he didn't see that as a "problem" but rather as a charming fact of nature that must be accommodated somehow by civil timekeeping.

Seidelmann said that Gernot Winkler had privately admitted to him that perhaps they made a mistake when they calibrated the definition of the *SI* second against the ephemeris second. If the second had been defined closer to the length of the mean solar second of the late 20th century, then the need for leap seconds might have been put off. McCarthy added that these types of discussions might have been postponed for perhaps a century, although Storz noted that the need for adjustments would likely have been needed much sooner due to decadal fluctuations in Earth rotation. Seidelmann concluded his remarks by offering thanks to the chairmen for coordinating a venue whereby such discussions could take place.

Frank Reed also expressed his appreciation for the colloquium organization and his pleasure with meeting all the attendees. Reed said that no matter how this issue turned out, some people would perceive the decision as a small group of know-it-all scientists messing with something that is important to people's daily lives. That could turn into what he called a "Pluto moment," where people get angry and waste a lot of time defending one position or another. While the public may find a one-second adjustment entertaining, Reed said a proposed one-hour adjustment would be "tyranny". He offered a historical example: the Soviet Union called "permanent daylight-saving time" (one-hour ahead year-round) "decree time". He cited the Solzhenitsyn novel where one character in the gulag says to another:[6]

> "Since then it's been decreed that the sun is highest at one o'clock." The other wonders, "Who decreed that?" And the first answers, "The Soviet government."

So at some point messing with daily time will seem to ordinary people like an authoritarian power that no small group should have. People complain enough about daylight-saving time; they will complain more if it becomes a bigger amount of time. Tyson remarked that this level of difference might not occur for a millennium, but Reed replied that people would talk about it as if time were already broken. It is an issue that could get ordinary people riled if they are not educated on the issues. That's the "Pluto moment." While Reed knew many people who feel that keeping the Sun aligned with noon o'clock is an important cosmological issue (in the old sense of that word), he felt that other operational time scales (TT, TAI, *etc.*) are aesthetically appealing in their own physical way, if perhaps not as cosmologically fundamental as the rotation angle of the Earth. Tyson remarked that the Pluto analogy seemed apt and well made, but only to a point: there is no Disney cartoon character involved and it is not something memorized from a young age, so there may be less personal investment in either type of time. Seaman said that Earth-rotation time apparently needed a mascot—perhaps a fluffy bunny rabbit.

Mark Storz said he learned a lot from the colloquium and thanked the chairmen. Storz commented that it was eye-opening to him that astronomers might use something like an "ephemeris longitude", or else perhaps shift cataloged right ascensions, in order to address some problems with the decoupling of civil timekeeping from Earth rotation. However, he noted that stop-gap measures do not appear to be viable options for the space-surveillance community. He had hoped by attending this colloquium that he would be introduced to some proposed workarounds that US Air Force Space Command could leverage, but now he is convinced that his organization will have to procure the funds to fix everything correctly if the proposal goes forward. One particularly important issue resulting from one of Seaman's talks was the necessity of tools to assist in identifying where code is likely to break in order to identify the scale of the problems within various program offices.

Seaman suggest an investigation of strategies employed prior to Y2K might be fruitful due to the similarity of the problem, also noting that some systems simply introduced "pivot points" in code rather than truly fixing the systems large scale. Storz commented that he is already heading up an Air Force Space Command working group to consider the discontinuation of *status-quo* UTC. This group needs to audit their system software as soon as possible to educate their leadership on the costs and risks. He is also contemplating sending an official letter to DoD leadership seeking guidance on how they are to fix the issues since it appears that some elements of the DoD may have already decided to favor UTC redefinition. He hopes that such formal communication will make some elements of the DoD better aware of the potential magnitude of problems this issue might cause. Finally, Storz noted that metrological models (both terrestrial weather and space weather) are tied to UT1 and this technology sector was not represented at this colloquium.

Seago shared a parting thought from David Simpson, who left the day's discussion early to address other professional obligations. Simpson had noted that he had not found any compelling reason for motivating a change in the definition of UTC, and thereby preferred the *status quo*.

Finally, John Seago shared that he did not see how humanity could usefully recouple civil timekeeping and Earth rotation once the two were officially decoupled. Based on these proceedings, it appeared that significant functionality and infrastructure was needed to maintain intercalary adjustments. If such infrastructure were phased out of civil timekeeping systems in the near term, the expense of reintroducing such infrastructure in the distant future seemed prohibitively burdensome. Alternate long-term proposals (leap minutes, leap hours, centennial adjustments, *etc.*) pushed the technicalities sufficiently far into the future such that the recoupling would not be pragmatically addressed when the declared time comes. For that reason, schemes involving very infrequent intercalary adjustments cannot be credibly presumed to work once leap seconds are formally abolished.

Malys added that if the differences between Earth rotation and civil time were allowed to accumulate to such a degree that a minute or more needed to be introduced, the disruption caused by a large adjustment would likely spawn many legal complications. A leap second, although possibly disruptive to some systems now, seems much more manageable compared to a more noticeable amount of time. Otherwise, financial transactions and other economic activity might have to be suspended, adding even more to the cost. Seago agreed, also adding that the fundamental issues involving leap seconds do not change by kicking the can down the road. Events occurring during a leap minute would still need to be tagged in some unconventional way as is required for a leap second, except that now people will be inconvenienced by re-building and testing specialized hardware and software to be used once in a lifetime versus a leap second introduced once every few years. It would seem to be a prohibitively expensive option that each generation would likely defer to the next. Kaplan added that safety-of-life systems, such as medical and transport systems, cannot be turned off for a minute or an hour or restarted in order to adjust a clock.

Reed offered that local offsets to civil time could be legislated much the way daylight-saving time is legislated today. Terrett noted that approach causes significant complications for people living close the International Dateline. Seago wondered if this legislative approach might foster a changeover of the civil day occurring at a time other than midnight o'clock, introducing another fundamental issue. Seaman remarked that the legislative approach would move what is now a common international standard into perhaps a thousand local time-zone decisions across different nations, states, and provinces, promoting historical chaos. Reed replied that such decisions are already being made with regard to zone times. Seago offered that the historic records of such decisions would be far-flung and harder to accurately reconstruct, perhaps conflicting with a desire to maintain very accurate civil timekeeping over long intervals.

REFERENCES

[1] Wilkins, G.A. (2009), "A Personal History of the Royal Greenwich Observatory at Herstmonceux Castle, 1948-1990, Volume 1—Narrative." Collections of the Royal Greenwich Observatory Archives, Scientific Manuscripts Collections of the Department of Manuscripts and University Archives in the Cambridge University Library, p. 130. (URL: http://www.lib.cam.ac.uk/deptserv/manuscripts/RGO_history/RGO_GAW-1948-1990_v1.pdf)

[2] Essen, L. (1968), "Time Scales." *Metrologia*, Vol. 4, pp. 161-165.

[3] Quinn, T. (2011) "Time, the SI and the Metre Convention." *Metrologia*, Vol. 48, S121-S124.

[4] Steele, J. (1999), "Report of the 14th Meeting of the Consultative Committee on Time and Frequency (CCTF), BIPM, Sevres, 20-22 April." Report on the URSI Commission A: Scientific Activities for the Period 1997-1999 citing a circular letter by Dr. Terry Quinn, BIPM Director, "Time scales for satellite navigation and electronic communication systems." 7 June 1999. (available via URL http://ursiweb.intec.ugent.be/A_97-99.htm)

[5] Palmer, W. (1972), "Standard Time." Journal of Navigation, Vol. 25, No. 4, pp. 535-36.

[6] Solzhenitsyn, A. (1962), *One Day in the Life of Ivan Denisovich*. English translation by H.T. Willetts, Farrar, Straus and Giroux, 2005.

Special Session:
THE LONGWOOD GARDENS ANALEMMATIC SUNDIAL

AAS 11-682

THE LONGWOOD GARDENS ANALEMMATIC SUNDIAL

P. Kenneth Seidelmann[*]

The analemmatic sundial at Longwood Gardens was originally built in 1939, based on the design of a sundial at the Cathedral of Brou, France, which was originally built in the early 16th century. The Longwood sundial is 24 by 37 feet with a gnomon movable along an analemma. The hour markers are moveable for daylight saving time. However, the sundial did not tell time correctly. The U.S. Naval Observatory was contacted by the director of maintenance at Longwood Gardens about the problem. After measuring the sundial and computing a solar ephemeris for the location, computations were made to determine the correct locations for the gnomon. Analemmas very different from the current ones were determined for the morning and afternoon hours. After contact with a historian in Brou, France, it was determined that the sundial there had been rebuilt twice, and the current design did not tell the correct time either. That inaccurate sundial design from France was revised by computer technology to give an analemmatic sundial that told mean solar time directly, including the correction for the equation of time.

INTRODUCTION

Horizontal sundials are perhaps the most common type of dial. Traditionally, these dials have a flat horizontal face and a fixed gnomon or *style* (the index casting the shadow) directed along the local meridian. The style is often triangular, with the upper edge angled according to the latitude of the dial. This type of dial tells time by comparing the direction of the style's shadow with directional rulings along the horizontal dial face.

The classical *analemmatic sundial* is a less-common type of dial, usually horizontal, with a *vertical* rod or pole as the gnomon which must be repositioned along the dial's meridian line according to the time of year in order for the dial to indicate the intended time. This kind of dial is read from the intersection of the gnomon's shadow with points on a graduated ellipse marking off the time of day. The analemmatic sundial gets its name from the fact that the word *analemma* originally referred to a type of orthographic projection technique which was useful for dividing the elliptical dial into equal solar hours.[1]

Most sundials, including analemmatic dials, intrinsically indicate local solar time. To realize local clock time from them, corrections must be applied for daylight-saving (summer) time, the difference between local time and standard time, and the difference between mean and apparent solar time over the course of the year. This last correction is commonly known today as the *equa-*

---

[*] Research Professor, Astronomy Department, University of Virginia, P.O. Box 400325, Charlottesville, Virginia 22904, U.S.A.

tion of time, which ironically, when plotted against solar declination is also known as an *analemma*. However, analemmatic sundials do not necessarily have a correction for the equation of time built into their design for indicating mean time.

THE LONGWOOD GARDENS SUNDIAL

The Longwood Analemmatic Sundial is 24 by 37 feet in size with morning and afternoon analemmas, along which two gnomons are moved daily. The hour markers are movable to change for daylight saving time. It was built in 1939 based on the design of the historic sundial at the cathedral of Brou in France. Noon observations were made for six years prior to the completion of the sundial. In 1946 du Pont commissioned work on repositioning the hour markers of the sundial. It was originally calculated for 11 am and 1 pm, but not for other hours.

In 1968, Art Jarvela was the director of maintenance for Longwood Gardens. His job description included a long detailed list of the standard type of maintenance that would be expected at a large complex facility such as Longwood Gardens. However, the last line of the job description said "Fix the sundial." That was one thing he did not know how to do. Hence, he contacted the US Naval Observatory and talked about the problem with Ralph Haupt, Assistant Director of the Nautical Almanac Office. Art and Ralph measured the sundial in detail.[2]

Ralph Haupt presented me with the problem, with some suggestions concerning the computations to be made. I computed a solar ephemeris of the apparent azimuth and altitude for the sundial location for the year 1962, as an average non leap year. Then I computed an hourly location for the gnomon on the analemma and a position between each pair of hours. The positions were very different from the measured positions on the current analemmas. There was no single position that would give the correct time for the morning or afternoon hours. I realized that the true positions for an accurate reading required that the gnomon be moved hourly along a helical path around the analemmas for both the morning and afternoon. At noon the major axis position was most important, and in the early morning or late afternoon the minor axis position was most important. So the x positions were weighted by the cosine of the solar azimuth and the y positions by the sine of the azimuth. Since the number of hours of daylight varies during the year, there is a slight offset in the smooth shape of the analemmas in the spring and fall. The average error for any day is less than one minute. The maximum error during a year is 2.71 minutes for 7 a.m. and 2.12 minutes for noon. The computed analemmas were then checked by determining the errors for all hours of sunlight during the year. Plots of the morning and afternoon analemmas were sent in April 1971 to Art Jarvela along with information concerning what we had done and the expected errors from this design, some details of which are given in the Appendix.

CHECKING THE SUNDIAL DESIGN AND COMPUTATIONS

When the new design and computations had been applied to the sundial, Raynor Duncombe (Director of the Nautical Almanac Office), Ralph Haupt, and I, together with our wives, travelled to Longwood Gardens in the summer of 1971 to check on the results. We arrived the afternoon of the day before we were to meet the people of Longwood Gardens, so we went to see the sundial late that afternoon. We discovered that they had painted the analemma on the surface, so it could be removed if it was wrong. The gnomons were in the correct positions for our arrival the next day. We adjusted them and took readings for an hour or so. Then we placed the gnomons back where they should be for the next day. The next day Art Jarvela and some staff of Longwood Gardens joined us in observing the readings of the sundial for a period of time. It was giving the readings and errors as we had calculated. After years of checking, the analemmas were engraved in stone in 1978.

Figure 1. The Analemmatic Sundial at Longwood Gardens.

THE SUNDIAL AT BROU

The Longwood Gardens sundial copied the historic sundial in front of the Cathedral of Brou at Bourg-en-Bresse, France, which was known to have been re-designed by Jerome Lalande in the mid-18th century. I was familiar with the name Lalande, because in 1795 Lalande had recorded two observations of a star that he noticed had moved between the days of the observations. He recorded the observations, but he did not follow up on them. They were observations of Neptune made some 70 years before the discovery of Neptune and with an offset of an unknown source from the current ephemeris of Neptune. Since Lalande was a good astronomer, it was assumed that the sundial in Brou told the correct time. If the Longwood sundial was originally a copy of the Brou sundial, why didn't the Longwood sundial tell the right time?

History of the Sundial in Brou

The methods of projection used to create an analemmatic sundial go back to Vitruvius and Ptolemy.[3] However the analemmatic sundial in Brou may have been the first of its type. There is some uncertainty concerning its construction date, either 1513 or 1532. It was originally built to determine the payments for the workmen building the cathedral. It was located on the cathedral walkways, so people walked on it and wore off the engraved markings indicating where the gnomon should be placed during the year. In 1756 Jerome Lalande rebuilt the sundial and described the theory and difficulties in a memoir in the French Academy of Science in 1758.[4] In 1644 Vanzelard, an expert in analemmatic sundials, described the Brou sundial.[5]

Figure 2. The Sundial at the Cathedral of Brou, France.

Current Sundial in Brou

I contacted the cathedral of Brou seeking an explanation as to why the copy of that sundial in Longwood Gardens did not tell the correct time. A historian at the cathedral responded and informed me that the sundial at Brou did not tell the correct time now either. He explained that Lalande's markings locating the gnomon had since worn away and that an amateur artisan had restored the gnomon markers with one of his own design in 1902. The new feature was that the dates were only located on a surrounding figure-eight analemma, while only the months were on the meridian line. The more precise graduations on the surrounding curve led users to think that the gnomon belonged on the decorative analemma curve instead of on the meridian, which resulted in an inaccurate dial. The historian requested that I send a letter to the cathedral supporting the restoration of the sundial in Brou so that it would tell the correct time as with the Lalande design.

ANALEMMATIC SUNDIALS

Analemmatic sundials have become increasingly common in the 20th century in public spaces, where a person usually acts as the vertical gnomon. Early and more well-known examples of analemmatic sundials also exist in Dijon, Montpellier, and Avignon, France, Vienna, Austria, and

Basel, Switzerland. These examples all have the gnomon on the meridian line, the equation of time correction is applied separately. This would be like the original versions of the sundial at Brou. So an amateur artisan designed a sundial that would not tell the correct time. However, the application of computer technology and accurate computations resulted in a correction to that analemmatic sundial design so it would tell mean solar time directly, including the correction for the equation of time.[1]

ACKNOWLEDGEMENTS

The author would like to acknowledge John H. Seago for providing some editorial suggestions and for transcribing a revision of the author's unpublished typescript "Analemmatic Sundials" as an appendix to this manuscript.

APPENDIX[*]

General Information

The analemmatic sundial differs from other sundials, because the gnomon, the index casting the shadow, is moved periodically so the sundial will indicate the correct time. This contrasts with a fixed gnomon type where corrections are made to the readings. It will be shown that the accuracy achievable will depend on the frequency with which the gnomon is moved. It is possible that analemmatic sundials are rather old, although the evidence is not convincing. Vitruvius in his book on architecture refers to "people familiar with the analemma" and the theoretical conception of this type of sundial shows similarity to the graphical methods used in works such as "Treatise of the Analemma" by Ptolemy in the second century A.D.[3] One of the oldest and most famous analemmatic sundials in existence is in front of the Cathedral of Brou at Bourg-en-Bresse, France. The sundial dates from the epoch of construction of the Cathedral, about 1506. It was remade by Lalande in 1756, after he conceived the theory of the analemmatic sundial which he published in *Memoires de l'Academie do Sciences* 1757. This sundial has dimensions of 11.18 meters (36.68 feet) by 9.09 meters (29.82 feet) according to René R.J. Rohr in *Les Cadrans solaires* edited by Gauthier-Villars 1965, which is a source of information on various types of sundials. There is a similar sundial in a park in the city of Dijon. The largest analemmatic sundial in the United States is located at Longwood Gardens, Kennett Square, Pennsylvania, and has the dimensions 37.223 feet by 23.832 feet. The layout and calculations for this sundial are the subject of the remainder of this appendix.

Theory of Construction

An analemmatic sundial will have the shape of an ellipse (except at the Earth's poles, where it will be a circle, and along the equator where it will be a straight line). The major axis is oriented East-West while the minor axis is North-South. The ratio of the minor axis to the major axis of the ellipse should equal the sine of the latitude of the location, or

$$\frac{\text{minor axis}}{\text{major axis}} = \sin(\varphi) \qquad (1)$$

---

[*] This appendix is based on the unpublished typescript "Analemmatic Sundials" by P. Kenneth Seidelmann, c. 1970.

where φ is the geodetic latitude. Longwood Gardens is located at 39° 52' 22.2" north latitude, so the sine of the latitude is 0.6410858. The ratio of the minor to the major axis is 0.64159, so the layout of the ellipse is quite accurate.[*]

The positions of the hour markers along the ellipse can be determined from the expressions:

$$X = R\sin(AH)$$
$$Y = R\sin(\varphi)\cos(AH) \qquad (2)$$

where $2R$ equals the major axis, AH is the hour angle of the Sun, X is measured along the major axis, and Y is measured along the minor axis. For a sundial located on a standard time meridian, one whose longitude is an exact multiple of fifteen degrees, the hour markers are determined for each hour (HR) by evaluating the hour angle of the Sun (AH) in degrees from $AH = 15 \times (HR - 12)$ and then determining X and Y from the above expressions. 'For sundials not located on a standard meridian, the deviation (DEV) in degrees from the standard time meridian must be determined from

$$DEV = \text{Longitude of Sundial - Longitude of Standard Time Meridian.} \qquad (3)$$

Then the hour angles must be determined from

$$AH = 15 \times (HR - 12) - DEV. \qquad (4)$$

when the theoretical positions of the hour markers for Longwood Gardens are thus calculated, they can be compared to the actual measured positions. Table 1 gives the actual positions, the theoretical positions, and the differences in feet. The largest difference is less than one half of an inch.

The Analemma

The analemma (representing the *equation of time*) is defined by Webster to be a graduated scale of the sun's declination and of the equation of time for each day of the year, drawn across the torrid zone on a terrestrial globe.[6] For most sundials, the analemma is shaped like the figure eight and marked for the months of the year to provide a correction due to the equation of time, i.e. the difference between the position of the true Sun and that of the mean Sun which is the basis of our time system.

A point on the analemma can be determined from the equation

$$x - X = (y - Y)\tan(A) \qquad (5)$$

where (x, y) are the coordinates of the point on the analemma, (X, Y) are the coordinates of the hour marker, and A is the azimuth of the Sun at that time. For any particular time there is one equation and two unknowns.

To determine the analemma for Longwood Gardens, an apparent solar ephemeris was prepared (at the time on magnetic tape) with the apparent azimuth A and altitude of the Sun for the

---

[*] The dimensions given above for the sundial at Bourg indicate a ratio of minor axis to major axis of 0.81305, which would be for a sundial at latitude 54°24' N. Bourg is located at approximately 46°14' N. latitude, whose sine is 0.72216. Since this is about the ratio of the minor to major axis of the analemmatic sundial in Fig. 94 of the book by Rohr, the dimensions published in that book probably contain a typographic mistake.

coordinates, longitude 75° 40' 29.2" W., latitude 39° 52' 22.2" N., for each hour of the year 1962 (selected as an average year midway between leap years), with the beginning of the Besselian year close to the beginning of the calendar year. Using the azimuth of the Sun and the actual hour markers for Longwood Gardens' sundial, a solution was made for the position of the gnomon for each successive pair of equations (5) for the hours 6 a.m. to 6 p.m. In other words, for January 1 at 6 a.m. EST, the azimuth of the Sun, A_6, and the hour markers for 6 a.m., (X_6, Y_6) were substituted into the equation (5) giving

$$x - X_6 = (y - Y_6)\tan(A_6) \tag{6}$$

and likewise for 7 a.m. giving the expression

$$x - X_7 = (y - Y_7)\tan(A_7) \tag{7}$$

These two equations were solved for x and y to determine the required position of the gnomon for the sundial to read the correct time during that one hour period. This was then done for each successive pair of hours between 6 a.m. and 6 p.m. The process was followed for each day of the year. While the analemmatic sundial will give its most accurate readings if the gnomon is uniquely positioned for each pair of hours, this will complicate the appearance of the analemma and will require someone to move the gnomon each hour. The hourly positions of the gnomon are a helical curve, whose daily average is shaped like the figure 8. Likewise, the curve through the values for any one hour is shaped like the figure 8. This lack of a discrete value satisfying all the equations (5) for a given day prevents the determination from a least squares solution.

Therefore, some attempts were made to determine an average position for the gnomon for each day. A straight average of values for an entire day and for a.m. and p.m. hours separately was attempted along with weighted averages for the entire day. It was found that the most successful solution was to determine a weighted average for an analemma for the a.m., separate from that for the p.m. The Longwood Garden sundial is designed for two analemmas which facilitates the use of a.m. and p.m. averages. Thus, the a.m. and p.m. values were combined separately with the values of the x coordinates weighted by the cosine of the azimuth of the Sun and the y coordinates weighted by the sine of the azimuth.

This method of weighting arises from the geometry of the sundial. At approximately 6 a.m. and 6 p.m. when the shadow is cast along the major X axis, the position of the gnomon along the X axis is relatively unimportant, while its position along the Y axis is very important. Likewise, at noon when the shadow is cast along the minor Y axis, the gnomon position in Y is much less important than its position in X. The hours entering the average were limited to the pairs of hours for which the Sun is above the horizon, specifically 7 a.m. to 5 p.m. If all the hours that the Sun is above the horizon are included in the average, the number of values entering the average varies during the year causing discontinuities in the analemma curve.

With the average values of the analemma calculated for each half day, the azimuth of the hour markers from those positions could be calculated for each hour and compared to the azimuth of the apparent sun. The angular difference, divided by the difference in the azimuth of the Sun between the successive hours, gives the error in the sundial reading for that hour due to the use of the weighted average position for the gnomon. These error values in minutes were evaluated for each hour between 6 a.m. and 6 p.m. The average error for the hours included in the average for any day is less than one minute. The errors for any given time vary periodically during the year, building up to a maximum and then decreasing. The maximum values for times included in the average occurred at 7 in the a.m. curve and noon in the p.m. curve. The maximum for times not included in the averages occurred at 6 in the a.m. curve and 5 in the p.m. curve. The maxima during the year for each of these times are tabulated in Table 2.

Conclusions

The analemmas of the Longwood Gardens sundial represent the average a.m. or p.m. gnomon positions for each day. The exact positions of the gnomon would describe a helical curve around the a.m. and p.m. averages. A weighted average can be used with reasonable accuracy as long as the a.m. and p.m. analemmas are separate. The average error for any day is less than one minute. The maximum error for times included in the average is -2.71 minutes for 7 a.m. and -2.12 minutes for noon.

Table 1. Actual v. Theoretical Positions for the
Longwood Gardens Hour Markers (feet)

Hour	Actual Positions X	Y	Theoretical Positions X	Y	Differences X	Y
6	-18.635	-0.125	-18.626	-0.141	0.009	-0.016
7	-18.057	2.974	-18.048	2.955	0.009	-0.019
8	-16.230	5.869	-16.241	5.849	-0.011	-0.020
9	-13.310	8.355	-13.326	8.344	-0.016	-0.011
10	-9.500	10.298	-9.503	10.271	-0.003	-0.027
11	-5.030	11.516	-5.033	11.498	-0.003	-0.018
12	-0.220	11.947	-0.219	11.941	0.001	-0.006
13	4.602	11.572	4.609	11.571	0.007	-0.001
14	9.114	10.404	9.123	10.412	0.009	0.008
15	13.020	8.542	13.016	8.543	-0.004	0.001
16	16.005	6.084	16.021	6.092	0.016	0.008
17	17.911	3.210	17.935	3.226	0.024	0.016
18	18.588	0.120	18.626	0.141	0.038	0.021

Table 2. Maximum Errors

Date	6 a.m.	7 a.m.	Noon	5 p.m.
Feb 10	-3.95	-1.80		
Feb 25			-2.12	
Mar 11				-0.65
May 14			0.77	
May 17	2.36	1.33		
May 18				-0.81
Jul 17				-2.30
Jul 21	-4.35	-2.71		
Jul 25			-2.05	
Oct 18	3.55	1.20		
Oct 20			1.79	
Nov 3				1.40

REFERENCES

[1] Sawyer, III, Frederick W., "On Analemmas, Mean Time and the Analemmatic Sundial." *Bulletin of the British Sundial Society*, June 1994, 94(2), 2-6 and *Bulletin of the British Sundial Society*, Feb 1995, 95(1), 39-44.

[2] Seidelmann, P. Kenneth, "A Design for an Analemmatic Standard-Time Sundial." *Sky & Telescope*, December 1975, 50(6), 368-369.

[3] Rohr, René R.J., *Sundials: History, Theory, and Practice*. (Translation by Gabriel Godin of *Les Cadrans solaires*, Gauthier-Villars, 1965) University of Toronto Press, Toronto, Ontario, 1970 p. 100.

[4] LaLande, Joseph Jerome, Probleme de Gnomonique. Tracer un Cadran analemmatique azimutal, horizontal, elliptique, don't le style soit une ligne vertical indefinite, Histoire de l'Academie Royale des Sciences, Anne MDCCLVII, Paris, 1757.

[5] Vaulezard, Traite de l'Origine, Demonstration, Construction et Usage du Quadrant Analematique, Paris, 1644.

[6] Webster's Revised Unabridged Dictionary, published 1913 by the C. & G. Merriam Co. Springfield, Mass. under the direction of Noah Porter, D.D., LL.D.

APPENDICES

Appendix A

ABOUT THE CONTRIBUTORS

Steve Allen tracked asteroids in the Summer Science Program before proceeding through the California Institute of Technology to the University of California Santa Cruz, where he is now a programmer/analyst for UCO/Lick Observatory. As a recognized researcher on 20th-century timekeeping, he makes routine use of Lick's considerable historical library. Mr. Allen wrote precision metrology software that enabled the figuring of Keck secondary mirrors and other aspheres in instrument cameras. He designed and maintains the real-time readout code for Keck CCD mosaics and the precision milling code for Keck spectrograph slitmasks. He oversees several web/database applications needed for ongoing operation of the two observatories. He is a member of the IAU *Flexible Image Transport System* (FITS) working group and co-author of the FITS MIME (RFC 4047) and World Coordinate System papers. (Sessions 6 and 7)

Florent Deleflie developed with Pierre Exertier a theory of mean orbital motion expressed in non-singular elements for eccentricity for his Ph.D. In 2005, he was appointed an astronomer with Observatoire de la Côte d'Azur, directing the French Official Analysis Center of the International Laser Ranging Service, which provides daily solutions of Earth-orientation parameters and station coordinates based on post-fit residuals analysis of geodetic satellite trajectories. He also organized the Virtual Observatory (called OV-GAFF), establishing a cooperative with all colleagues in France to routinely produce space-geodetic data-based results, particularly terrestrial and celestial reference frame and long-wavelength gravity field information. In 2010, Dr. Deleflie joined IMCCE at Paris Observatory to apply mean-orbit theory to the problem of space debris using astronomical images, leveraging many years of research by colleagues working on the small solar-system bodies. He developed a model based on the official software provided by CNES in the framework of the French Space Act, which is used to design appropriate disposal orbits for spacecraft before launch. (Session 7)

Wolfgang R. Dick, physicist, received his Ph.D. in astronomy in 1989. He worked in photographic astrometry from 1982 to 1991 at the Institute of Astrophysics, Potsdam, and from 1991 to 1992 at the Institute of Astronomy of Bonn University. From 1992 to 2000 he analyzed Satellite Laser Ranging at the Institut fuer Angewandte Geodaesie/Bundesamt fuer Kartographie und Geodaesie (BKG) at Potsdam. Since 2001, Dr. Dick has been at the IERS Central Bureau (hosted by BKG in Frankfurt am Main), working on IERS publications, websites, the data center, and public relations. He is a member of IAU Commissions 8, 19, 31, and 41, and was a member of the IAU Working Group on the Definition of UTC (2000-2006). (Session 3)

Fr. Paul Gabor, S.J., Ph.D., studied particle physics at Charles University in Prague, Czech Republic (M.Sc.) and joined the Jesuits in 1995. He studied philosophy in Cracow, Poland, theology in Paris, France, where he also obtained his Ph.D. in astrophysics in 2009. At the Vatican Observatory, his primary field is astronomical instrumentation but he also teaches history and philosophy of astronomy at the University of Arizona in Tucson. Timekeeping has been one of the constant interests of papal and Jesuit astronomers since Fr. Christoph Clavius, S.J., and the Gregorian calendar reform of 1582. (Session 2)

Daniel Gambis, Ph.D., is an astronomer at the Paris Observatory and Head of the Earth Orientation Center of the International Earth Rotation and Reference Systems Service (IERS). His research activity explores the use of geodetic techniques for monitoring Earth-orientation and the excitation of Earth-rotation variability due to geophysical processes, including mass transport within atmosphere and ocean, solar activity, and earthquakes. He is a member of IAU Division I Commissions 19 (Rotation of the Earth) and 31 (Time). (Sessions 3, 4, and 7).

Danny Hillis is Co-Chairman and Chief Technology Officer of Applied Minds, Inc., a research and development company creating a range of new products and services in software, entertainment, electronics, biotechnology and mechanical design. Considered a legendary designer of computer architecture, Dr. Hillis is an inventor, scientist, author, and engineer. He pioneered the concept of parallel computers that is now the basis for most supercomputers. He holds over 100 U.S. patents, covering parallel computers, disk arrays, forgery prevention methods, and various electronic and mechanical devices. Previously, he was vice president of research and development at Walt Disney Imagineering, co-founder of Thinking Machines Corp., and an adjunct professor at the MIT Media Lab. Dr. Hillis has also worked as a consultant to many companies developing technology-related business strategies, and has served on the Presidential Information Technology Advisory Committee. He earned his Ph.D. in computer science from MIT. (Session 2)

George H. Kaplan, Ph.D., retired from the U.S. Naval Observatory in 2005, having worked as an astronomer there since 1971 in a number of research and management positions. He earned his Ph.D. in astronomy from the University of Maryland in 1985. His work at USNO involved a wide variety of projects related to positional astronomy, including planetary orbit computations, Earth rotation measurements, radio and optical interferometry, astrometry of Jupiter's satellites, binary star motions, and the mathematics of celestial navigation. He specialized in algorithm development and software tools. Currently he is a part-time contractor to USNO, serving as a consultant to the Astronomical Applications Department. He is a member of the American Astronomical Society, the American Association for the Advancement of Science, the Institute of Navigation, the International Astronomical Union, and Sigma Xi. He is currently serving as president of IAU Commission 4, Ephemerides. (Session 4)

Stephen Malys holds a BS from The Pennsylvania State University and an MS in geodetic science from The Ohio State University. During his 30-year career in the federal government he has contributed to advancements in the geodetic exploitation of satellite systems, including the Navy's TRANSIT system and the Navstar Global Positioning System (GPS). He has played leading roles in the implementation of refinements to the global coordinate system known as the World Geodetic System 1984 (WGS 84) and improvements to the GPS Precise Positioning Service. He has authored or co-authored more than 23 technical papers dealing with accuracy analysis of GPS and geodetic applications of TRANSIT and GPS. He is a member of the editorial advisory board for the journal *GPS Solutions*. He was selected as a Science and Technology Fellow by the Director of National Intelligence in 2007. (Session 5)

Dennis D. McCarthy studied astronomy at Case Institute of Technology and the University of Virginia and received his Ph.D. in 1972. He is currently President of International Astronomical Union (IAU) Division 1 on Fundamental Astronomy and is a co-author of the book *Time: from Earth Rotation to Atomic Physics*. He served as Head of the Earth Orientation Department and Director of the Directorate of Time of the U.S. Naval Observatory (USNO), developing the USNO's VLBI program and the use of GPS observations for the determination of Earth orientation. He has served as President of the IAU Commission on Time, President of the IAU Commission on the Rotation of the Earth, Chairman of the IAU Working Group on Nutation, Chairman of the IAU Working Group on the Definition of UTC, and Secretary of the International Association of Geodesy Section on Geodynamics. He has also served as Chairman of the Directing Board of the National Earth Orientation Service, the head of the International Earth Rotation Service (IERS) Product Center for Conventions, and as a member of the Directing Board of the IERS. (Session 3)

Frank Reed teaches celestial navigation at various locations around Chicago, Illinois and at the Treworgy Planetarium at the Mystic Seaport Museum in Mystic, Connecticut. He has practiced celestial navigation for over thirty years and has extensive experience in the history, practice, and theoretical basis of traditional celestial navigation. He has organized three bi-annual *Navigation Weekend* seminars at Mystic Seaport, delivering presentations primarily focused on lunar distance navigation historically and in present practice. Mr. Reed is the head cartographer, developer, and owner of the *Centennia Historical Atlas* software. Mr. Reed holds a Bachelor of Arts in Physics from Wesleyan University, Middletown, Connecticut, his high-honors thesis being on hydrogen spectra in strongly curved space-time. (Session 4)

Arnold Rots started out as a radio astronomer with expertise in spectral line aperture synthesis techniques. He worked at the University of Groningen, the Netherlands Foundation for Radio Astronomy (WSRT), the National Radio Astronomy Observatory (Green Bank and VLA), and the Tata Institute of Fundamental Research. After switching to X-ray astronomy, he worked at the Rossi X-ray Timing Explorer where he wrote the timing code, performed the astronomical clock calibration, and worked on pulsars. Currently, he is archive astrophysicist at the Chandra X-ray Center. He is the author of the IVOA Space-Time Coordinates metadata standard and lead author of the FITS WCS Time standard (in preparation). (Session 6)

Denis Savoie, Ph.D., is the head of the Astronomy Department at the Palais de le découverte science museum in Paris. He is an associated researcher at the SYRTE at the Paris Observatory, and is a corresponding member of the International Academy of History of Science. He is a renowned specialist on sundials, from both a theoretical and historical perspective, and was the president of the Sundial Commission of the French Astronomical Society (established 1887) from 1990 to 2010. He has written numerous papers about gnomonics, has made celebrated sundials, and has published several books valued as reference works on dialing. (Session 4)

John H. Seago studied aerospace engineering at the University of Texas at Austin with specialization in orbital mechanics. As an astrodynamics engineer at Analytical Graphics, Inc. (AGI), he pursues research interests related to astrodynamics, orbit determination, and statistical inference. Prior to joining AGI in 2006, he was a space-systems engineer at Honeywell Technology Solutions, Inc., supporting NASA and DoD activities utilizing high precision timekeeping and Earth rotation in projects related to precision orbit determination, space surveillance, and remote sensing. (Sessions 1 and 2)

Rob Seaman studied astronomy and physics at Villanova University and the University of Massachusetts. As Five College Observer in Residence at the Wyoming Infrared Observatory he used the long, cold nights to port the Image Reduction and Analysis Facility to System V Unix. Since joining the IRAF group at the National Optical Astronomy Observatory in 1988, Mr. Seaman has tackled diverse projects of CCD data acquisition, image processing and archiving, astronomical image and table compression, data and metadata standards, and complex observing modes such as heliocentric time series cadencing. He was Y2K remediation lead for IRAF and chaired the IVOA celestial transient event working group (http://voevent.org). His current position is senior software systems engineer in the Science Data Management group of the NOAO System Science Center. (Sessions 1 and 6)

P. Kenneth Seidelmann received his Ph.D. in Dynamical Astronomy in 1968 from the University of Cincinnati. After military service as a Research and Development Coordinator at the U. S. Army Missile Command from 1963 to 1965, he joined the Nautical Almanac Office of the U. S. Naval Observatory. In February 1976 he was named Director of the Nautical Almanac Office and in September 1990 he became director of the Orbital Mechanics Department. In June 1994 Dr. Seidelmann became Director of the Directorate of Astrometry involving three departments dealing with astrometry and astronomical data. In 2000 he retired from the USNO to become a research professor in the Astronomy Department of the University of Virginia. He is coauthor of two books, *Fundamentals of Astrometry* and *Time: from Earth Rotation to Atomic Physics*, and editor of the *Explanatory Supplement to the Astronomical Almanac*. Minor planet 3217 is named *Seidelmann* in his honor. (Session 4 and Special Session)

David Simpson, Ph.D., is a physicist at the NASA Goddard Space Flight Center in Greenbelt, Maryland, doing space physics research with the Cassini mission to Saturn and supporting the upcoming Magnetospheric Multiscale Mission. Prior to that he worked on the flight software of the Hubble Space Telescope's DF-224 and 486 computers, along with the on-board computers of several other NASA spacecraft. He has authored on-board computer software to compute UTC and spacecraft and planetary ephemerides. Dr. Simpson has also served as an adjunct professor of physics at Prince George's Community College in Maryland since 1991. (Session 5)

Steven E. Slojkowski studied physics at the University of Chicago and Rutgers University. He has been employed at the Flight Dynamics Facility of Goddard Space Flight Center since 1997. He has participated in prelaunch planning, launch and early orbit operations, and on-orbit operations for NASA missions in the low-Earth-orbit, geostationary-orbit, Lagrange-point, and Lunar regimes, specializing particularly in the areas of orbit determination and prediction (discussant).

Mark F. Storz worked in the Space Control Analysis Branch at Air Force Space Command from 1993 to 2004 and now is Chief of the Force Enhancement Analysis Branch. He is currently responsible for analysis of GPS, communications satellites, weather satellites, and space-based intelligence, surveillance and reconnaissance. (Session 5)

David Terrett is a software engineer in the space science department of Rutherford Appleton Laboratory. He has been writing software for astronomy research since about 1980, initially applications for science data analysis and, for the last 15 years, control software for ground-based telescopes and instrumentation. Projects he has worked on include the control systems for the Gemini telescopes, the Large Binocular Telescope and the VISTA survey telescope and instruments for Gemini and the European Southern Observatory. (Session 7)

Neil deGrasse Tyson is an astrophysicist with the American Museum of Natural History, where he also serves as the Frederick P. Rose director of the Hayden Planetarium. In addition to his research publications on galactic structure, he is the author of ten books, most recently *Space Chronicles: Facing the Ultimate Frontier* which critically assesses the past, present, and future of space exploration. A frequent commentator on the universe for leading television news programs, talk shows, and documentaries, Tyson has just begun work as on camera host and executive editor for the 21st century version of Carl Sagan's Cosmos. Tyson is the recipient of fourteen honorary doctorates and the NASA Distinguished Public Service Medal. His contributions to the public appreciation of the cosmos have been recognized by the International Astronomical Union in their official naming of asteroid 13123 *Tyson*. (discussant)

Appendix B

PUBLICATIONS OF THE AMERICAN ASTRONAUTICAL SOCIETY

Following are the principal publications of the American Astronautical Society:

JOURNAL OF THE ASTRONAUTICAL SCIENCES (1954 -   )
Published quarterly and distributed by AAS Business Office, 6352 Rolling Mill Place, Suite #102, Springfield, Virginia 22152-2352. Back issues available from Univelt, Inc., P.O. Box 28130, San Diego, California 92198.

SPACE TIMES (1986 -   )
Published bi-monthly and distributed by AAS Business Office, 6352 Rolling Mill Place, Suite #102, Springfield, Virginia 22152-2352.

AAS NEWSLETTER (1962 - 1985)
Incorporated in Space Times.

ASTRONAUTICAL SCIENCES REVIEW (1959 -1962)
Incorporated in Space Times. Back issues still available from Univelt, Inc., P.O. Box 28130, San Diego, California 92198.

ADVANCES IN THE ASTRONAUTICAL SCIENCES (1957 -   )
Proceedings of major AAS technical meetings. Published and distributed for the American Astronautical Society by Univelt, Inc., P.O. Box 28130, San Diego, California 92198.

SCIENCE AND TECHNOLOGY SERIES (1964 -   )
Supplement to Advances in the Astronautical Sciences. Proceedings and monographs, most of them based on AAS technical meetings. Published and distributed for the American Astronautical Society by Univelt, Inc., P.O. Box 28130, San Diego, California 92198.

AAS HISTORY SERIES (1977 -   )
Supplement to Advances in the Astronautical Sciences. Selected works in the field of aerospace history under the editorship of R. Cargill Hall. Published and distributed for the American Astronautical Society by Univelt, Inc., P.O. Box 28130, San Diego, California 92198.

AAS MICROFICHE SERIES (1968 - 1999)
Supplement to Advances in the Astronautical Sciences. Consists principally of technical papers not included in the hard-copy volume. Published and distributed for the American Astronautical Society by Univelt, Inc., P.O. Box 28130, San Diego, California 92198.

Subscriptions to the *Journal of the Astronautical Sciences* and the *Space Times* should be ordered from the AAS Business Office. Back issues of the *Journal* and all books and microfiche should be ordered from Univelt, Incorporated.

ADVANCES IN THE ASTRONAUTICAL SCIENCES SERIES
(1957-   )

ISSN 0065-3438,

LIBRARY OF CONGRESS CARD NO. 57-43769

Proceedings of Major AAS Technical Meetings

Vol. 1 Third Annual AAS Meeting, Dec. 6-7, 1956, New York, NY, 1957, 184p., ed. Norman V. Petersen, Microfiche only, $20 (ISBN 0-87703-002-2)

Vol. 2 Fourth Annual AAS Meeting, Jan. 29-31, 1958, New York, NY, 1958, 440p., eds. Norman V. Petersen, Horace Jacobs, Microfiche only, $20 (ISBN 0-87703-003-0)

Vol. 3 First Western National AAS Meeting, Aug. 18-19, 1958 530p., eds. Norman V. Petersen, Horace Jacobs, Microfiche only, $20 (ISBN 0-87703-004-9)

Vol. 4 Fifth Annual AAS Meeting, Dec. 27-31, 1958, Washington, D.C., 1959, 462p., ed. Horace Jacobs, Microfiche only, $20 (ISBN 0-87703-005-7)

Vol. 5 Second Western National AAS Meeting, Aug. 4-5, 1959, Los Angeles, CA, 1960, 364p., ed. Horace Jacobs, Microfiche only, $20 (ISBN 0-87703-006-3)

Vol. 6 Sixth Annual AAS Meeting, Jan. 18-21, 1960, New York, NY, 1961, 968p., eds. Horace Jacobs and Eric Burgess, Hard Cover $45 (ISBN 0-87703-007-3)

Vol. 7 Third Western National AAS Meeting, Aug. 4-5, 1960, Seattle, WA, 1961, 464p., eds. Horace Jacobs and Eric Burgess, Microfiche only, $20 (ISBN 0-87703-008-1)

Vol. 8 Seventh Annual AAS Meeting, Jan. 16-18, 1961, Dallas, TX, 1963, 602p., ed. Horace Jacobs, Microfiche only, $20 (ISBN 0-87703-009-X)

Vol. 9 Fourth Western Regional AAS Meeting, Aug. 1-3, 1961, San Francisco, CA, 1963, 910p., ed. Eric Burgess, Hard Cover $45 (ISBN 0-87703-010-3)

Vol. 10 Manned Lunar Flight (AAS/AAAS Symposium) Dec. 19, 1961, Denver, CO, 1963, 310p., eds. George W. Morgenthaler and Horace Jacobs, Hard Cover $35 (ISBN 0-87703-011-1)

Vol. 11 Eighth Annual AAS Meeting, Jan. 16-18, 1962, Washington, D.C., 1963, 808p., ed. Horace Jacobs, Hard Cover $45 (ISBN 0-87703-012-X)

Vol. 12 Scientific Satellites - Mission and Design (AAS/AAAS Symposium), Dec. 27, 1962, Philadelphia, PA, 1963, 262p., ed. Irving E. Jeter, Hard Cover $25 (ISBN 0-87703-013-8)

Vol. 13 Interplanetary Missions, 9th Annual AAS Meeting, Jan. 15-17, 1963, Los Angeles, CA, 1963, 690p., ed. Eric Burgess, Hard Cover $45 (ISBN 0-87703-014-6)

Vol. 14 Second AAS Symposium on Physical and Biological Phenomena under Zero G Conditions, Jan. 18, 1963, Los Angeles, CA, 1963, 382p., eds. Elliot T. Benedikt and Robert W. Halliburton, Hard Cover $30 (ISBN 0-87703-015-4)

Vol. 15 Exploration of Mars Symposium, Jun. 6-7, 1963, Denver, CO, 1963, 634p., ed. George W. Morgenthaler, Hard Cover $45 (ISBN 0-87703-016-2)

Vol. 16 Space Rendezvous, Rescue, and Recovery Symposium, Sept. 10-12, 1963, Edwards, CA 1963, 1408p., ed. Norman V. Petersen, Hard Cover, Part 1, 1028p., $45 (ISBN 0-87703-017-0); Part 2, 380p., $30 (ISBN 0-87703-018-9)

Vol. 17 Bioastronautics - Fundamental and Practical Problems (AAS/AAAS Symposium), Dec. 30, 1963, Cleveland, OH, 1964, 128p., ed. William C. Kaufman, Microfiche only, $10 (ISBN 0-87703-019-7)

Vol. 18 Lunar Flight Programs, 10th Annual AAS Meeting, May 4-7, 1964, New York, NY, 1964, 630p., ed. Ross Fleisig, Hard Cover $45 (ISBN 0-87703-020-0)

Vol. 19 Unmanned Exploration of the Solar System Symposium, Feb. 8-10, 1965, Denver, CO, 1965, 1000p., eds. George W. Morgenthaler, Robert G. Morra, Hard Cover $45 (ISBN 0-87703-021-9)

Vol. 20 Post Apollo Exploration, 11th Annual AAS Meeting, May 3-6, 1965, Chicago, IL, 1966, 1220p., ed. Francis Narin, Microfiche only, Part I, 572p., $30 (ISBN 0-87703-022-7); Part 2, 648p., $35 (ISBN 0-87703-023-5)

Vol. 21 Practical Space Applications Symposium, Feb. 21-23, 1966, San Diego, CA, 1967, 508p., ed. Lawrence L. Kavanau, Microfiche Only $40 (ISBN 0-87703-024-3)

Vol. 22 The Search for Extraterrestrial Life, 12th Annual AAS Meeting, May 23-25, 1966, Anaheim, CA, 1967, 388p., ed. James S. Hanrahan, Microfiche only $30 (ISBN 0-87703-025-1); Microfiche Suppl. (Vol. 1 AAS Microfiche Series) $12 (ISBN 0-87703-132-0)

Vol. 23 Commercial Utilization of Space, 13th Annual AAS Meeting, May 1-3, 1967, Dallas, TX, 1968, 512p., eds. J. Ray Gilmer, Alfred M. Mayo, Ross C. Peavey, Hard Cover (ISBN 0-87703-026-X); plus Microfiche Suppl. (Vol. 3 AAS Microfiche Series) $60 (ISBN 0-87703-216-5)

Vol. 24 Exploitation of Space for Experimental Research, 14th Annual AAS Meeting, May 13-15, 1968, Dedham, MA, 1968, 363p., ed. Harry Zuckerberg, Hard Cover $30 (ISBN 0-87703-027-8)

Vol. 25 Advanced Space Experiments, Sept. 16-18, 1968, Ann Arbor, MI, 1969, 530p., eds. O. Lyle Tiffany and Eugene M. Zaitzeff, Hard Cover $40 (ISBN 0-87703-028-6)

Vol. 26 Planning Challenges of the 70?s in Space, 15th Annual AAS Meeting, Jun. 17-20, 1969, Denver, CO, 1970, 470p., eds. George W. Morgenthaler and Robert G. Morra, Hard Cover $35 (ISBN 0-87703-053-7); Microfiche Suppl. (Vol. 14 AAS Microfiche Series) $20 (ISBN 0-87703-130-4)

Vol. 27/28 Space Stations (v27) and Space Shuttles and Interplanetary Missions (v28), 16th Annual AAS Meeting, Jun. 8-10, 1970, Anaheim, CA, 1970, Vol. 27, eds. Lewis Larmore and Robert L. Gervais, 606p., Hard Cover $45 (ISBN 0-87703-054-5); Vol. 28, eds. Lewis Larmore and Robert L. Gervais, 488p., Hard Cover $35 (ISBN 0-87703-055-3)

Vol. 29 The Outer Solar System, 17th Annual AAS Meeting, Jun. 28-30, 1971, Seattle, WA, 1971, ed. Juris Vagners, Part 1 (on Microfiche Only), 618p., $40 (ISBN 0-87703-059-6); Part 2, Hard Cover, 740p., $45 (ISBN 0-87703-060-X)

Vol. 30 International Congress of Space Benefits, 19th Annual AAS Meeting, Jun. 19-21, 1973, Dallas, TX, 1974, 528p., ed. Francis S. Johnson, Hard Cover $40 (ISBN 0-87703-065-0)

Vol. 31 The Skylab Results, 20th Annual AAS Meeting, Aug. 20-22, 1974, Los Angeles, CA, 1975, 1174p., eds. William C. Schneider and Thomas E. Hanes, Microfiche only (ISBN 0-87703-072-3); Plus Microfiche Suppl. (Vol. 22 AAS Microfiche Series) $60 (ISBN 0-87703-143-6)

Vol. 32 Space Shuttle Missions of the 80's, 21st Annual AAS Meeting, Aug. 26-28, 1975, Denver, CO, 1977, 1364p., eds. William J. Bursnall, George W. Morgenthaler, Gerald E. Simonson, Hard Cover, Part 1, 598p., $40 (ISBN 0-87703-078-2); Hard Cover, Part 2, 766p., $55 (ISBN 0-87703-087-1); Microfiche Suppl. (Vol. 25 AAS Microfiche Series $65 (ISBN 0-87703-133-9)

Vol. 33 AAS/AIAA Astrodynamics Conference, July 28-30, 1975, Nassau, Bahamas, 1976, 390p., eds. William F. Powers, Herbert E. Rauch, Byron D. Tapley, Carmelo E. Velez, Hard Cover $35 (ISBN 0-87703-079-0); Microfiche Suppl. (Vol. 26 AAS Microfiche Series) $40 (ISBN 0-87703-142-8)

Vol. 34 Apollo Soyuz Mission Report, 1977, 336p., ed. Chester M. Lee, Hard Cover $95 (ISBN 0-87703-089-8)

Vol. 35 The Bicentennial Space Symposium - New Themes for Space: Mankind?s Future Needs and Aspirations, 22nd AAS Meeting, Oct. 6-8, 1976, Washington, D.C., 1977, 242p., ed. William C. Schneider, Hard Cover $25 (ISBN 0-87703-090-1)

Vol. 36 The Industrialization of Space, 23rd Annual AAS Meeting, Oct. 18-20, 1977, San Francisco, CA, 1978, 1160p., eds. Richard A. Van Patten, Paul Siegler, Edward V.B. Stearns, Hard Cover, Part 1, 610p., $55 (ISBN 0-87703-094-4); Hard Cover, Part 2, 550p., $45 (ISBN 0-87703-095-2); Microfiche Suppl. (Vol. 28 AAS Microfiche Series) $15 (ISBN 0-87703-121-5)

Vol. 37 Space Shuttle and Spacelab Utilization, What are the Near-Term and Long-Term Benefits for Mankind?, 16th Goddard Memorial Symposium, 24th Annual AAS Meeting, March 8-10, 1978, Washington, D.C., 1978, 865p., eds. George W. Morgenthaler and Manfred Hollstein, Hard Cover, Part 1, 400p., $40 (ISBN 0-87703-096-0); Hard Cover, Part 2, 465p., $45 (ISBN 0-87703-097-9)

Vol. 38 The Future U.S. Space Program, 25th Anniversary Conference, Oct. 20 - Nov. 2, 1978, Houston, TX, 1979, 880p., eds. Richard S. Johnston, Albert Naumann, Jr., Clay W. G. Fulcher, Hard Cover, Part 1, 444p., $45 (ISBN 0-87703-098-7); Hard Cover, Part 2, 436p., $40 (ISBN 0-87703-099-5); Microfiche Suppl. (Vol. 30 AAS Microfiche Series) $15 (ISBN 0-87703-129-0)

Vol. 39 Guidance and Control 1979, Feb. 24-28, 1979, Keystone, CO, 1979, 492p., ed. Robert D. Culp, Hard Cover $45 (ISBN 0-87703-100-2); Microfiche Suppl. (Vol. 31 AAS Microfiche Series) $10 (ISBN 0-87703-128-2)

Vol. 40 AAS/AIAA Astrodynamics Conference, Jun. 25-27, 1979, Provincetown, MA, 1980, 996p., eds. Paul A. Penzo, Bernard Kaufman, Louis Friedman, Richard Battin, Hard Cover, Part 1, 494p., $45 (ISBN 0-87703-107-X); Soft Cover $35 (ISBN 0-87703-108-8); Hard Cover, Part 2, 502p., $45 (ISBN 0-87703-109-6); Soft Cover $35 (ISBN 0-87703-110-X); Microfiche Suppl. (Vol. 32 AAS Microfiche Series) $20 (ISBN 0-87703-139-8)

Vol. 41 Space Shuttle: Dawn of an Era, 26th Annual AAS Meeting, Oct. 29-Nov. 1, 1979, Los Angeles, CA, 1980, 980p., eds. William F. Rector, III and Paul A. Penzo, Hard Cover, Part 1, 452p., $45 (ISBN 0-87703-111-8); Soft Cover $35 (ISBN 0-87703-112-6); Hard Cover, Part 2, 528p., $55 (ISBN 0-87703-113-4); Soft Cover $40 (ISBN 0-87703-114-2); Microfiche Suppl. (Vol. 33 AAS Microfiche Series) $10 (ISBN 0-87703-136-3)

Vol. 42 Guidance and Control 1980, Feb. 17-21, 1980, Keystone, CO, 1980, 738p., ed. Louis A. Morine, Hard Cover $60 (ISBN 0-87703-137-1); Soft Cover $45 (ISBN 0-87703-138-X)

Vol. 43 Shuttle/Spacelab - The New Transportation System and its Utilization, (3rd DGLR/AAS Symposium), Apr. 28-30, 1980, Hannover, Germany, 1981, 342p., eds. Dietrich E. Koelle and George V. Butler, Hard Cover $45 (ISBN 0-87703-144-4); Soft Cover $35 (ISBN 0-87703-146-0)

Vol. 44 Space-Enhancing Technological Leadership, 27th Annual AAS Meeting, Oct. 20-23, 1980, Boston, MA, 1981, 580p., ed. Lawrence P. Greene, Hard Cover $65 (ISBN 0-87703-147-9); Soft Cover $50 (ISBN 0-87703-148-7); Microfiche Suppl. (Vol. 35 AAS Microfiche Series) $10 (ISBN 0-87703-164-9)

Vol. 45 Guidance and Control 1981, Jan. 31-Feb. 4, 1981, Keystone, CO, 1981, 506p., ed. Edward J. Bauman, Hard Cover $60 (ISBN 0-87703-150-9); Soft Cover $50 (ISBN 0-87703-151-7); Microfiche Suppl. (Vol. 36 AAS Microfiche Series) $15 (ISBN 0-87703-156-8)

Vol. 46 AAS/AIAA Astrodynamics Conference, Aug. 3-5, 1981, North Lake Tahoe, NV, 1982, 1124p., eds. Alan L. Friedlander, Paul J. Cefola, Bernard Kaufman, Walt Williamson, G.T. Tseng, Hard Cover, Part 1, 552p., $55 (ISBN 0-87703-159-2); Soft Cover $45 (ISBN 0-87703-160-6); Hard Cover, Part 2, 572p., $55 (ISBN 0-87703-161-4); Soft Cover $45 (ISBN 0-87703-162-2); Microfiche Suppl. (Vol. 37 AAS Microfiche Series) $40 (ISBN 0-87703-163-0)

Vol. 47 Leadership in Space - For Benefits on Earth, 28th Annual AAS Meeting, Oct. 26-29, 1981, San Diego, CA, 1982, 310p., ed. William F. Rector, III, Hard Cover $45 (ISBN 0-87703-168-1); Soft Cover $35 (ISBN 0-87703-169-X)

Vol. 48 Guidance And Control 1982, Jan. 30 -Feb. 3, 1982, Keystone, CO, 1982, 558p., eds. Robert D. Culp, Edward J. Bauman, W. E. Dorroh, Jr., Hard Cover $65 (ISBN 0-87703-170-3); Soft Cover $50 (ISBN 0-87703-171-1); Microfiche Suppl. (Vol. 38 AAS Michrofiche Series) $10 (ISBN 0-87703-180-0)

Vol. 49 Spacelab, Space Platforms, and the Future, Fourth AAS/DGLR Symposium and 20th Goddard Memorial Symposium, Mar. 17-19, 1982, Greenbelt, MD, 1982, 502p., eds. Peter M. Bainum, Dietrich E. Koelle, Hard Cover $55 (ISBN 0-87703-174-6); Soft Cover $45 (ISBN 0-87703-175-4); Microfiche Suppl. (Vol. 42 AAS Microfiche Series) $15 (ISBN 0-87703-181-9)

Vol. 50 Proceedings on an International Symposium on Engineering Sciences and Mechanics, Dec. 29-31, Tainan, Taiwan, 1983, two parts, 1570p., eds. Han-Min Hsia, Richard W. Longman, You-Li Chou, Hard Cover $120 (ISBN 0-87703-176-2); Microfiche Suppl. (Vol. 43 AAS Microfiche Series) $10 (ISBN 0-87703-215-7)

Vol. 51 Guidance and Control 1983, Feb. 5-9, 1983, Keystone, CO, 1983, 494p., eds. Edward J. Bauman, Zubin W. Emsley, Hard Cover $60 (ISBN 0-87703-182-7); Soft Cover $50 (ISBN 0-87703-183-5); Microfiche Suppl. (Vol. 44 AAS Microfiche Series) $10 (ISBN 0-87703-214-9)

Vol. 52 Developing the Space Frontier, 29th Annual AAS Meeting, Oct. 25-27, 1982, Houston, TX, 1983, 436p., eds. Albert Naumann, Grover Alexander, Microfiche Only $45 (ISBN 0-87703-189-4)

Vol. 53 Space Manufacturing 1983, May 9-12, 1983, Princeton, NJ, 1983, 496p., eds. James D. Burke, April S. Whitt, On Microfiche Only $50 (ISBN 0-87703-188-6)

Vol. 54 AAS/AIAA Astrodynamics Conference, Aug. 22-25, 1983, Lake Placid, NY, 1984, two parts, 1370p., eds. G.T. Tseng, Paul J. Cefola, Peter M. Bainum, David A. Levinson, Hard Cover $120 (ISBN 0-87703-190-8); Soft Cover $90 (ISBN 0-87703-191-6); Microfiche Suppl. (Vol. 45 AAS Microfiche Series) $40 (ISBN 0-87703-192-4)

Vol. 55 Guidance and Control 1984, Feb. 4-8, 1984, Keystone, CO, 1984, 500p., eds. Robert D. Culp, Parker S. Stafford, Hard Cover $60 (ISBN 0-87703-199-1); Soft Cover $50 (ISBN 0-87703-200-9); Microfiche Suppl. (Vol. 48 AAS Microfiche Series $15 (ISBN 0-87703-201-7)

Vol. 56 From Spacelab to Space Station, Fifth DGLR/AAS Symposium, Oct. 3-5, 1984, Hamburg, Germany, 1985, 270p., eds. H. Stoewer, Peter M. Bainum, Microfiche Only $30 (ISBN 0-87703-209-2)

Vol. 57 Guidance and Control 1985, Feb. 2-6, 1985, Keystone, CO, 1985, 618p., eds. Robert D. Culp, Edward J. Bauman, Charles A. Cullian, Hard Cover $65 (ISBN 0-87703-211-4); Soft Cover $50 (ISBN 0-87703-212-2); Microfiche Suppl. (Vol. 50 AAS Microfiche Series) $15 (ISBN 0-87703-213-0)

Vol. 58 AAS/AIAA Astrodynamics Conference, Aug. 12-15, 1985, Vail, CO, 1986, two parts, 1556p., eds. Bernard Kaufman, Joseph J.F. Liu, Robert A. Calico, Felix R. Hoots, Hard Cover $140 (ISBN 0-87703-245-9); Soft Cover $110 (ISBN 0-87703-246-7); Microfiche Suppl. (Vol. 51 AAS Microfiche Series); $60 (ISBN 0-87703-247-5)

Vol. 59 Space Station Beyond IOC, 32nd Annual AAS Meeting, Nov 6-7, 1985, Los Angeles, CA, 1986, 188p., ed. M. Jack Friedenthal, Hard Cover $40 (ISBN 0-87703-252-1); Soft Cover $30 (ISBN 0-87703-253-X)

Vol. 60 Space Exploitation and Utilization, First AAS/JRS Symposium, Dec. 15-19, 1985, Honolulu, HI, 1986, 740p., eds. Gayle L. May, Peter M. Bainum, Kenji Ikeda, Tamiya Nomura, Tatsuo Yamanaka, Ryojiro Akiba, Hard Cover $70 (ISBN 0-87703-254-8); Soft Cover $55 (ISBn 0-87703-255-6); Microfiche Suppl. (Vol. 52 AAS Microfiche Series) $10 (ISBN 0-87703-256-4)

Vol. 61 Guidance and Control 1986, Feb. 1-5, 1986, Keystone, CO, 1986, 460p., eds. Robert D. Culp, John C. Durrett, Hard Cover $60 (ISBN 0-87703-257-2); Soft Cover $50 (ISBN 0-87703-258-0); Microfiche Suppl. (Vol. 53 AAS Microfiche Series) $10 (ISBN 0-87703-259-9)

Vol. 62 Tethers in Space, Proceedings of First International Conference on Tethers in Space (NASA & PSN Sponsors; AIAA, AAS, & AIDAA Co-Sponsors), Sept. 17-19, 1986, Arlington, VA, 1987, 784p., eds. Peter M. Bainum, Ivan Bekey, Luciano Guerriero, Paul A. Penzo, Hard Cover $80 (ISBN 0-87703-264-5); Soft Cover $70 (ISBN 0-87703-265-3)

Vol. 63 Guidance and Control 1987, Jan. 31 -Feb. 4, 1987, Keystone, CO, 1987, 638p., eds. Robert D. Culp, Terry J. Kelly, Hard Cover $75 (ISBN 0-87703-268-8); Soft Cover $60 (ISBN 0-87703-269-6)

Vol. 64 Aerospace Century XXI, 33rd AAS Annual Meeting, Oct. 26-29, 1986, Boulder, CO, 1987, all three parts, Hard Cover $225 (ISBN 0-87703-276-9); Soft Cover $180 (ISBN 0-87703-277-7); Part I, Space Missions and Policy, 686p., eds. George W. Morgenthaler, Gayle L. May, Hard Cover $75 (ISBN 0-87703-279-3); Soft Cover $60 (ISBN 0-87703-282-3); Part II, Space Flight Technologies, 608p., eds. George W. Morgenthaler, W. Kent Tobiska, Hard Cover $75 (ISBN 0-87703-280-7); Soft Cover $60 (ISBN 0-87703-283-1); Part III, Space Sciences, Applications, and Commercial Developments, 724p., eds. George W. Morgenthaler, Jean N. Koster, Hard Cover $75 (ISBN 0-87703-281-5); Soft Cover $60 (ISBN 0-87703-284-X); Microfiche Suppl. (Vol. 54 AAS Microfiche Series) $25 (ISBN 0-87703-278-5)

Vol. 65 AAS/AIAA Astrodynamics Conference, Aug. 10-13, 1987, Kalispell, MT, 1988, two parts, 1774p., eds. John K. Soldner, Arun K. Misra, Robert E. Lindberg, Walton Williamson, Hard Cover $180 (ISBN 0-87703-285-8); Soft Cover $150 (ISBN 0-87703-286-6); Microfiche Suppl. (Vol. 55 AAS Microfiche Series); $70 (ISBN 0-87703-287-4)

Vol. 66 Guidance and Control 1988, Jan. 30 -Feb. 3, 1988, Keystone, CO, 1988, 576p., eds. Robert D. Culp, Paul L. Shattuck, Hard Cover $75 (ISBN 0-87703-288-2); Soft Cover $60 (ISBN 0-87703-289-0); Microfiche Suppl. (Vol. 56 AAS Microfiche Series) $10 (ISBN 0-87703-290-4)

Vol. 67 Space - A New Community of Opportunity, 34th AAS Annual Meeting, Nov. 3-5, 1987, Houston, TX, 1989, 472p., eds. William G. Straight, Henry N. Bowes, Hard Cover $70 (ISBN 0-87703-297-1); Soft Cover $55 (ISBN 0-87703-298-X)

Vol. 68 Guidance and Control 1989, Feb. 4-8, 1989, Keystone, CO, 1989, 708p., eds. Robert D. Culp, Robert A. Lewis, Hard Cover $85 (ISBN 0-87703-299-8); Soft Cover $70 (ISBN 0-87703-300-5)

Vol. 69 Orbital Mechanics and Mission Design, Apr. 24-27, 1989, Greenbelt, MD, 1989, 862p., ed. Jerome Teles, Hard Cover $95 (ISBN 0-87703-311-0); Microfiche Suppl. (Vol. 57 AAS Microfiche Series) $10 (ISBN 0-87703-313-7)

Vol. 70 The 21st Century in Space, 35th AAS Annual Meeting, Oct. 24-26, 1988, St. Louis, MO, 1990, 446p., ed. George V. Butler, Hard Cover $90 (ISBN 0-87703-314-5); Soft Cover $75 (ISBN 0-87703-315-3); Microfiche Suppl. (Vol. 58 AAS Microfiche Series) $10 (ISBN 0-87703-316-1)

Vol. 71 AAS/AIAA Astrodynamics Conference, Aug. 7-10, 1989, Stowe, VT, 1990, two parts, 1472p., eds. Catherine L. Thornton, Ronald J. Proulx, John E. Prussing, Felix R. Hoots, Hard Cover $200 (ISBN 0-87703-317-X); Soft Cover $170 (ISBN 0-87703-318-8); Microfiche Suppl. (Vol. 59 AAS Microfiche Series) $50 (ISBN 0-87703-319-6)

Vol. 72 Guidance and Control 1990, Feb. 3-7, 1990, Keystone, CO, 1990, 676p., eds. Robert D. Culp, Arlo D. Gravseth, Hard Cover $95 (ISBN 0-87703-320-X); Soft Cover $80 (ISBN 0-87703-321-8)

Vol. 73 Space Utilization and Applications in the Pacific, Third (PISSTA) AAS/JRS/CSA Symposium, Nov. 6-8, 1989, Los Angeles, CA, 1990, 764p., eds. Peter M. Bainum, Gayle L. May, Tatsuo Yamanaka, Yang Jiachi, Hard Cover $95 (ISBN 0-87703-325-0); Soft Cover $80 (ISBN 0-87703-326-9)

Vol. 74 Guidance and Control 1991, Feb. 2-6, 1991, Keystone, CO, 1991, 730p., eds. Robert D. Culp, James P. McQuerry, Hard Cover $120 (ISBN 0-87703-334-X); Soft Cover $90 (ISBN 0-87703-335-8)

Vol. 75 AAS/AIAA Spaceflight Mechanics Meeting, Feb. 11-13, 1991, Houston, TX, 1991, two parts, 1354p., eds. John K. Soldner, Arun K. Misra, Lester L. Sackett, Richard Holdaway, Hard Cover $220 (ISBN 0-87703-338-2); Soft Cover $190 (ISBN 0-87703-339-0); Microfiche Suppl. (Vol. 62 AAS Microfiche Series) $50 (ISBN 0-87703-340-4)

Vol. 76 AAS/AIAA Astrodynamics Conference, Aug. 19-22, 1991, Durango, CO, 1992, three parts, 2590p., eds. Bernard Kaufman, Kyle T. Alfriend, Ronald L. Roehrich, Robert R. Dasenbrock, Hard Cover $390 (ISBN 0-87703-347-1); Microfiche Suppl. (Vol. 63 AAS Microfiche Series) $75 (ISBN 0-87703-348-X)

Vol. 77 International Space Year (ISY) in the Pacific Basin, Fourth ISCOPS (formerly PISSTA) AAS/JRS/CSA Symposium, Nov. 17-20, 1991, Kyoto, Japan, 1992, 798p., eds. Peter M. Bainum, Gayle L. May, Makoto Nagatomo, Yoshiaki Ohkami, Yang Jiachi, Hard Cover $120 (ISBN 0-87703-351-X); Soft Cover $90 (ISBN 0-87703-352-8)

Vol. 78 Guidance and Control 1992, Feb. 2-6, 1992, Keystone, CO, 1992, 754p., eds. Robert D. Culp, Richard P. Zietz, Hard Cover $120 (ISBN 0-87703-353-6); Soft Cover $90 (ISBN 0-87703-354-4); Microfiche Suppl. (Vol. 64 AAS Microfiche Series) $20 (ISBN 0-87703-355-2)

Vol. 79 AAS/AIAA Spaceflight Mechanics Meeting, Feb. 24-26, 1992, Colorado Springs, CO, 1992, two parts, 1312p., eds. Roger E. Diehl, Ralph G. Schinnerer, Walton E. Williamson, Daryl G. Boden, Hard Cover $240 (ISBN 0-87703-358-7); Microfiche Suppl. (Vol. 65 AAS Microfiche Series) $40 (ISBN 0-87703-359-5)

Vol. 80 Space Business Opportunities, 37th and 38th AAS Annual Meetings, Nov. 5-7, 1990, Dec. 3-5, 1991, Los Angeles, CA, 1992, 380p., eds. Wayne J. Esser, Don K. Tomajan, Hard Cover $90 (ISBN 0-87703-360-9); Soft Cover $70 (ISBN 0-87703-361-7); Microfiche Suppl. (Vol. 61 AAS Microfiche Series) $50 (ISBN 0-87703-331-5); (Vol. 66 AAS Microfiche Series) $60 (ISBN 0-87703-362-5)

Vol. 81 Guidance and Control 1993, Feb. 6-10, 1993, Keystone, CO, 1993, 648p., eds. Robert D. Culp, George Bickley, Hard Cover $120 (ISBN 0-87703-365-X); Soft Cover $90 (ISBN 0-87703-366-8); Microfiche Suppl. (Vol. 67 AAS Microfiche Series) $20 (ISBN 0-87703-367-6)

Vol. 82 AAS/AIAA Spaceflight Mechanics Meeting, Feb. 22-24, 1993, Pasadena, CA, 1993, two parts, 1454p., eds. Robert G. Melton, Lincoln J. Wood, Roger C. Thompson, Stuart J. Kerridge, Hard Cover $240 (ISBN 0-87703-368-4); Microfiche Suppl. (Vol. 68 AAS Microfiche Series) $15 (ISBN 0-87703-369-2)

Vol. 83 Dynamics of Space Tether Systems (English Language Edition), 1993, 508p., by Vladimir V. Beletsky, Evgenii M. Levin, Hard Cover $120 (ISBN 0-87703-370-6); Soft Cover $90 (ISBN 0-87703-371-4)

Vol. 84 AAS/GSFC International Symposium on Spaceflight Dynamics, Apr. 26-30, 1993, Greenbelt, MD, 1993, two parts, 1450p., eds. Jerome Teles, Mina V. Samii, Hard Cover $240 (ISBN 0-87703-378-1); Microfiche Suppl. (Vol. 69 AAS Microfiche Series) $10 (ISBN 0-87703-379-X)

Vol. 85 AAS/AIAA Astrodynamics Conference, Aug. 16-19, 1993, Victoria, British Columbia, Canada, 1994, three parts, 2750p., eds. Arun K. Misra, Vinod J. Modi, Richard Holdaway, Peter M. Bainum, Hard Cover $390 (ISBN 0-87703-380-3); Microfiche Suppl. (Vol. 70 AAS Microfiche Series) $30 (ISBN 0-87703-381-1)

Vol. 86 Guidance and Control 1994, Feb. 2-6, 1994, Keystone, CO, 1994, 700p., eds. Robert D. Culp, Ronald D. Rausch, Hard Cover $120 (ISBN 0-87703-384-6); Soft Cover $90 (ISBN 0-87703-385-4)

Vol. 87 AAS/AIAA Spaceflight Mechanics Meeting, Feb. 14-16, 1994, Cocoa Beach, FL, 1994, two parts, 1272p., eds. John E. Cochran, Jr., Charles D. Edwards, Jr., Stephen J. Hoffman, Richard Holdaway, Hard Cover $240 (ISBN 0-87703-386-2)

Vol. 88 Guidance and Control 1995, Feb. 1-5, 1995, Keystone, CO, 1995, 600p. eds. Robert D. Culp, James D. Medbery, Hard Cover $120 (ISBN 0-87703-399-4); Soft Cover $90 (ISBN 0-87703-400-1)

Vol. 89 AAS/AIAA Spaceflight Mechanics Meeting, Feb. 13-16, 1995, Albuquerque, NM, 1995, two parts, 1774p., eds. Ronald J. Proulx, Joseph J. F. Liu, P. Kenneth Seidelmann, Salvatore Alfano, Hard Cover $280 (ISBN 0-87703-401-X); Microfiche Suppl. (Vol. 71 AAS Microfiche Series) $15 (ISBN 0-87703-402-8)

Vol. 90 AAS/AIAA Astrodynamics Conference, Aug. 14-17, 1995, Halifax, Nova Scotia, Canada, 1996, two parts, 2270p., eds. K. Terry Alfriend, I. Michael Ross, Arun K. Misra, C. Fred Peters, Hard Cover $290 (ISBN 0-87703-407-9); Microfiche Suppl. (Vol. 72 AAS Microfiche Series) $15 (ISBN 0-87703-408-7)

Vol. 91 Strengthening Cooperation in the 21st Century, Sixth ISCOPS (formerly PISSTA) AAS/JRS/CSA Symposium, Dec. 6-8, 1995, Marina Del Rey, CA, 1996, 1154p., eds. Peter M. Bainum, Gayle L. May, Yoshiaki Ohkami, Kuninori Uesugi, Qi Faren, Li Furong, Hard Cover $145 (ISBN 0-87703-409-5)

Vol. 92 Guidance and Control 1996, Feb. 7-11, 1996, Breckenridge, CO, 1996, 744p. Eds. Robert D. Culp, Marv Odefey, Hard Cover $120 (ISBN 0-87703-412-5); Soft Cover $90 (ISBN 0-87703-413-3)

Vol. 93 AAS/AIAA Spaceflight Mechanics Meeting, Feb. 12-15, 1996, Austin, TX, 1996, two parts, 1776p., eds. G. Edward Powell, Robert H. Bishop, John B. Lundberg, Robert H. Smith, Hard Cover $280 (ISBN 0-87703-414-1); Microfiche Suppl. (Vol. 73 AAS Microfiche Series) $10 (ISBN 0-87703-415-X)

Vol. 94 Guidance and Control 1997, Feb. 5-9, 1997, Breckenridge, CO, 1997, 458p. Eds. Robert D. Culp, Stuart B. Wiens, Hard Cover $120 (ISBN 0-87703-430-3); Soft Cover $90 (ISBN 0-87703-431-1)

Vol. 95 AAS/AIAA Spaceflight Mechanics Meeting, Feb. 10-12, 1997, Huntsville, AL, 1997, two parts, 1168p., eds. Kathleen C. Howell, David A. Cicci, John E. Cochran, Jr., Thomas S. Kelso, Hard Cover $240 (ISBN 0-87703-432-X)

Vol. 96 Space Cooperation into the 21st Century, Seventh ISCOPS (formerly PISSTA) JRS/AAS/CSA Symposium, Jul. 15-18, 1997, Nagasaki, Japan, 1997, 1089p., eds. Peter M. Bainum, Gayle L. May, Makoto Nagatomo, Kuninori T. Uesugi, Fu Bing-chen, Zhang Hui, Hard Cover $145.00 (ISBN 0-87703-438-9)

Vol. 97 AAS/AIAA Astrodynamics Conference, Aug. 4-7, 1997, Sun Valley, ID, 1998, two parts, 2184p., eds. Felix R. Hoots, Bernard Kaufman, Paul J. Cefola, David B. Spencer, Hard Cover $310 (ISBN 0-87703-441-9)

Vol. 98 Guidance and Control 1998, Feb. 4-8, 1998, Breckenridge, CO, 1998, 706p. Eds. Robert D. Culp, David Igli, Hard Cover $120 (ISBN 0-87703-448-6); Soft Cover $90 (ISBN 0-87703-449-4)

Vol. 99 AAS/AIAA Spaceflight Mechanics Meeting, Feb. 9-11, 1998, Monterey, CA, 1998, two parts, 1638p., eds. Jay W. Middour, Lester L. Sackett, Louis A. D?Amario, Dennis V. Byrnes, Hard Cover $280 (ISBN 0-87703-450-8); Microfiche Suppl. (Vol. 78 AAS Microfiche Series) $10 (ISBN 0-87703-452-4)

Vol. 100 AAS/GSFC International Symposium on Space Flight Dynamics, May 11-15, 1998, Greenbelt, MD, 1998, two parts, 1092p., ed. Thomas H. Stengle, Hard Cover $250 (ISBN 0-87703-453-2)

Vol. 101 Guidance and Control 1999, Feb. 3-7, 1999, Breckenridge, CO, 1999, 528p. Eds. Robert D. Culp, Douglas Wiemer, Hard Cover $120 (ISBN 0-87703-456-7); Soft Cover $90 (ISBN 0-87703-457-5)

Vol. 102 AAS/AIAA Spaceflight Mechanics Meeting, Feb. 7-10, 1999, Breckenridge, CO, 1999, two parts, 1600p., eds. Robert H. Bishop, Robert D. Culp, Donald L. Mackison and Maria Evans, Hard Cover $280 (ISBN 0-87703-458-3)

Vol. 103 AAS/AIAA Astrodynamics Conference, Aug. 16-19, 1999, Girdwood, AK, 2000, three parts, 2724p., eds. Kathleen C. Howell, Felix R. Hoots, Bernard Kaufman, K. Terry Alfriend, Hard Cover $450 (ISBN 0-87703-467-2)

Vol. 104 Guidance and Control 2000, Feb. 2-6, 2000, Breckenridge, CO, 2000, 738p. Eds. Robert D. Culp, Eileen M. Dukes, Hard Cover $130 (ISBN 0-87703-468-0); Soft Cover $95 (ISBN 0-87703-469-9)

Vol. 105 AAS/AIAA Spaceflight Mechanics Meeting, Jan. 23-26, 2000, Clearwater, FL, 2000, two parts, 1704p., eds. Craig A. Kluever, Beny Neta, Christopher D. Hall, John M. Hanson Hard Cover $300 (ISBN 0-87703-470-2)

Vol. 106 The Richard H. Battin Astrodynamics Symposium, Mar. 20-21, 2000, College Station, TX, 2000, 492p., Eds. John L. Junkins, K. Terry Alfriend, Kathleen C. Howell, Hard Cover $130 (ISBN 0-87703-471-0); Soft Cover $95 (ISBN 0-87703-472-9)

Vol. 107 Guidance and Control 2001, Jan. 31 - Feb. 4, 2001, Breckenridge, CO, 2001, 738p. Eds. Robert D. Culp, Charles N. Schira, Hard Cover $140 (ISBN 0-87703-479-6); Soft Cover $100 (ISBN 0-87703-480-X)

Vol. 108 AAS/AIAA Spaceflight Mechanics Meeting, Feb. 11-15, 2001, Santa Barbara, CA, 2001, two parts, 2174p., eds. Louis A. D'Amario, Lester L. Sackett, Daniel J. Scheeres, Bobby G. Williams, Hard Cover $330 (ISBN 0-87703-487-7)

Vol. 109 AAS/AIAA Astrodynamics Conference, Jul. 30 to Aug. 2, 2001, Quebec City, Quebec, Canada, 2002, three parts, 2592p., eds. David B. Spencer, Calina C. Seybold, Arun K. Misra, Ronald J. Lisowski, Hard Cover $450 (ISBN 0-87703-488-5)

Vol. 110 Space Development & Cooperation Among all Pacific Basin Countries, Ninth ISCOPS (formerly PISSTA) AAS/JRS/CSA Symposium, Nov. 14-16, 2001, Pasadena, CA, 2002, 444p., eds. Peter M. Bainum, Joan Johnson-Freese, Takashi Nakajima, Kuninori T. Uesugi, Hard Cover $130.00 (ISBN 0-87703-489-3); Soft Cover $100.00 (ISBN 0-87703-490-7)

Vol. 111 Guidance and Control 2002, Feb. 6-10, 2002, Breckenridge, CO, 2002, 544p. Eds. Robert D. Culp, Steven D. Jolly, Hard Cover $140 (ISBN 0-87703-491-5); Soft Cover $100 (ISBN 0-87703-492-3)

Vol. 112 AAS/AIAA Spaceflight Mechanics Meeting, Jan. 27-30, 2002, San Antonio, TX, 2002, two parts, 1570p., eds. Kyle T. Alfriend, Beny Neta, Kim Luu, Cheryl A. Hilton Walker, Hard Cover $330 (ISBN 0-87703-487-7)

Vol. 113 Guidance and Control 2003, Feb. 5-9, 2003, Breckenridge, CO, 2003, 662p. Eds. Ian J. Gravseth and Robert D. Culp, Hard Cover $150 (ISBN 0-87703-502-4); Soft Cover $110 (ISBN 0-87703-503-2)

Vol. 114 AAS/AIAA Spaceflight Mechanics Meeting, Feb. 9-13, 2003, Ponce, Puerto Rico, 2003, three parts, 2294p., plus a CD ROM, eds. Daniel J. Scheeres, Mark E. Pittelkau, Ronald J. Proulx and L. Alberto Cangahuala, Hard Cover plus a CD ROM $450 (ISBN 0-87703-504-0)

Vol. 115 The John L. Junkins Astrodynamics Symposium, May 23-24, 2003, College Station, TX, 2003, 542p., Eds. Srinivas Rao Vadali and Daniele Mortari, Hard Cover plus a CD ROM $150 (ISBN 0-87703-505-9); Soft Cover plus a CD ROM $110 (ISBN 0-87703-506-7)

Vol. 116 AAS/AIAA Astrodynamics Conference, Aug. 3-7, 2003, Big Sky, MT, 2004, three parts, 2746p., eds. Jean de Lafontaine, Alfred J. Treder, Mark T. Soyka and Jon A. Sims, Hard Cover plus a CD ROM $490 (ISBN 0-87703-509-1)

Vol. 117 Space Activities and Cooperation Contributing to All Pacific Basin Countries, Tenth ISCOPS (formerly PISSTA) JRS/AAS/CSA Symposium, Dec. 10-12, 2003, Tokyo, Japan, 2004, 1022p., eds. Peter M. Bainum, Li Furong and Takashi Nakajima, Hard Cover plus a CD ROM $180 (ISBN 0-87703-510-5)

Vol. 118 Guidance and Control 2004, Feb. 4-8, 2004, Breckenridge, CO, 2004, 684p. Eds. Jim D. Chapel and Robert D. Culp, Hard Cover $150 plus a CD ROM (ISBN 0-87703-511-3); Soft Cover plus a CD ROM $110 (ISBN 0-87703-512-1)

Vol. 119 AAS/AIAA Spaceflight Mechanics Meeting, Feb. 8-12, 2004, Maui, Hawaii, 2005, three parts, 3318p., plus a CD ROM, eds. Shannon L. Coffey, Andre P. Mazzoleni, K. Kim Luu, Robert A. Glover, Hard Cover plus a CD ROM $560 (ISBN 0-87703-515-6)

Vol. 120 AAS/AIAA Spaceflight Mechanics Meeting, Jan. 23-27, 2005, Copper Mountain, CO, 2005, two parts, 2152p., plus a CD ROM, eds. David A. Vallado, Michael J. Gabor, Prasun N. Desai, Hard Cover plus a CD ROM $390 (ISBN 0-87703-520-2)

Vol. 121 Guidance and Control 2005, Feb. 5-9, 2005, Breckenridge, CO, 2005, 618p. Eds. William Frazier and Robert D. Culp, Hard Cover $150 plus a CD ROM (ISBN 0-87703-521-0); Soft Cover plus a CD ROM $110 (ISBN 0-87703-522-9)

Vol. 122 The Malcolm D. Shuster Astronautics Symposium, June 12-15, 2005, Grand Island, NY, 2006, 628p., Eds. John L. Crassidis, F. Landis Markley, John L. Junkins, and Kathleen C. Howell, Hard Cover plus a CD ROM $150 (ISBN 0-87703-525-3); Soft Cover plus a CD ROM $110 (ISBN 0-87703-526-1)

Vol. 123 AAS/AIAA Astrodynamics Conference, Aug. 7-11, 2005, South Lake Tahoe, CA, 2006, three parts, plus a CD ROM, 2878p., eds. Bobby G. Williams, Louis A. D'Amario, Kathleen C. Howell and Felix R. Hoots, Hard Cover plus a CD ROM $520 (ISBN 0-87703-527-X)

Vol. 124 AAS/AIAA Spaceflight Mechanics Meeting, Jan. 22-26, 2006, Tampa, FL, 2006, two parts, 2282p., plus a CD ROM, eds. Srinivas Rao Vadali, L. Alberto Cangahuala, Paul W. Schumacher, Jr. and Jose J. Guzman, Hard Cover plus a CD ROM $430 (ISBN 0-87703-528-8)

Vol. 125 Guidance and Control 2006, Feb. 4-8, 2006, Breckenridge, CO, 2006, 624p. Eds. Steven D. Jolly and Robert D. Culp, Hard Cover $150 plus a CD ROM (ISBN 0-87703-531-8); Soft Cover plus a CD ROM $110 (ISBN 0-87703-532-6)

Vol. 126 Dynamic Analysis of Space Tether Missions, 2007, 462p., by Eugene M. Levin, Hard Cover $150 (ISBN 978-0-87703-537-4); Soft Cover $110 (ISBN 978-0-87703-538-1)

Vol. 127 AAS/AIAA Spaceflight Mechanics Meeting, Jan. 28 to Feb. 1, 2007, Sedona, AZ, 2007, two parts, 2230p., plus a CD ROM, eds. Maruthi R. Akella, James W. Gearhart, Robert H. Bishop and Alfred J. Treder, Hard Cover plus a CD ROM $430 (ISBN 978-0-87703-541-1)

Vol. 128 Guidance and Control 2007, Feb. 3-7, 2007, Breckenridge, CO, 2007, 1110p. Eds. Heidi E. Hallowell and Robert D. Culp, Hard Cover $215 plus a CD ROM (ISBN 978-0-87703-542-8)

Vol. 129 AAS/AIAA Astrodynamics Conference, Aug. 19-23, 2007, Mackinac Island, MI, 2008, three parts, plus a CD ROM, 2892p., eds. Ronald J. Proulx, Thomas F. Starchville, Jr., R. D. Burns, Daniel J. Scheeres, Hard Cover plus a CD ROM $520 (ISBN 978-0-87703-543-5)

Vol. 130 AAS/AIAA Spaceflight Mechanics Meeting, Jan. 27-31, 2008, Galveston, TX, 2008, two parts, 2190p., plus a CD ROM, eds. John H. Seago, Beny Neta, Thomas J. Eller and Frederic J. Pelletier, Hard Cover plus a CD ROM $430 (ISBN 978-0-87703-544-2)

Vol. 131 Guidance and Control 2008, Feb. 1-6, 2008, Breckenridge, CO, 2008, 818p. Eds. Michale E. Drews and Robert D. Culp, Hard Cover $190 plus a CD ROM (ISBN 978-0-87703-545-9)

Vol. 132 The F. Landis Markley Astronautics Symposium, June 29 to July 2, 2008, Cambridge, MD, 2008, 954p., Eds. John L. Crassidis, John L. Junkins, Kathleen C. Howell, Yaakov Oshman and Julie K. Thienel, Hard Cover plus a CD ROM $190 (ISBN 978-0-87703-548-0)

Vol. 133 Guidance and Control 2009, Jan. 31 to Feb. 4, 2009, Breckenridge, CO, 2009, 712p. Eds. Edward J. Friedman and Robert D. Culp, Hard Cover $190 plus a CD ROM (ISBN 978-0-87703-553-4)

Vol. 134 AAS/AIAA Spaceflight Mechanics Meeting, Feb. 8-12, 2009, Savannah, GA, 2009, three parts, 2496p., plus a CD ROM, eds. Alan M. Segerman, Peter C. Lai, Matthew P. Wilkins and Mark E. Pittelkau, Hard Cover plus a CD ROM $490 (ISBN 978-0-87703-554-1)

Vol. 135 AAS/AIAA Astrodynamics Conference, Aug. 9-13, 2009, Pittsburgh, PA, 2010, three parts, plus a CD ROM, 2446p., eds. Anil V. Rao, T. Alan Lovell, F. Kenneth Chan, L. Alberto Cangahuala, Hard Cover plus a CD ROM $490 (ISBN 978-0-87703-557-2)

Vol. 136 AAS/AIAA Spaceflight Mechanics Meeting, Feb. 14-17, 2010, San Diego, CA, 2010, three parts, 2652p., plus a CD ROM, eds. Daniele Mortari, Thomas F. Starchville, Jr., Aaron J. Trask and James K. Miller, Hard Cover plus a CD ROM $510 (ISBN 978-0-87703-560-2)

Vol. 137 Guidance and Control 2010, Feb. 5-10, 2010, Breckenridge, CO, 2010, 892p. ed. Shawn C. McQuerry, Hard Cover plus a CD ROM $190 (ISBN 978-0-87703-561-9)

Vol. 138 Applications of Space Technology for Humanity, Twelfth ISCOPS (formerly PISSTA) AAS/JRS/CSA Symposium, July 27-30, 2010, Montreal, Canada, 2010, 762p., eds. Peter M. Bainum, Arun K. Misra, Yasuhiro Morita and Zhang Chi, Hard Cover plus a CD ROM $190 (ISBN 978-0-87703-562-6)

Vol. 139 The Kyle T. Alfriend Astrodynamics Symposium, May 18-19, 2010, Monterey, CA, 2011, 544p., Eds. Shannon L. Coffey, John L. Junkins, K. Kim Luu, I. Michael Ross, Chris Sabol and Paul W. Schumacher, Jr., Hard Cover plus CD ROM $150 (ISBN 978-0-87703-565-7); CD ROM Only $100 (ISBN 978-0-87703-566-4)

Vol. 140 AAS/AIAA Spaceflight Mechanics Meeting, Feb. 13-17, 2011, New Orleans, LA, 2011, three parts, 2622p., plus a CD ROM, eds. Moriba K. Jah, Yanping Guo, Angela L. Bowes and Peter C. Lai, Hard Cover plus CD ROM $510 (ISBN 978-0-87703-569-5); CD ROM Only $330 ((ISBN 978-0-87703-570-1)

Vol. 141 Guidance and Control 2011, Feb. 4-9, 2011, Breckenridge, CO, 2011, 830p. ed. Kyle B. Miller, Hard Cover plus CD ROM $190 (ISBN 978-0-87703-571-8); CD ROM Only $125 (ISBN 978-0-87703-572-5)

Order from Univelt, Inc., P.O. Box 28130, San Diego, California 92198
(Web Site: http://www.univelt.com) STANDING ORDERS ACCEPTED

SCIENCE AND TECHNOLOGY SERIES (1964-   )

ISSN 0278-4017

A Supplement to *Advances in the Astronautical Sciences*. Proceedings and monographs, most of them based on AAS technical meetings.

Vol. 1 Manned Space Reliability Symposium, Jun. 9, 1964, Anaheim, CA, 1964, 112p., ed. Paul Horowitz, Hard Cover $20 (ISBN 0-87703-029-4)

Vol. 2 Towards Deeper Space Penetration (AAS/AAAS Symposium), Dec. 29, 1964, Montreal, Canada, 1964, 182p., ed. Edward R. Van Driest, Hard cover $20 (ISBN 0-87703-030-8)

Vol. 3 Orbital Hodograph Analysis, 1965, 150p., ed. Samuel P. Altman, Hard Cover $20 (ISBN 0-87703-031-6)

Vol. 4 Scientific Experiments for Manned Orbital Flight, 3rd Goddard Memorial Symposium, Mar. 18-19, 1965, Washington, D.C., 1965, 372p., ed. Peter C. Badgley, Hard Cover $30 (ISBN 0-87703-032-4)

Vol. 5 Physiological and Performance Determinants in Manned Space Systems (AAS/HFS Symposium, Apr. 14-15, 1965, Northridge, CA, 1965, 220p., ed. Paul Horowitz, Microfiche Only $20 (ISBN 0-87703-033-2)

Vol. 6 Space Electronics Symposium (AAS/AES Meeting), May 25-27, 1965, Los Angeles, CA, 1965, 404p., ed. Chung-Ming Wong, Hard Cover $30 (ISBN 0-87703-034-0)

Vol. 7 Theodore von Karman Memorial Seminar, May 12, 1965, Los Angeles, CA, 1966, 140p., ed. Shirley Thomas, Hard Cover $30 (ISBN 0-87703-035-9)

Vol. 8 Impact of Space Exploration on Society, Aug. 18-20, 1965, San Francisco, CA, 1966, 382p., ed. William E. Frye, Hard Cover $30 (ISBN 0-87703-036-7)

Vol. 9 Recent Developments in Space Flight Mechanics, (AAS/AAAS Symposium), Dec. 29, 1965, Berkeley, CA, 1966, 280p., ed. Paul B. Richards, Hard Cover $25 (ISBN 0-87703-037-5)

Vol. 10 Space in the Fiscal Year 2001, 4th Goddard Memorial Symposium, Mar. 15-16, 1966, Washington, D.C., 1967, 458p., eds. Eugene B. Konecci, Maxwell W. Hunter, II, Robert F. Trapp, Hard Cover $35 (ISBN 0-87703-038-3)

Vol. 11 Space Flight Specialist Conference, Jul. 6-8, 1966, Denver, CO, 1967, 618p., ed. Maurice L. Anthony, Microfiche Only (ISBN 0-87703-039-1); Plus Microfiche Suppl. (Vol. 2 AAS Microfiche Series) $60 (ISBN 0-87703-221-1)

Vol. 12 Management of Aerospace Programs Conference, Nov. 16-18, 1966, Columbia, MO, 1967, 392p., ed. Walter K. Johnson, Hard Cover $30 (ISBN 0-87703-040-5)

Vol. 13 Physics of the Moon (AAS/AAAS Symposium), Dec. 29, 1966, Washington, D.C., 1967, 260p., ed. S. Fred Singer, Hard Cover $25 (ISBN 0-87703-041-3)

Vol. 14 Interpretation of Lunar Probe Data, Sept. 17, 1966, Huntington Beach, CA, 1967, 270p., ed. Jack Green, Hard Cover $25 (ISBN 0-87703-042-1)

Vol. 15 Future Space Program and Impact on Range and Network Development Symposium, Mar. 22-24, 1967, Las Cruces, NM, 1967, 588p., ed. George W. Morgenthaler, Hard Cover $40 (ISBN 0-87703-043-X)

Vol. 16 Voyage to the Planets, 5th Goddard Memorial Symposium, Mar. 14-15, 1967, Washington, D.C., 1968, 184p., ed. S. Fred Singer, Hard Cover $20 (0-87703-044-8)

Vol. 17 Use of Space Systems for Planetary Geology and Geophysics Symposium, May 25-27, 1967, Boston, MA, 1968, 623p., ed. Robert D. Enzmann, Hard Cover $45 (ISBN 0-87703-045-6); Microfiche Suppl. (Vol. 5 AAS Microfiche Series) $15 (ISBN 0-87703-135-5)

Vol. 18 Technology and Social Progress, 6th Goddard Memorial Symposium, Mar. 12-13, 1968, Washington, D.C., 1969, 170p., ed. Philip K. Eckman, Hard Cover $20 (ISBN 0-87703-046-4)

Vol. 19 Exobiology - The Search for Extraterrestrial Life (AAS/AAAS Symposium) Dec. 30, 1967, New York, NY, 1969, 184p., eds. Martin M. Freundlich, Bernard W. Wagner, Hard Cover $20 (ISBN 0-87703-047-2)

Vol. 20 Bioengineering and Cabin Ecology (AAS/AAAS Symposium) Dec. 30, 1968, Dallas, TX, 1969, 162p., ed. William Cassidy, Hard Cover $20 (ISBN 0-87703-048-0)

Vol. 21 Reducing the Cost of Space Transportation, 7th Goddard Memorial Symposium, Mar. 4-5, 1969, Washington, D.C., 1969, 264p., ed. George K. Chacko, Microfiche only $25 (ISBN 0-87703-049-9)

Vol. 22 Planning Challenges of the 70?s in the Public Domain, 15th Annual AAS Meeting, Jun. 17-20, 1969, Denver, CO, 1970, 504p., eds. William J. Burnsnall, George K. Chacko, George W. Morgenthaler, Hard Cover $40 (ISBN 0-87703-050-2); Microfiche Suppl. (Vol. 13 AAS Microfiche Series) $20 (ISBN 0-87703-131-2); See also Vols. 15-17, AAS Microfiche Series

Vol. 23 Space Technology and Earth Problems Symposium, Oct. 23-25, 1969, Las Cruces, NM, 1970, 418p., ed. C. Quentin Ford, Hard Cover $35 (ISBN 0-87703-051-0); Microfiche Suppl. (Vol. 12 AAS Microfiche Series) $20 (ISBN 0-87703-134-7)

Vol. 24 Aerospace Research and Development, Jul. 14, 1966, Holloman AFB, NM, 1970, 500p., ed. Ernst A. Steinhoff, Microfiche Only $40 (ISBN 0-87703-052-9)

Vol. 25 Geological Problems in Lunar and Planetary Research, Feb. 17-18, 1969, Huntington Beach, CA, 1971, 750p., ed. Jack Green, Hard Cover $45 (ISBN 0-87703-056-1)

Vol. 26 Technology Utilization Ideas for the 70s and Beyond, Oct. 30, 1970, Winrock, AR, 1971, 312p., eds. Fred W. Forbes, Paul Dergarabedian, Microfiche only $30 (ISBN 0-87703-057-X)

Vol. 27 International Cooperation in Space Operations and Exploration, 9th Goddard Memorial Symposium, Mar. 11, 1971, Washington, D.C. 1971, 194p., ed. Michael Cutler, Hard Cover $20 (ISBN 0-88703-058-8)

Vol. 28 Astronomy from a Space Platform (AAS/AAAS Symposium) Dec. 27-28, 1971, Philadelphia, PA, 1972, 416p., eds. George W. Morgenthaler, Howard D. Greyber, Hard Cover $35 (ISBN 0-87703-061-8)

Vol. 29 Space Technology Transfer to Community and Industry, 10th Goddard Memorial Symposium, 18th Annual AAS Meeting, Mar. 13-14, 1972, Washington, D.C., 1972, 196p., eds. Ralph H. Tripp, John K. Stotz, Jr., Hard Cover $20 (ISBN 0-87703-062-6); on Microfiche $15

Vol. 30 Space Shuttle Payloads (AAS/AAAS Symposium) Dec. 27-28, 1972, Washington, D.C., 1973, 532p., eds. George W. Morgenthaler, William J. Bursnall, Hard Cover $40 (ISBN 0-87703-063-4)

Vol. 31 The Second Fifteen Years in Space, 11th Goddard Memorial Symposium, Mar. 8-9, 1973, Washington, D.C., 1973, 212p., ed. Saul Ferdman, Hard Cover $25 (ISBN 0-87703-064-2)

Vol. 32 Health Care Systems Conference, Nov. 21-22, 1972, Dallas, TX, 1974, 265p., ed. Eugene B. Konecci, Hard Cover $25 (ISBN 0-87703-067-7)

Vol. 33 Orbital International Laboratory, 3rd and 4th IAF/OIL Symposia, Oct. 5-6, 1970, Constance, Germany, Sept. 24-25, 1971, Brussels, Belgium, 1974, 322p., ed. Ernst A. Steinhoff, Hard Cover $30 (ISBN 0-87703-068-5)

Vol. 34 Management and Design of Long-Life Systems, Apr. 24-26, 1973, Denver, CO, 1974, 198p., ed. Harris M. Schurmeier, Hard Cover $20 (ISBN 0-87703-069-3)

Vol. 35 Energy Delta, Supply vs. Demand, (AAS/AAAS Symposium) Feb. 25-27, 1974, San Francisco, CA, 1975, 2nd Printing 1976, 604p., eds. George W. Morgenthaler, Aaron N. Silver, Hard Cover $35 (ISBN 0-87703-070-7); Soft Cover $25 (ISBN 0-87703-082-0); on Microfiche $20

Vol. 36 Skylab and Pioneer Report, 12th Goddard Memorial Symposium, Mar. 8, 1974, Washington, D.C., 1975, 160p., eds. Philip H. Bolger, Paul B. Richards, Hard Cover $20 (ISBN 0-87703-071-5)

Vol. 37 Space Rescue and Safety 1974, 7th International IAA Symposium, Sept. 30 - Oct. 5, 1974, Amsterdam, Netherlands, 1975, 294p., ed. Philip H. Bolger, Microfiche Only $25 (ISBN 0-87703-073-1)

Vol. 38 Skylab Science Experiments, (AAS/AAAS Symposium) Feb. 28, 1974, San Francisco, CA, 1976, 274p., eds. George W. Morgenthaler, Gerald E. Simonson, Microfiche only $20 (ISBN 0-87703-074-X)

Vol. 39 Environmental Control and Agri-Technology, 1976, 346p., ed. Eugene B. Konecci, Microfiche only $20 (ISBN 0-87703-075-8)

Vol. 40 Future Space Activities, 13th Goddard Memorial Symposium, Apr. 11, 1975, Washington, D.C., 1976, 182p., ed. Carl H. Tross, Microfiche only $20 (ISBN 0-87703-076-6)

Vol. 41 Space Rescue and Safety 1975, 8th International IAA Symposium, Sept. 21-27, 1975, Lisbon, Portugal, 1976, 230p., ed. Philip H. Bolger, Hard Cover $25 (ISBN 0-87703-077-4)

Vol. 42 The End of an Era in Space Exploration, From International Rivalry to International Cooperation, 1976, 216p., by J.C.D. Blaine, Hard Cover $25 (ISBN 0-87703-084-7); without volume number (ISBN 0-87703-080-4)

Vol. 43 The Eagle Has Returned, Part I, International Space Hall of Fame Dedication Conference, Oct. 5-9, 1976, Alamogordo, NM, 1976, 370p., ed. Ernst. A. Steinhoff, Hard Cover $30 (ISBN 0-87703-086-3)

Vol. 44 Satellite Communications in the Next Decade, 14th Goddard Memorial Symposium, Mar. 12, 1976, Washington, D.C., 1977, 188p., ed. Leonard Jaffe, Hard Cover $20 (ISBN 0-87703-088-X)

Vol. 45 The Eagle Has Returned, Part 2, International Space Hall of Fame Dedication Conference, Oct. 5-9, 1976, Alamogordo, NM, 1977, 454p., ed. Ernst A. Steinhoff, Hard Cover $35 (ISBN 0-87703-092-8)

Vol. 46 Export of Aerospace Technology, 15th Goddard Memorial Symposium, Mar. 31 - Apr. 1, 1977, Washington, D.C., 1978, 174p., ed. Carl H. Tross, Hard Cover $20 (ISBN 0-87703-093-6)

Vol. 47 Handbook of Soviet Lunar and Planetary Exploration, 1979, 276p., by Nicholas L. Johnson, Microfiche Only $25 (ISBN 0-87703-105-3)

Vol. 48 Handbook of Soviet Manned Space Flight, 2nd Edition, 1988, 474p., by Nicholas L. Johnson, Hard Cover $60 (ISBN 0-87703-115-0); Soft Cover $45 (ISBN 0-87703-116-9)

Vol. 49 Space - New Opportunities for International Ventures, 17th Goddard Memorial Symposium, Mar. 28-30, 1979, Washington, D.C., 1980, 300p., ed. William C. Hayes, Jr., Hard Cover $35 (ISBN 0-87703-124-X); Soft Cover $25 (ISBN 0-87703-125-8); see also Vol. 2 AAS History Series

Vol. 50 Remember the Future - The Apollo Legacy, Jul. 20-21, 1979, San Francisco, CA, 1980, 218p., ed. Stan Kent, Hard Cover $25 (ISBN 0-87703-126-6); Soft Cover $15 (ISBN 0-87703-127-4)

Vol. 51 Commercial Operations in Space 1980-2000, 18th Goddard Memorial Symposium, Mar. 27-28, 1980, Washington, D.C., 1981, 214p., eds. John L. McLucas, Charles Sheffield, Hard Cover $30 (ISBN 0-87703-140-1); Soft Cover $20 (ISBN 0-87703-141-X); Microfiche Suppl. (Vol. 34 AAS Microfiche Series) $10 (ISBN 0-87703-165-7); see also Vols. 2 and 3, AAS History Series

Vol. 52 International Space Technical Applications, 19th Goddard Memorial Symposium, Mar. 26-27, 1981, Washington, D.C., 1981, 186p., eds. Andrew Adelman, Peter M. Bainum, Hard Cover $30 (ISBN 0-87703-152-5); Soft Cover $20 (ISBN 0-87703-153-3); see also Vol. 5, AAS History Series

Vol. 53 Space in the 1980?s and Beyond, 17th European Space Symposium, Jun. 4-6, 1980, London, England, 1981, 302p., ed. Peter M. Bainum, Hard Cover $40 (ISBN 0-87703-154-1); Soft Cover $30 (ISBN 0-87703-155-X)

Vol. 54 Space Safety and Rescue 1979-1981 (with abstracts 1976-1978), Proceedings of symposia of the International Academy of Astronautics held in conjunction with the 30th, 31st, and 32nd International Astronautical Federation Congresses, Munich, Germany, 1979, Tokyo, Japan, 1980, and Rome, Italy, 1981, 1983, 456p., ed. Jeri W. Brown, Hard Cover $45 (ISBN 0-87703-177-0); Soft Cover $35 (ISBN 0-87703-178-9); Microfiche Suppl. (Vols. 39-41 AAS Microfiche Series) $39 (ISBN 0-87703-222-X); (ISBN 0-87703-223-8); (ISBN 0-87703-224-6)

Vol. 55 Space Applications at the Crossroads, 21st Goddard Memorial Symposium, Mar. 24-25, 1983, Greenbelt, MD, 1983, 308p., eds. John H. McElroy, E. Larry Heacock, Hard Cover $45 (ISBN 0-87703-186-X); Soft Cover $35 (ISBN 0-87703-187-8)

Vol. 56 Space: A Developing Role for Europe, 18th European Space Symposium, Jun. 6-9, 1983, London, England, 1984, 278p., eds. Len J. Carter, Peter M. Bainum, Hard Cover $45 (ISBN 0-87703-193-2); Soft Cover $35 (ISBN 0-87703-194-0); Microfiche Suppl. (Vol. 46 AAS Microfiche Series) $15 (ISBN 0-87703-195-9)

Vol. 57 The Case for Mars, Apr. 29 - May 2, 1981, Boulder, CO, 1984, Second Printing 1987, 348p., ed. Penelope J. Boston, on Microfiche only $25 (ISBN 0-87703-198-3)

Vol. 58 Space Safety and Rescue 1982-1983, Proceedings of the International Academy of Astronautics held in conjunction with the 33rd and 34th International Astronautical Congresses, Paris, France, Sept. 27 - Oct. 2, 1982, and Budapest, Hungary, Oct. 10-15, 1983, 1984, 378p., ed. Gloria W. Heath, Hard Cover $50 (ISBN 0-87703-202-5); Soft Cover $40 (ISBN 0-87703-203-3)

Vol. 59 Space and Society - Challenges and Choices, April 14-16, 1982, University of Texas at Austin, 1984, 442p., eds. Paul Anaejionu, Nathan C. Goldman, Philip J. Meeks, Hard Cover $55 (ISBN 0-87703-204-1); Soft Cover $35 (ISBN 0-87703-205-X)

Vol. 60 Permanent Presence - Making It Work, 22nd Goddard Memorial Symposium, Mar. 15-16, 1984, Greenbelt, MD, 1985, 190p., ed. Ivan Bekey, Hard Cover $40 (ISBN 0-87703-207-6); Soft Cover $30 (ISBN 0-87703-208-4)

Vol. 61 Europe/United States Space Activities - With a Space Propulsion Supplement, 23rd Goddard Memorial Symposium/19th European Space Symposium, Mar. 27-29, 1985, Greenbelt, MD, 31st Annual AAS Meeting, Oct. 22-24, 1984, Palo Alto, CA, 1985, 442p., eds. Peter M. Bainum, Friedrich von Bun, Hard Cover $55 (ISBN 0-87703-217-3); Soft Cover $45 (ISBN 0-87703-218-1)

Vol. 62 The Case for Mars II, July 10-14, 1984, Boulder, CO, 1985, 730p., ed. Christopher P. McKay, Hard Cover $60 (ISBN 0-87703-219-1); Soft Cover $40 (ISBN 0-87703-220-3)

Vol. 63 Proceedings of 4th International Conference on Applied Numerical Modeling, Dec. 27-29, 1984, Tainan, Taiwan, 1986, 800p., ed. Han-Min Hsia, You-Li Chou, Shu-Yi Wang, Sheng-Jii Hsieh, Hard Cover $70 (ISBN 0-87703-242-4)

Vol. 64 Space Safety and Rescue 1984-1985, Proceedings of the International Academy of Astronautics held in conjunction with the 35th and 36th International Astronautical Congresses, Lausanne, Switzerland, Oct. 7-13, 1984, and Stockholm, Sweden, Oct. 7-12, 1985, 1986, 400p., ed. Gloria W. Heath, Hard Cover $55 (ISBN 0-87703-248-3); Soft Cover $45 (ISBN 0-87703-249-1)

Vol. 65 The Human Quest in Space, 24th Goddard Memorial Symposium, Mar. 20-21, 1986, Greenbelt, MD, 1987, 312p., ed. Gerald L. Burdett, Gerald A. Soffen, Hard Cover $55 (ISBN 0-87703-262-9); Soft Cover $45 (ISBN 0-87703-263-7)

Vol. 66 Soviet Space Programs 1980-1985, 1987, 298p., by Nicholas L. Johnson, Microfiche Only $45 (ISBN 0-87703-266-1)

Vol. 67 Low-Gravity Sciences, Seminar Series 1986, University of Colorado at Boulder, 290p., ed. Jean N. Koster, Hard Cover $55 (ISBN 0-87703-270-X); Soft Cover $45 (ISBN 0-87703-271-8)

Vol. 68 Proceedings of the Fourth Annual L5 Space Development Conference, Apr. 25-28, 1985, Washington, D.C., 1987, 268p., ed. Frank Hecker, Hard Cover $50 (ISBN 0-87703-272-6); Soft Cover $35 (ISBN 0-87703-273-4)

Vol. 69 Visions of Tomorrow: A Focus on National Space Transportation Issues, 25th Goddard Memorial Symposium, Mar. 18-20, 1987, Greenbelt, MD, 1987, 338p., ed. Gerald A. Soffen, Hard Cover $55 (ISBN 0-87703-274-2); Soft Cover $45 (ISBN 0-87703-275-0)

Vol. 70 Space Safety and Rescue 1986-1987, Proceedings of the International Academy of Astronautics held in conjunction with the 37th and 38th International Astronautical Congresses, Innsbruck, Austria, Oct. 4-11, 1986, and Brighton, England, Oct. 11-16, 1987, 1988, 360p., ed. Gloria W. Heath, Hard Cover $55 (ISBN 0-87703-291-2); Soft Cover $45 (ISBN 0-87703-292-0)

Vol. 71 The NASA Mars Conference, Jul. 21-23, 1986, Washington, D.C., 1988, 570p., ed. Duke B. Reiber, Hard Cover $50 (ISBN 0-87703-293-9); Soft Cover $30 (ISBN 0-87703-294-7)

Vol. 72 Working in Orbit and Beyond: The Challenges for Space Medicine, Jun. 20-21, 1987, Washington, D.C., 1989, 188p., ed. David Lorr, Victoria Garshnek, Hard Cover $45 (ISBN 0-87703-295-5); Soft Cover $35 (ISBN 0-87703-296-3)

Vol. 73 Technology and the Civil Future in Space, 26th Goddard Memorial Symposium, Mar. 16-18, 1988, Greenbelt, MD, 1989, 246p., ed. Leonard A. Harris, Hard Cover $50 (ISBN 0-87703-301-3); Soft Cover $35 (ISBN 0-87703-302-1)

Vol. 74 The Case for Mars III: Strategies for Exploration - General Interest and Overview, July 18-22, 1987, Boulder, CO, 1989, 744p., ed. Carol Stoker, Hard Cover $75 (ISBN 0-87703-303-X); Soft Cover $55 (0-87703-304-8)

Vol. 75 The Case for Mars III: Strategies for Exploration - Technical, July 18-22, 1987, Boulder, CO, 1989, 646p., ed. Carol Stoker, Hard Cover $70 (ISBN 0-87703-305-6); Soft Cover $50 (ISBN 0-87703-306-4)

Vol. 76 Global Environmental Change: The Role of Space in Understanding Earth, 27th Goddard Memorial Symposium, Mar. 8-10, 1989, Washington, D.C., 1990, 178p., ed. Richard G. Johnson, Hard Cover $50 (ISBN 0-87703-322-6); Soft Cover $40 (ISBN 0-87703-323-4); Microfiche Suppl. (Vol. 60 AAS Microfiche Series) $10 (ISBN 0-87703-324-2)

Vol. 77 Space Safety and Rescue 1988 - 1989, Proceedings of the International Academy of Astronautics held in conjunction with the 39th and 40th International Astronautical Congresses, Bangalore, India, Oct. 8-15, 1988, and Málaga, Spain, Oct. 7-12, 1989, 1990, 500p., ed. Gloria W. Heath, Hard Cover $70 (ISBN 0-87703-327-7); Soft Cover $55 (ISBN 0-87703-328-5)

Vol. 78 Leaving the Cradle: Human Exploration of Space in the 21st Century, 28th Goddard Memorial Symposium, Mar. 14-16, 1990, Washington, D.C., 1991, 348p., ed. Thomas O. Paine, Hard Cover $70 (ISBN 0-87703-336-6); Soft Cover $55 (ISBN 0-87703-337-4)

Vol. 79 Space Safety and Rescue 1990, Proceedings of the International Academy of Astronautics held in conjunction with the 41st International Astronautical Congress, Dresden, Germany, Oct. 6-12, 1990, 1991, 232p., ed. Gloria W. Heath, Hard Cover $65 (ISBN 0-87703-341-2); Soft Cover $50 (ISBN 0-87703-342-0)

Vol. 80 Prospects for Interstellar Travel, 1992, 390p., by John H. Mauldin, Hard Cover $50 (ISBN 0-87703-344-7); Soft Cover $27 (ISBN 0-87703-345-5)

Vol. 81 Humans and Machines in Space: The Vision, The Challenge, The Payoff, 29th Goddard Memorial Symposium, Mar. 14-15, 1991, Washington, D.C., 1992, 204p., ed. Bradley Johnson, Gayle L. May, Paula Korn, Hard Cover $50 (ISBN 0-87703-356-0); Soft Cover $35 (ISBN 0-87703-357-9)

Vol. 82 Space Safety and Rescue 1991, Proceedings of the International Academy of Astronautics held in conjunction with the 42nd International Astronautical Congress, Montreal, Canada, Oct. 5-11, 1991, 1993, 270p., ed. Gloria W. Heath, Hard Cover $65 (ISBN 0-87703-372-2); Soft Cover $50 (ISBN 0-87703-373-0)

Vol. 83 Space: A Vital Stimulus to Our National Well-Being, 31st Goddard Memorial Symposium, March 9-10, 1993, Arlington, Virginia, and World Space Programs and Fiscal Reality, 30th Goddard Memorial Symposium, April 9-10, 1992, Alexandria, Virginia, 1994, 334p., ed. Gayle L. May, Saunders B. Kramer, Paula Korn, Leonard David, Barbara Sprungman, Hard Cover $70 (ISBN 0-87703-389-7); Soft Cover $50 (ISBN 0-87703-390-0)

Vol. 84 Space Safety and Rescue 1992, Proceedings of the International Academy of Astronautics held in conjunction with the World Space Congress, Washington, D.C., Aug. 28 to Sept. 5, 1992, 1994, 372p., ed. Gloria W. Heath, Hard Cover $70 (ISBN 0-87703-391-9); Soft Cover $55 (ISBN 0-87703-392-7)

Vol. 85 Civil Space in the Clinton Era, 32nd Goddard Memorial Symposium, March 1-2, 1994, Crystal City, Virginia, and Partners in Space . . . 2001, 41st Annual Meeting, November 14-16, 1994, Crystal City, Virginia, 1995, 292p., ed. Donald R. McConathy, Paula Korn, Hard Cover $70 (ISBN 0-87703-397-8); Soft Cover $50 (ISBN 0-87703-398-6)

Vol. 86 Strategies for Mars: A Guide to Human Exploration, 1996, 644p, ed. Carol R. Stoker, Carter Emmart, Hard Cover $70 (ISBN 0-87703-405-2); Soft Cover $45 (ISBN 0-87703-406-0)

Vol. 87 Space Safety and Rescue 1993, Proceedings of the International Academy of Astronautics held in conjunction with the 44th International Astronautical Congress, Graz, Austria, Oct. 16-22, 1993, 1996, 344p., ed. Gloria W. Heath, Hard Cover $70 (ISBN 0-87703-410-9); Soft Cover $50 (ISBN 0-87703-411-7)

Vol. 88 Space Safety and Rescue 1994, Proceedings of the International Academy of Astronautics held in conjunction with the 45th International Astronautical Congress, Jerusalem, Israel, Oct. 9-14, 1994, 1996, 326p., ed. Gloria W. Heath, Hard Cover $70 (ISBN0-87703-416-8); Soft Cover $50 (ISBN 0-87703-417-6)

Vol. 89 The Case for Mars IV: The International Exploration of Mars?Mission Strategy and Architectures, June 4-8, 1990, Boulder, CO, 1997, 790p., ed. Thomas R. Meyer, Hard Cover $80 (ISBN 0-87703-418-4); Soft Cover $55 (ISBN 0-87703-419-2)

Vol. 90 The Case for Mars IV: The International Exploration of Mars?Considerations for Sending Humans, June 4-8, 1990, Boulder, CO, 1997, 502p., ed. Thomas R. Meyer, Hard Cover $70 (ISBN 0-87703-420-6); Soft Cover $55 (ISBN 0-87703-421-4)

Vol. 91 From Imagination to Reality: Mars Exploration Studies of the Journal of the British Interplanetary Society (Part I, Precursors and Early Piloted Missions), 1997, 388p., ed. Robert M. Zubrin, Hard Cover $70 (ISBN 0-87703-426-5); Soft Cover $45 (ISBN 0-87703-427-3)

Vol. 92 From Imagination to Reality: Mars Exploration Studies of the Journal of the British Interplanetary Society (Part II, Base Building, Colonization and Terraformation), 1997, 376p., ed. Robert M. Zubrin, Hard Cover $70 (ISBN 0-87703-428-1); Soft Cover $45 (ISBN 0-87703-429-X)

Vol. 93 Space Safety and Rescue 1995, Proceedings of the International Academy of Astronautics held in conjunction with the 46th International Astronautical Congress, Jerusalem, Israel, Oct. 2-6, 1995, 1997, 482p., ed. Gloria W. Heath, Hard Cover $80 (ISBN 0-87703-416-8); Soft Cover $55 (ISBN 0-87703-417-6)

Vol. 94 Fluid and Electrolyte Regulation in Spaceflight, 1998, 238p., Carolyn S. Leach Huntoon, Anatoliy I. Grigoriev, Yuri V. Natochin, Hard Cover $60 (ISBN 0-87703-442-7); Soft Cover $40 (ISBN 0-87703-443-5)

Vol. 95 Space Safety and Rescue 1996, Proceedings of the International Academy of Astronautics held in conjunction with the 47th International Astronautical Congress, Beijing, China, Oct. 7-11, 1996, 1998, 362p., ed. Gloria W. Heath, Hard Cover $75 (ISBN 0-87703-446-X); Soft Cover $50 (ISBN 0-87703-447-8)

Vol. 96 Space Safety and Rescue 1997, Proceedings of the International Academy of Astronautics held in conjunction with the 48th International Astronautical Congress, Turin, Italy, Oct. 6-10, 1997, 1999, 412p., ed. Gloria W. Heath, Hard Cover $80 (ISBN 0-87703-454-0); Soft Cover $55 (ISBN 0-87703-455-9)

Vol. 97 The Case for Mars V, May 26-29, 1993, Boulder, CO, 2000, 564p., ed. Penelope J. Boston, Hard Cover $80 (ISBN 0-87703-459-1); Soft Cover $55 (ISBN 0-87703-460-5)

Vol. 98 The Case for Mars VI: Making Mars an Affordable Destination, July 17-20, 1996, Boulder, CO, 2000, 578p., ed. Kelly R. McMillen, Hard Cover $80 (ISBN 0-87703-461-3); Soft Cover $55 (ISBN 0-87703-462-1)

Vol. 99 Space Safety and Rescue 1998, Proceedings of the International Academy of Astronautics held in conjunction with the 49th International Astronautical Congress, Melbourne, Australia, Sept. 28 - Oct. 2, 1998, 2000, 410p., ed. Macgregor S. Reid and Walter Flury, Hard Cover $80 (ISBN0-87703-463-X); Soft Cover $55 (ISBN 0-87703-464-8)

Vol. 100 Space Debris 1999, Proceedings of the International Academy of Astronautics held in conjunction with the 50th International Astronautical Federation Congress, Amsterdam, The Netherlands, Oct. 2-8, 1999, 2001, 494p., ed. Joerg Bendisch, Hard Cover $90 (ISBN 0-87703-473-7); Soft Cover $65 (ISBN 0-87703-474-5)

Vol. 101 Space Access and Utilization Beyond 2000, Based on, but not limited to, the Symposium held at the Annual General Assembly of the American Association for the Advancement of Science, Washington, D.C., Feb. 18, 2000, 234p., ed. Yoji Kondo, Charles Sheffield and Frederick C. Bruhweiler, Hard Cover $70 (ISBN 0-87703-475-3); Soft Cover $50 (ISBN 0-87703-476-1)

Vol. 102 Space Safety, Rescue and Quality 1999-2000, Proceedings of the International Academy of Astronautics held in conjunction with the 50th and 51st International Astronautical Federation Congresses, Amsterdam, The Netherlands, Oct. 2-8, 1999; Rio de Janeiro, Brazil, Oct. 2-6, 2000, 2001, 286p., eds. Macgregor S. Reid and Manola Romero, Hard Cover $80 (ISBN 0-87703-483-4); Soft Cover $55 (ISBN 0-87703-484-2)

Vol. 103 Space Debris 2000, Proceedings of the International Academy of Astronautics held in conjunction with the 51st International Astronautical Federation Congress, Rio de Janeiro, Brazil, Oct. 2-6, 2000, 2001, 368p., ed. Joerg Bendisch, Hard Cover $90 (ISBN 0-87703-485-0); Soft Cover $65 (ISBN 0-87703-486-9)

Vol. 104 ISOLATION: NASA Experiments in Closed-Environment Living - Advanced Human Life Support Enclosed System Final Report, 2002, 470p., eds. Helen W. Lane, Richard L. Sauer and Daniel L. Feeback, Hard Cover $90 (ISBN 0-87703-493-1); Soft Cover $65 (ISBN 0-87703-494-X)

Vol. 105 Space Debris 2001, Proceedings of the International Academy of Astronautics held in conjunction with the 52nd International Astronautical Federation Congress, Toulouse, France, Oct. 1-5, 2001, 2002, 306p., ed. Joerg Bendisch, Hard Cover $90 (ISBN 0-87703-496-6); Soft Cover $65 (ISBN 0-87703-497-4)

Vol. 106 Space Safety, Rescue and Quality 2001, Proceedings of the International Academy of Astronautics held in conjunction with the 52nd International Astronautical Federation Congress, Toulouse, France, Oct. 1-5, 2001, 2003, 224p., eds. Macgregor S. Reid and Manola Romero, Hard Cover $80 (ISBN 0-87703-500-8); Soft Cover $55 (ISBN 0-87703-501-6)

Vol. 107 Martian Expedition Planning, Based on the British Interplanetary Society Symposium held Feb. 24, 2003, London, England, plus invited papers, 2004, 518p., ed. Charles S. Cockell, Hard Cover plus a CD ROM $90 (ISBN 0-87703-507-5); Soft Cover plus a CD ROM $65 (ISBN 0-87703-508-3)

Vol. 108 Proceedings of the International Lunar Conference 2003 / International Lunar Exploration Working Group 5 – ILC2003 / ILEWG 5, held Nov. 16-22, 2003, Hawaii Island, Hawaii, 2004, 602p., eds. Steve M. Durst, Charles T. Bohannan, Christopher G. Thomason, Michael R. Cerney, Leilehua Yuen, Hard Cover plus a CD ROM $110 (ISBN 0-87703-513-X); Soft Cover plus a CD ROM $85 (ISBN 0-87703-514-8)

Vol. 109 Space Debris and Space Traffic Management 2003, Proceedings of the International Academy of Astronautics Symposium held in conjunction with the 54th International Astronautical Congress, Bremen, Germany, Sept. 29 – Oct. 3, 2003, 2004, 648p., ed. Joerg Bendisch, Hard Cover plus a CD ROM $140 (ISBN 0-87703-516-4); Soft Cover plus a CD ROM $110 (ISBN 0-87703-517-2)

Vol. 110 Space Debris and Space Traffic Management 2004, Proceedings of the International Academy of Astronautics Symposium held in conjunction with the 55th International Astronautical Congress, Vancouver, British Columbia, Canada, Oct. 4-8, 2004, 2004, 444p., ed. Joerg Bendisch, Hard Cover plus a CD ROM $140 (ISBN 0-87703-523-7); Soft Cover plus a CD ROM $110 (ISBN 0-87703-524-5)

Vol. 111 Mars Analog Research, Based on invited papers, 2006, 356p., ed. Jonathan D. A. Clarke, Hard Cover plus a CD ROM $90 (ISBN 0-87703-529-6); Soft Cover plus a CD ROM $65 (ISBN 0-87703-530-X)

Vol. 112 Space Debris and Space Traffic Management 2005, Proceedings of the International Academy of Astronautics Symposium held in conjunction with the 56th International Astronautical Congress, Fukuoka, Japan, Oct. 17-21, 2005, 2006, 414p., ed. Joerg Bendisch, Hard Cover plus a CD ROM $140 (ISBN 0-87703-533-4); Soft Cover plus a CD ROM $110 (ISBN 0-87703-534-2)

Vol. 113 Decoupling Civil Timekeeping from Earth Rotation, Proceedings of a Colloquium Exploring Implications of Redefining Coordinated Universal Time (UTC), Exton, Pennsylvania, Oct. 5–7, 2011, 2011, 408p., eds. John H. Seago, Robert L. Seaman, Steven L. Allen, Hard Cover plus a CD ROM $120 (ISBN 978-0-87703-575-6); CD ROM Only $80 (ISBN 978-0-87703-576-3)

Order from Univelt, Inc., P.O. Box 28130, San Diego, California 92198 (Web Site: http://www.univelt.com) **STANDING ORDERS ACCEPTED**

AAS HISTORY SERIES

Vol. 1 Two Hundred Years of Flight in America: A Bicentennial Survey, Edited by Eugene M. Emme, 1977, 326p, Third Printing 1981, Hard Cover $35 (ISBN 0-87703-091-X); Soft Cover $25 (ISBN 0-87703-101-0); special price for classroom text or bulk purchase.

Vol. 2 Twenty-Five Years of the American Astronautical Society: Historical Reflections and Projections, 1954-1979, Edited by Eugene M. Emme, 1980, 248p, Hard Cover $25 (ISBN 0-87703-117-7); Soft Cover $15 (ISBN 0-87703-118-5).

Vol. 3 Between Sputnik and the Shuttle: New Perspectives on American Astronautics, 1957-1980, Edited by Frederick C. Durant, III, 1981, 350p, Hard Cover $40 (ISBN 0-87703-145-2); Soft Cover $30 (ISBN 0-87703-149-9).

Vol. 4 The Endless Space Frontier: A History of the House Committee on Science and Astronautics, By Ken Hechler, Abridged and edited by Albert E. Eastman, 1982, 460p, Hard Cover $45 (ISBN 0-87703-157-6); Soft Cover $35 (ISBN 0-87703-158-4).

Vol. 5 Science Fiction and Space Futures: Past and Present, Edited by Eugene M. Emme, 1982, 278p, Hard Cover $35 (ISBN 0-87703-172-X); Soft Cover $25 (ISBN 0-87703-173-8).

Vol. 6 First Steps Toward Space, Edited by Frederick C. Durant, III and George S. James, 1986, 318p, Hard Cover $45 (ISBN 0-87703-243-2); Soft Cover $35 (ISBN 0-87703-244-0).

Vol. 7 History of Rocketry and Astronautics, Edited by R. Cargill Hall, 1986, Part I, 250p, Part II, 502p, sold as a set, Hard Cover $100 (ISBN 0-87703-260-2); Soft Cover $80 (ISBN 0-87703-261-0).

Vol. 8 History of Rocketry and Astronautics, Edited by Kristan R. Lattu, 1989, 368p, Hard Cover $50 (ISBN 0-87703-307-2); Soft Cover $35 (ISBN 0-87703-308-0).

Vol. 9 History of Rocketry and Astronautics, Edited by Frederick I. Ordway, III, 1989, 330p, Hard Cover $50 (ISBN 0-87703-309-9); Soft Cover $35 (ISBN 0-87703-310-2).

Vol. 10 History of Rocketry and Astronautics, Edited by Å. Ingemar Skoog, 1990, 330p, Hard Cover $60 (ISBN 0-87703-329-3); Soft Cover $40 (ISBN 0-87703-330-7)

Vol. 11 History of Rocketry and Astronautics, Edited by Roger D. Launius, 1994, 236p, Hard Cover $60 (ISBN 0-87703-382-X); Soft Cover $40 (ISBN 0-87703-383-8).

Vol. 12 History of Rocketry and Astronautics, Edited by John L. Sloop, 1991, 252p, Hard Cover $60 (ISBN 0-87703-332-3); Soft Cover $40 (ISBN 0-87703-333-1).

Vol. 13 History of Liquid Rocket Engine Development in the United States 1955-1980, Edited by Stephen E. Doyle, 1992, 176p, Hard Cover $50 (ISBN 0-87703-349-8); Soft Cover $35 (ISBN 0-87703-350-1).

Vol. 14 History of Rocketry and Astronautics, Edited by Tom D. Crouch, Alex M. Spencer, 1993, 222p, Hard Cover $50 (ISBN 0-87703-374-9); Soft Cover $35 (ISBN 0-87703-375-7).

Vol. 15 History of Rocketry and Astronautics, Edited by Lloyd H. Cornett, Jr., 1993, 452p, Hard Cover $60 (ISBN 0-87703-376-5); Soft Cover $40 (ISBN 0-87703-377-3).

Vol. 16 Out From Behind the Eight-Ball: A History of Project Echo, by Donald C. Elder, 1995, 176p, Hard Cover $50 (ISBN 0-87703-387-0); Soft Cover $30 (ISBN 0-87703-388-9).

Vol. 17 History of Rocketry and Astronautics, Edited by John Becklake, 1995, 480p, Hard Cover $60 (ISBN 0-87703-395-1); Soft Cover $40 (ISBN 0-87703-396-X).

Vol. 18 Organizing for the Use of Space: Historical Perspectives on a Persistent Issue, Edited by Roger D. Launius, 1995, 234p, Hard Cover $60; Soft Cover $40

Vol. 19 History of Rocketry and Astronautics, Edited by J. D. Hunley, 1997, 318p, Hard Cover $60 (ISBN 0-87703-422-2); Soft Cover $40 (ISBN 0-87703-423-0).

Vol. 20 History of Rocketry and Astronautics, Edited by J. D. Hunley, 1997, 344p, Hard Cover $60 (ISBN 0-87703-424-9); Soft Cover $40 (ISBN 0-87703-425-7).

Vol. 21 History of Rocketry and Astronautics, Edited by Philippe Jung, 1997, 368p, Hard Cover $60 (ISBN 0-87703-439-7); Soft Cover $40 (ISBN 0-87703-440-0).

Vol. 22 History of Rocketry and Astronautics, Edited by Philippe Jung, 1998, 418p, Hard Cover $60 (ISBN 0-87703-444-3); Soft Cover $40 (ISBN 0-87703-445-1).

Vol. 23 History of Rocketry and Astronautics, Edited by Donald C. Elder and Christophe Rothmund, 2001, 566p, Hard Cover $85 (ISBN 0-87703-477-X); Soft Cover $60 (ISBN 0-87703-478-8).

Vol. 24 The Origins and Technology of the Advanced Extravehicular Space Suit, by Gary L. Harris, 2001, 558p, Hard Cover $85 (ISBN 0-87703-481-8); Soft Cover $60 (ISBN 0-87703-482-6).

Vol. 25 History of Rocketry and Astronautics, Edited by Hervé Moulin and Donald C. Elder, 2003, 370p, Hard Cover $85 (ISBN 0-87703-498-2); Soft Cover $60 (ISBN 0-87703-499-0).

Vol. 26 History of Rocketry and Astronautics, Edited by Donald C. Elder and George S. James, 2005, 430p, Hard Cover $95 (ISBN 0-87703-518-0); Soft Cover $70 (ISBN 0-87703-519-9).

Vol. 27 History of Rocketry and Astronautics, Edited by Kerrie Dougherty and Donald C. Elder, 2007, 416p, Hard Cover $95 (ISBN 978-0-87703-535-0); Soft Cover $70 (ISBN 978-0-87703-536-7).

Vol. 28 History of Rocketry and Astronautics, Edited by Frank H. Winter, 2007, 560p, Hard Cover $95 (ISBN 978-0-87703-539-8); Soft Cover $70 (ISBN 978-0-87703-540-4).

Vol. 29 Space Shuttle Main Engine: The First Twenty Years and Beyond, by Robert E. Biggs, 2008, 270p, Hard Cover $70 (ISBN 978-0-87703-546-6); Soft Cover $50 (ISBN 978-0-87703-547-3)

Vol. 30 History of Rocketry and Astronautics, Edited by Otfrid G. Liepack, 2009, 346p, Hard Cover $80 (ISBN 978-0-87703-549-7); Soft Cover $70 (ISBN 978-0-87703-550-3).

Vol. 31 History of Rocketry and Astronautics: IAA History Symposia 1967–2000 Abstracts and Index, Edited by Hervé Moulin 2009, 386p, Hard Cover plus a CD ROM $95 (ISBN 978-0-87703-551-0); Soft Cover plus a CD ROM $80 (ISBN 978-0-87703-552-7).

Vol. 32 History of Rocketry and Astronautics, Edited by Christophe Rothmund, 2010, 482p, Hard Cover $95 (ISBN 978-0-87703-555-8); Soft Cover $75 (ISBN 978-0-87703-556-5).

Vol. 33 History of Rocketry and Astronautics, Edited by Michael L. Ciancone, 2010, 568p, Hard Cover $95 (ISBN 978-0-87703-558-9); Soft Cover $75 (ISBN 978-0-87703-559-6).

Vol. 34 History of Rocketry and Astronautics, Edited by Otfrid G. Liepack, 2011, 576p, Hard Cover $95 (ISBN 978-0-87703-563-3); Soft Cover $75 (ISBN 978-0-87703-564-0).

Vol. 35 History of Rocketry and Astronautics, Edited by Å. Ingemar Skoog, 2011, 558p, Hard Cover $95 (ISBN 978-0-87703-567-1); Soft Cover $75 (ISBN 978-0-87703-568-8).

Vol. 36 History of Rocketry and Astronautics, Edited by Emily D. Springer, 2011, 316p, Hard Cover $95 (ISBN 978-0-87703-573-2); Soft Cover $75 (ISBN 978-0-87703-574-9).

Order from Univelt, Incorporated, P.O. Box 28130, San Diego, California 92198 (Web Site: http://www.univelt.com) STANDING ORDERS ACCEPTED

INDEX

INDEX TO ALL AMERICAN ASTRONAUTICAL SOCIETY PAPERS AND ARTICLES 1954 - 1992

This index is a numerical/chronological index (which also serves as a citation index) and an author index. (A subject index volume will be forthcoming.)

It covers all articles that appear in the following:
- *Advances in the Astronautical Sciences* (1957 - 1992)
- *Science and Technology Series* (1964 -1992)
- *AAS History Series* (1977 - 1992)
- *AAS Microfiche Series* (1968 - 1992)
- *Journal of the Astronautical Sciences* (1954 -September 1992)
- *Astronautical Sciences Review* (1959 - 1962)

If you are in aerospace you will want this excellent reference tool which covers the first 35 years of the Space Age.

Numerical/Chronological/Author Index in three volumes,

Ordered as a set:
- Library Binding (all three volumes) $120.00;
- Soft Cover (all three volumes) $90.00.

Ordered by individual volume:
- Volume I (1954 - 1978) Library Binding $40.00; Soft Cover $30.00;
- Volume II (1979 - 1985/86) Library Binding $60.00; Soft Cover $45.00;
- Volume III (1986 - 1992) Library Binding $70.00; Soft Cover $50.00.

**Order from Univelt, Inc., P.O. Box 28130, San Diego, California 92198.
Web Site: http://www.univelt.com**

NUMERICAL INDEX

VOLUME 113	**SCIENCE AND TECHNOLOGY SERIES**, *DECOUPLING CIVIL TIMEKEEPING FROM EARTH ORBIT*, 2011 (Proceedings of a Colloquium Exploring Implications of Redefining Coordinated Universal Time (UTC), October 5–7, 2011, Exton, Pennsylvania
AAS 11-660	The Colloquium on Decoupling Civil Timekeeping from Earth Rotation, John H. Seago, Robert L. Seaman and Steven L. Allen
AAS 11-661	Systems Engineering for Civil Timekeeping, Rob Seaman
AAS 11-662	Legislative Specifications for Coordinating with Universal Time, John H. Seago, P. Kenneth Seidelmann and Steve Allen
AAS 11-663	The Heavens and Timekeeping, Symbolism and Expediency, Paul Gabor
AAS 11-664	Leap Seconds in Literature, John H. Seago
AAS 11-665	Time in the 10,000-Year Clock, Danny Hillis, Rob Seaman, Steve Allen and Jon Giorgini
AAS 11-666	Using UTC to Determine the Earth's Rotation Angle, Dennis D. McCarthy
AAS 11-667	The IERS, the Leap Second, and the Public, Wolfgang R. Dick
AAS 11-668	Results from the 2011 IERS Earth Orientation Center Survey about a Possible UTC Redefinition, Daniel Gambis, Gérard Francou and Teddy Carlucci
AAS 11-669	Traditional Celestial Navigation and UTC, Frank E. Reed
AAS 11-670	The Consequences of Decoupling UTC on Sundials, Denis Savoie and Daniel Gambis
AAS 11-671	Time Scales in Astronomical and Navigational Almanacs, George H. Kaplan
AAS 11-672	Issues Concerning the Future of UTC, P. Kenneth Seidelmann and John H. Seago
AAS 11-673	UTC and the Hubble Space Telescope Flight Software, David G. Simpson
AAS 11-674	Computation Errors in Look Angle and Range Due to Redefinition of UTC, Mark F. Storz
AAS 11-675	Proposal for the Redefinition of UTC: Influence on NGA Earth Orientation Predictions and GPS Operations, Stephen Malys
AAS 11-676	UTC at the Harvard-Smithsonian Center for Astrophysics (CFA) and Environs, Arnold H. Rots

AAS 11-677 An Inventory of UTC Dependencies for IRAF, Rob Seaman

AAS 11-678 Telescope Systems at Lick Observatory and Keck Observatory,
Steven L. Allen

AAS 11-679 Automating Retrieval of Earth Orientation Predictions, David L. Terrett

AAS 11-680 Dissemination of UT1-UTC through the use of Virtual Observatory
Florent Deleflie, Daniel Gambis, Christophe Barache and Jérome Berthier

AAS 11-681 Timekeeping System Implementations: Options for the *Pontifex Maximus*
Steven L. Allen

AAS 11-682 The Longwood Gardens Analemmatic Sundial, P. Kenneth Seidelmann

AUTHOR INDEX[*]

Allen, Steven L., AAS 11-660, S&T v113, pp3-12; AAS 11-662, S&T v113, pp29-50; AAS 11-665, S&T v113, pp79-96; AAS 11-678, S&T v113, pp289-297; AAS 11-681, S&T v113, pp325-334

Barache, Christophe, AAS 11-680, S&T v113, pp317-324

Berthier, Jérome, AAS 11-680, S&T v113, pp317-324

Carlucci, Teddy, AAS 11-668, S&T v113, pp123-179

Deleflie, Florent, AAS 11-680, S&T v113, pp317-324

Dick, Wolfgang R., AAS 11-667, S&T v113, pp117-122

Francou, Gérard, AAS 11-668, S&T v113, pp123-179

Gabor, Paul, AAS 11-663, S&T v113, pp53-64

Gambis, Daniel, AAS 11-668, S&T v113, pp123-179; AAS 11-670, S&T v113, pp195-199; AAS 11-680, S&T v113, pp317-324

Giorgini, Jon, AAS 11-665, S&T v113, pp79-96

Hillis, Danny, AAS 11-665, S&T v113, pp79-96

Kaplan, George H., AAS 11-671, S&T v113, pp201-214

Malys, Stephen, AAS 11-675, S&T v113, pp265-270

McCarthy, Dennis D., AAS 11-666, S&T v113, pp105-116

Reed, Frank E., AAS 11-669, S&T v113, pp183-195

Rots, Arnold H., AAS 11-676, S&T v113, pp273-276

Savoie, Denis, AAS 11-670, S&T v113, pp195-199

Seago, John H., AAS 11-660, S&T v113, pp3-12; AAS 11-662, S&T v113, pp29-50; AAS 11-664, S&T v113, pp65-78; AAS 11-672, S&T v113, pp215-233

Seaman, Robert L., AAS 11-660, S&T v113, pp3-12; AAS 11-661, S&T v113, pp15-27; AAS 11-665, S&T v113, pp79-96; AAS 11-677, S&T v113, pp277-287

Seidelmann, P. Kenneth, AAS 11-662, S&T v113, pp29-50; AAS 11-672, S&T v113, pp215-233; AAS 11-682, S&T v113, pp351-359

Simpson, David G., AAS 11-673, S&T v113, pp237-248

Storz, Mark F., AAS 11-674, S&T v113, pp249-263

Terrett, David L., AAS 11-679, S&T v113, pp309-315

---

[*] For each author the AAS paper number is given. The page numbers refer to Volume 113, AAS *Science and Technology Series*.